T0325718

Environmental Data Analysis with MATLAB®

Environmental Data Analysis with MATLAB®

Second Edition

William Menke
Professor of Earth and Environmental Sciences, Columbia University

Joshua Menke
Software Engineer, JOM Associates

AMSTERDAM · BOSTON · HEIDELBERG · LONDON
NEW YORK · OXFORD · PARIS · SAN DIEGO
SAN FRANCISCO · SINGAPORE · SYDNEY · TOKYO
Academic Press is an imprint of Elsevier

Academic Press is an imprint of Elsevier
32 Jamestown Road, London NW1 7BY, UK
525 B Street, Suite 1800, San Diego, CA 92101–4495, USA
50 Hampshire Street, 5th Floor, Cambridge, MA 02139, USA
The Boulevard, Langford Lane, Kidlington, Oxford OX5 1GB, UK

Library of Congress Cataloging-in-Publication Data
A catalog record for this book is available from the Library of Congress

British Library Cataloguing in Publication Data
A catalogue record for this book is available from the British Library

ISBN: 978-0-12-804488-9

For information on all Academic Press publications
visit our website at https://www.elsevier.com/

  Working together
to grow libraries in
developing countries

www.elsevier.com • www.bookaid.org

Publisher: Glyn Jones
Acquisition Editor: Graham Nisbet
Editorial Project Manager: Susan Ikeda
Production Project Manager: Poulouse Joseph
Designer: Matthew Limbert

Typeset by SPi Books

FOR
H1

Contents

For additional information on the topics covered in the book, visit the
companion site: http://store.elsevier.com/9780128044889

Preface

The first question that I ask an environmental science student who comes seeking my advice on a data analysis problem is *Have you looked at your data?* Very often, after some beating around the bush, the student answers, *Not really.* The student goes on to explain that he or she loaded the data into an analysis package provided by an advisor, or downloaded off the web, and it didn't work. The student tells me, *Something is wrong with my data!* I then ask my second question: *Have you ever used the analysis package with a dataset that did work?* After some further beating around the bush, the student answers, *Not really.* At this point, I offer two pieces of advice. The first is to spend time getting familiar with the dataset. Taking into account what the student has been able to tell me, I outline a series of plots, histograms, and tables that will help him or her prise out its general character and some of its nuances. Second, I urge the student to create several simulated datasets with properties similar to those expected of the data and run the data analysis package on them. The student needs to make sure that he or she is operating it correctly and that it returns the right answers. We then make an appointment to review the student's progress in a week or two. Very often the student comes back reporting, *The problem wasn't at all what I thought it was!*

Then the real works begins, either to solve the problem or if the student has already solved it—which often he or she has—to get on with the data analysis.

Environmental Data Analysis with MATLAB® is organized around two principles. The first is that real proficiency in data analysis requires analyzing realistic data on a computer, and not merely working through ultra-simplified examples with pencil and paper. The second is that the skills needed to perform data analysis are best learned in a series of steps that alternate between theory and application and that start simple but rapidly expand as one's toolkit of skills grows. The real world puts many impediments in the way of analyzing data—errors of all sorts, missing information, inconvenient units of measurements, inscrutable data formats, and more. Consequently, real proficiency is as much about confidence and experience as it is about formal knowledge of techniques. This book teaches a core set of techniques that are widely applicable across all of Environmental Science, and it reinforces them by leading the student through a series of case studies on real-world data that has both the richness and the blemishes inherent in real-world things.

Two fundamental themes are used to tie together many different data analysis techniques:

The first is that measurement *error* is a fundamental aspect of observation and experiment. Error has a profound influence on the way that knowledge is distilled from data. We use probability theory to develop the concept of *covariance*, the key tool for quantifying error. We then show how covariance propagates through a chain of calculations leading to a result that possesses uncertainty. Dealing with that uncertainty is as important a part of data analysis as arriving at the result, itself. From Chapter 3, where it is introduced, through the book's end, we are always returning to the idea of the propagation of error.

The second is that many problems are special cases of a *linear model* linking the observations to the knowledge that we aspire to derive from them. Measurements of the world around us create data, numbers that describe the results of observations and experiments. But measurements, in and of themselves, are of little utility. The purpose of data analysis is to distill them down to a few significant and insightful *model parameters*. We develop the idea of the linear model in Chapter 4 and in subsequent chapters show that very many, seemingly different data analysis techniques are special cases of it. These include curve fitting, Fourier analysis, filtering, factor analysis, empirical function analysis and interpolation. While their uses are varied, they all share a common structure, which when recognized makes understanding them easier. Most important, covariance propagates through them in nearly identical ways.

As the title of this book implies, it relies very heavily on *MatLab* to connect the theory of data analysis to its practice in the real world. *MatLab*, a commercial product of *The MathWorks, Inc.*, is a popular scientific computing environment that fully supports data analysis, data visualization, and data file manipulation. It includes a scripting language through which complicated data analysis procedures can be developed, tested, performed, and archived. *Environmental Data Analysis with MATLAB*® makes use of scripts in three ways. First, the text includes many short scripts and excerpts from scripts that illustrate how particular data analysis procedures are actually performed. Second, a set of complete scripts and accompanying datasets is provided as a companion to the book. They implement all of the book's figures and case studies. Third, each chapter includes recommended homework problems that further develop the case studies. They require existing scripts to be modified and new scripts to be written.

Environmental Data Analysis with MATLAB® is a relatively short book that is appropriate for a one-semester course at the upper-class undergraduate and graduate level. It requires a working knowledge of calculus and linear algebra, as might be taught in a two-semester undergraduate calculus course. It does *not* require any knowledge of differential equations or more advanced forms of applied mathematics. Students with some familiarity with the practice of environmental science and with its underlying issues will be better able to put examples in context, but detailed knowledge of the science is not required. The book is self-contained; it can be read straight through, and profitably, even by someone with no access to *MatLab*. But it is meant to be used in a setting where students are actively using *MatLab* both as an aid to studying (i.e., by reproducing the case studies described in the book) and as a tool for completing the recommended homework problems.

Environmental Data Analysis with MATLAB® uses eight exemplary environmental science datasets:

Air temperature,
Chemical composition of sea floor samples,
Earthquake magnitudes,

Ground level ozone concentration,
Precipitation,
Sea surface temperature,
Stream flow, and
Water quality.

Most datasets are used in several different contexts and in several different places in the text. They are used both as a test bed for particular data analysis techniques and to illustrate how data analysis can lead to important insights about the environment.

Chapter 1, *Data Analysis with MatLab*, is a brief introduction to *MatLab* as a data analysis environment and scripting language. It is meant to teach *barely enough* to enable the reader to understand the *MatLab* scripts in the book and to begin to start using and modifying them. While *MatLab* is a fully featured programming language, *Environmental Data Analysis with MATLAB®* is not a book on computer programming. It teaches scripting mainly by example and avoids long discussions on programming theory.

Chapter 2, *A First Look at Data*, leads students through the steps that, in our view, should be taken when first confronted with a new dataset. Time plots, scatter plots, and histograms, as well as simple calculations, are used to examine the data with the aim both of understanding its general character and spotting problems. We take the position that *all data*sets have problems—errors, data gaps, inconvenient units of measurement, and so forth. Such problems should not scare a person away from data analysis! The chapter champions the use of the *reality check*—checking that observations of a particular parameter really have the properties that we know it must possess. Several case study datasets are introduced, including a hydrograph from the Neuse River (North Carolina, USA), air temperature from Black Rock Forest (New York), and chemical composition from the floor of the Atlantic Ocean.

Chapter 3, *Probability and What It Has to Do with Data Analysis,* is a review of probability theory. It develops the techniques that are needed to understand, quantify, and propagate measurement error. Two key themes introduced in this chapter and further developed throughout the book are that error is an unavoidable part of the measurement process and that error in measurement propagates through the analysis to affect the conclusions. Bayesian inference is introduced in this chapter as a way of assessing how new measurements improve our state of knowledge about the world.

Chapter 4, *The Power of Linear Models*, develops the theme that making inferences from data occurs when the data are distilled down to a few parameters in a quantitative model of a physical process. An integral part of the process of analyzing data is developing an appropriate quantitative model. Such a model links to the questions that one aspires to answer to the parameters upon which the model depends, and ultimately, to the data. We show that many quantitative models are *linear* in form and, thus, are very easy to formulate and manipulate using the techniques of linear algebra. The method of least squares, which provides a means of estimating model parameters from data, and a rule for propagating error are introduced in this chapter.

Chapter 5, *Quantifying Preconceptions*, argues that we usually know things about the systems that we are studying that can be used to supplement actual observations. Temperatures often lie in specific ranges governed by freezing and boiling points.

Chemical gradients often vary smoothly in space, owing to the process of diffusion. Energy and momentum obey conservation laws. The methodology through which this prior information can be incorporated into the models is developed in this chapter. Called generalized least squares, it is applied to several substantial examples in which prior information is used to fill in data gaps in datasets.

Chapter 6, *Detecting Periodicities*, is about spectral analysis, the procedures used to represent data as a superposition of sinusoidally varying components and to detect periodicities. The key concept is the Fourier series, a type of linear model in which the data are represented by a mixture of sinusoidally varying components. The chapter works to make the student completely comfortable with the Discrete Fourier Transform (DTF), the key algorithm used in studying periodicities. Theoretical analysis and a practical discussion of *MatLab's* DFT function are closely interwoven.

Chapter 7, *The Past Influences the Present*, focuses on using past behavior to predict the future. The key concept is the *filter*, a type of linear model that connects the past and present states of a system. Filters can be used both to quantify the physical processes that connect two related sets of measurements and to predict their future behavior. We develop the *prediction error filter* and apply it to hydrographic data, in order to explore the degree to which stream flow can be predicted. We show that the filter has many uses in addition to prediction; for instance, it can be used to explore the underlying processes that connect two related types of data.

Chapter 8, *Patterns Suggested by Data*, explores linear models that characterize data as a mixture of a few significant patterns, whose properties are determined by the data, themselves (as contrasted to being imposed by the analyst). The advantage to this approach is that the patterns are a distillation of the data that bring out features that reflect the physical processes of the system. The methodology, which goes by the names, *factor analysis* and *empirical orthogonal function (EOF)* analysis, is applied to a wide range of data types, including chemical analyses and images of sea surface temperature (SST). In the SST case, the strongest pattern is the El Niño climate oscillation, which brings immediate attention to an important instability in the ocean–atmosphere system.

Chapter 9, *Detecting Correlations Among Data*, develops techniques for quantifying correlations within datasets, and especially within and among time series. Several different manifestations of correlation are explored and linked together: from probability theory, covariance; from time series analysis, cross-correlation; and from spectral analysis, coherence. The effect of smoothing and band-pass filtering on the statistical properties of the data and its spectra is also discussed.

Chapter 10, *Filling in Missing Data*, discusses the interpolation of one and two dimensional data. Interpolation is shown to be yet another special case of the linear model. The relationship between interpolation and the gap-filling techniques developed in Chapter 5 are shown to be related to different approaches for implementing prior information about the properties of the data. Linear and spline interpolation, as well as kriging, are developed. Two-dimensional interpolation and Delaunay triangulation, a critical technique for organizing two-dimensional data, are explained. Two dimensional Fourier transforms, which are also important in many two-dimensional data analysis scenarios, are also discussed.

Chapter 11, *'Approximate' is not a pejorative word*, explores the use of approximations to make data analysis simpler, faster and more adaptable. Taylor's theorem is derived and used to linearize nonlinear functions. The method is used to create small-number approximations, to estimate the variance of nonlinear functions and to solve nonlinear least squares problems. The gradient of the error is identified as the key quantity controlling the solutions and its further application to the gradient method is explored. The lookup table is introduced as an approximate method of evaluating functions of one or two variables and applied to speeding up iterative calculations. Finally, the artificial neural network is developed as an approximation technique that shares the adaptability of a lookup table while providing smoothness and enhanced flexibility. The properties of several simple network designs are illustrated and the back-propagation algorithm for training them is derived. The power of a neural network is demonstrated by predicting the non-linear response of a river network to precipitation.

Chapter 12, *Are My Results Significant?*, returns to the issue of measurement error, now in terms of *hypothesis testing*. It concentrates on four important and very widely applicable statistical tests—those associated with the statistics, Z, χ^2, t, and F. Familiarity with them provides a very broad base for developing the practice of *always* assessing the significance of *any* inference made during a data analysis project. We also show how empirical distributions created by *bootstrapping* can be used to test the significance of results in more complicated cases.

Chapter 13, *Notes*, is a collection of technical notes that supplement the discussion in the main text.

William Menke
July, 2015

Advice on scripting for beginners

For many of you, this book will be your first exposure to scripting, the process of instructing *MatLab* what to do to your data. Although you will be learning something new, many other tasks of daily life will have already taught you relevant skills. Scripts are not so different than travel directions, cooking recipes, carpentry and building plans, and tailoring instructions. Each is in pursuit of a specific goal, a final product that has value to you. Each has a clear starting place and raw materials. And each requires a specific, and often lengthy, set of steps that need to be seen to completion in order to reach the goal. Put the skills that you have learned in these other arenas of life to use!

As a beginner, you should approach scripting as you would approach giving travel directions to a neighbor. Always focus on the goal. Where does the neighbor want to go? What analysis products do you want *MatLab* to produce for you? With a clear goal in mind, you will avoid the common pitfall of taking a drive that, while scenic, goes nowhere in particular. While *MatLab* can make pretty plots and interesting tables, you should not waste your valuable time creating any that does not support your goal.

When starting a scripting project, think about the information that you have. How did you get from point A to point B, the last time that you made the trip? Which turns should you point out to your neighbor as particularly tricky? Which aspects of the script are likely to be the hardest to get right? It is these parts on which you want to focus your efforts.

Consider the value of good landmarks. They let you know when you are on the right road (you will pass a firehouse about halfway) and when you have made the wrong turn (if you go over a bridge). And remember that the confidence-building value of landmarks is just as important as is error detection. You do not want your neighbor to turn back, just because the road seems longer than expected. Your *MatLab* scripts should contain landmarks, too. Any sort of output, such as a plot, that enables you to judge whether or not a section of a script is working is critical. You do not want to spend time debugging a section of your script that already works. Make sure that every script that you write has landmarks.

Scripts relieve you from the tedium of repetitive data analysis tasks. A finished script is something in which you can take pride, for it is a tool that has the potential for helping you in your work for years to come.

Joshua Menke
February, 2011

1 Data analysis with *MatLab*

1.1   Why *MatLab*?

Data analysis requires computer-based computation. While a person can learn much of the *theory* of data analysis by working through short pencil-and-paper examples, he or she cannot become proficient in the *practice* of data analysis that way—for reasons both good and bad. Real datasets, which are almost always too large to handle manually, are inherently richer and more interesting than stripped-down examples. They have more to offer, but an expanded skill set is required to successfully tackle them. In particular, a new kind of judgment is required for selecting the analysis technique that is right for the problem at hand. These are good reasons. Unfortunately, the practice of data analysis is littered with bad reasons, too, most of which are related to the very steep learning curve associated with using computers. Many practitioners of data analysis find that they spend rather too many frustrating hours solving computer-related problems that have very little to do with data analysis, *per se*. That's bad, especially in a classroom setting where time is limited and where frustration gets in the way of learning.

One approach to dealing with this problem is to conduct all the data analysis within a single software environment—to *limit the damage*. Frustrating software problems will still arise, but fewer than if data were being shuffled between several different environments. Furthermore, in a group setting such as a classroom, the memory and experience of the group can help individuals solve commonly encountered problems. The trick is to select a single software environment that is capable of supporting *real* data analysis.

The key decision is whether to go with a spreadsheet or a scripting language-type software environment. Both are viable environments for computer-based data analysis. Stable implementations of both are available for most types of computers from commercial software developers at relatively modest prices (and especially for those eligible for student discounts). Both provide support for the data analysis itself, as well as associated tasks such as loading and writing data to and from files and plotting them on graphs. Spreadsheets and scripting languages are radically different in approach, and each has advantages and disadvantages.

In a spreadsheet-type environment, typified by *Microsoft Excel*, data are presented as one or more *tables*. Data are manipulated by selecting the rows and columns of a table and operating on them with functions selected from a menu and with formulas entered into the cells of the table itself. The immediacy of a spreadsheet is both its greatest advantage and its weakness. You see the data and all the intermediate results as you manipulate the table. You are, in a sense, touching the data, which gives you a great sense of what the data are like. More of a problem, however, is keeping track of what you did in a spreadsheet-type environment, as is transferring useful procedures from one spreadsheet-based dataset to another.

In a scripting language, typified by *The MathWorks MatLab*, data are presented as one or more *named variables* (in the same sense that the "c" and "d" in the formula $c = \pi d$ are named variables). Data are manipulated by typing formulas that create new variables from old ones and by running *scripts*, that is, sequences of formulas stored in a file. Much of data analysis is simply the application of well-known formulas to novel data, so the great advantage of this approach is that the formulas that you type usually have a strong similarity to those printed in a textbook. Furthermore, scripts provide a way of both documenting the sequence of formulas used to analyze a particular dataset and transferring the overall data analysis procedure from one dataset to another. The main disadvantage of a scripting language environment is that it hides the data within the variable—not absolutely, but a conscious effort is nonetheless needed to display it as a table or as a graph. Things can go badly wrong in a script-based data analysis scheme without the practitioner being aware of it. Another disadvantage is that the parallel between the syntax of the scripting language and the syntax of standard mathematical notation is nowhere near perfect. One needs to learn to translate one into the other.

While both spreadsheets and scripting languages have *pros* and *cons*, our opinion is that, on balance, a scripting language wins out, at least for the data analysis scenarios encountered in Environmental Science. In our experience, these scenarios often require a long sequence of data manipulation steps before a final result is achieved. Here, the self-documenting aspect of the script is paramount. It allows the practitioner to review the data processing procedure both as it is being developed and years after it has been completed. It provides a way of communicating *what you did*, a process that is at the heart of science.

We have chosen *MatLab*, a commercial software product of *The MathWorks, Inc.* as our preferred software environment for several reasons, some having to do with its designs and others more practical. The most persuasive design reason is that its syntax fully supports both linear algebra and complex arithmetic, both of which are important in data analysis. Practical considerations include the following: it is a long-lived and stable product, available since the mid 1980s; implementations are available for most commonly used types of computers; its price, especially for students, is fairly modest; and it is widely used, at least in university settings.

1.2 Getting started with *MatLab*

We cannot walk you through the installation of *MatLab*, for procedures vary from computer to computer and quickly become outdated, anyway. Furthermore, we will avoid discussion of the appearance of *MatLab* on your computer screen, because its Graphical User Interface has evolved significantly over the years and can be expected to continue to do so. We will assume that you have successfully installed *MatLab* and that you can identify the Command Window, the place where *MatLab* formula and commands are typed.

You might try typing

```
date
```

in this window. If *MatLab* responds by displaying the current date, you're on track!

All the *MatLab* commands that we use are in *MatLab* scripts that are provided as a companion to this book. This one is named `eda01_01` and is in a *MatLab* script file (*m-file*, for short) named `eda01_01.m` (conventionally, m-files have file names that end with ".m"). In this case, the script is pretty boring, as it contains just this one command, `date`, together with a couple of comment lines (which start with the character "%"):

```
% eda01_01
% displays the current date
date                                    (MatLab eda01_01)
```

After you install *MatLab*, you should copy the `eda` folder, provided with this book, to your computer's file system. Put it in some convenient and easy-to-remember place that you are not going to accidentally delete!

1.3 Getting organized

Files proliferate at an astonishing rate, even in the most trivial data management project. You start with a file of data, but then write m-scripts, each of which has its own file. You will usually output final results of the data analysis to a file, and you may well output intermediate results to files, too. You will probably have files containing graphs and files containing notes as well. Furthermore, you might decide to analyze

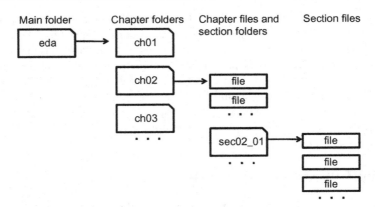

Figure 1.1 Folder (directory) structure used for the files accompanying this book.

your data in several different ways, so you may have several versions of some of these files.

A practitioner of data analysis may find that a little organization up front saves quite a bit of confusion down the line.

As data analysis scenarios vary widely, there can be no set rule regarding organization of the files associated with them. The goal should simply be to create a system of folders (directories), subfolders (sub-directories), and file names that are sufficiently systematic so that files can be located easily and they are not confused with one another. Predictability in both the pattern of filenames and in the arrangement of folders and subfolders is an extremely important part of the design.

By way of example, the files associated with this book are in a three-tiered folder/subfolder structure modeled on the chapter and section format of the book itself (Figure 1.1). Most of the files, such as the m-files, are in the chapter folders. However, some chapters have longish case studies that use a large number of files, and in those instances, section folders are used. Folder and file names are systematic. The chapter folder names are always of the form chNN, where NN is the chapter number. The section folder names are always of the form secNN_MM, where NN is the chapter number and MM is the section number. We have chosen to use leading zeros in the naming scheme (for example, ch01) so that filenames appear in the correct order when they are sorted alphabetically (as when listing the contents of a folder).

1.4 Navigating folders

The *MatLab* command window supports a number of commands that enable you to navigate from folder to folder, list the contents of folders, and so on. For example, when you type

```
pwd
```

(for "print working directory") in the Command Window, *MatLab* responds by displaying the name of the current folder. Initially, this is almost invariably the wrong folder, so you will need to `cd` (for "change directory") to the folder where you want to be—the `ch01` folder in this case. The pathname will, of course, depend on where you copied the `eda` folder, but will end in `eda/ch01`. On our computer, typing

```
cd c:/menke/docs/eda/ch01
```

does the trick. If you have spaces in your pathname, just surround it with single quotes:

```
cd 'c:/menke/my docs/eda/ch01'
```

You can check if you are in the right folder by typing `pwd` again. Once in the `ch01` folder, typing

```
eda01_01
```

will run the `eda01_01` m-script, which displays the current date. You can move to the folder above the current one by typing

```
cd ..
```

and to one below it by giving just the folder name. For example, if you are in the `eda` folder you can move to the `ch01` folder by typing

```
cd ch01
```

Finally, the command `dir` (for "directory"), lists the files and subfolders in the current directory.

```
dir
```
                                                                              (*MatLab* eda01_02)

1.5 Simple arithmetic and algebra

The *MatLab* commands for simple arithmetic and algebra closely parallel standard mathematical notation. For instance, the command sequence

```
a = 3.5;
b = 4.1;
c = a+b;
c
```
                                                                              (*MatLab* eda01_03)

evaluates the formula $c = a + b$ for the case $a = 3.5$ and $b = 4.1$ to obtain $c = 7.6$. Only the semicolons require explanation. By default, *MatLab* displays the value of every formula typed into the Command Window. A semicolon at the end of the formula suppresses the display. Hence, the first three lines, which end with semicolons, are evaluated but not displayed. Only the final line, which lacks the semi-colon, causes *MatLab* to print the final result, c.

A perceptive reader might have noticed that the m-script could have been made shorter by one line, simply by omitting the semicolon in the formula, c = a + b. That is,

```
a = 3.5;
b = 4.1;
c = a+b
```

However, we recommend *against* such cleverness. The reason is that many intermediate results will need to be temporarily displayed and then un-displayed in the process of developing and debugging a long m-script. When this is accomplished by adding and then deleting the semicolon at the end of a functioning—and important—formula in the script, the formula can be inadvertently damaged by deleting one or more extra characters. Editing a line of the code that has no function other than displaying a value is *safer*.

Note that *MatLab* variables are *static*, meaning that they persist in *MatLab's Workspace* until you explicitly delete them or exit the program. Variables created by one script can be used by subsequent scripts. At any time, the value of a variable can be examined, either by displaying it in the Command Window (as we have done above) or by using the spreadsheet-like display tools available through *MatLab's* Workspace Window. The persistence of *MatLab* variables can sometimes lead to scripting errors, as described in Note 1.1.

The four commands discussed above can be run as a unit by typing `eda01_03`. Now open the m-file `eda01_03` in *MatLab*, using the File/Open menu. *MatLab* will bring up a text-editor type window. First save it as a new file, say `myeda01_03`, edit it in some simple way, say by changing the `3.5` to `4.5`, save the edited file, and run it by typing `myeda01_03` in the Command Window. The value of `c` that is displayed will have changed appropriately.

A somewhat more complicated *MatLab* formula is

$$c = \sqrt{a^2 + b^2} \quad \text{with } a = 3 \text{ and } b = 4$$

```
a = 3;
b = 4;
c = sqrt(a^2 + b^2);
c
```
                                                                        (*MatLab* eda01_04)

Note that the *MatLab* syntax for a^2 is `a^2` and that the square root is computed using the function, `sqrt()`. This is an example of *MatLab's* syntax differing from standard mathematical notation.

A final example is

$$c = \sin\frac{n\pi(x - x_0)}{L} \quad \text{with } n = 2, \ x = 3, \ x_0 = 1, \ L = 5$$

```
n = 2;  x = 3;  x0 = 1;  L = 5;
c = sin(n*pi*(x-x0)/L);
c
```
                                                                        (*MatLab* eda01_05)

Note that several formulas separated by semicolons can be typed on the same line. Variables, such as `x0` and `pi` above, can have names consisting of more than one character, and can contain numerals as well as letters (although they must start with a letter). *MatLab* has a variety of predefined mathematical constants, including `pi`, which is the usual mathematical constant, π.

1.6 Vectors and matrices

Vectors and matrices are fundamental to data analysis both because they provide a convenient way to organize data and because many important operations on data can be very succinctly expressed using *linear algebra* (that is, the algebra of vectors and matrices).

Vectors and matrices are very easy to define in *MatLab*. For instance, the quantities

$$\mathbf{r} = [2 \quad 4 \quad 6] \quad \text{and} \quad \mathbf{c} = \begin{bmatrix} 1 \\ 3 \\ 5 \end{bmatrix} = [1 \quad 3 \quad 5]^T \quad \text{and} \quad \mathbf{M} = \begin{bmatrix} 1 & 2 & 3 \\ 4 & 5 & 6 \\ 7 & 8 & 9 \end{bmatrix}$$

are defined with the following commands:

```
r = [2, 4, 6];
c = [1, 3, 5]';
M = [ [1, 4, 7]', [2, 5, 8]', [3, 6, 9]' ];        (MatLab eda01_06)
```

Note that the column-vector, c, is created by first defining a row vector, [1, 3, 5], and then converting it to a column vector by taking its transform, which in *MatLab* is indicated by a single quote. Note, also, that the matrix, **M**, is being constructed from a "row vector of column vectors".

Although *MatLab* allows both column-vectors and row-vectors to be defined with ease, our experience is that using both creates serious opportunities for error. A formula that requires a column-vector will usually yield incorrect results if a row-vector is substituted into it, and vice-versa. Consequently, we adhere to a protocol where all vectors defined in this book are column vectors. Row vectors are created when needed—and as close as possible to where they are used in the script—by transposing the equivalent column vector. We also adhere to the convention that vectors have lower-case names and matrices have upper-case names (or, at least, names that start with an upper-case letter).

1.7 Multiplication of vectors of matrices

MatLab performs all multiplicative operations with ease. For example, suppose column vectors **a** and **b**, and matrices **M** and **N** are defined as

$$\mathbf{a} = \begin{bmatrix} 1 \\ 3 \\ 5 \end{bmatrix} \quad \text{and} \quad \mathbf{b} = \begin{bmatrix} 2 \\ 4 \\ 6 \end{bmatrix} \quad \text{and} \quad \mathbf{M} = \begin{bmatrix} 1 & 0 & 2 \\ 0 & 1 & 0 \\ 2 & 0 & 1 \end{bmatrix} \quad \text{and} \quad \mathbf{N} = \begin{bmatrix} 1 & 0 & -1 \\ 0 & 2 & 0 \\ -1 & 0 & 3 \end{bmatrix}$$

Then,

$$s = \mathbf{a}^T \mathbf{b} = \begin{bmatrix} 1 \\ 3 \\ 5 \end{bmatrix}^T \begin{bmatrix} 2 \\ 4 \\ 6 \end{bmatrix} = [1 \quad 3 \quad 5] \begin{bmatrix} 2 \\ 4 \\ 6 \end{bmatrix} = 2 \times 1 + 3 \times 4 + 5 \times 6 = 44$$

$$\mathbf{T} = \mathbf{ab}^T = \begin{bmatrix} 1 \\ 3 \\ 5 \end{bmatrix} \begin{bmatrix} 2 \\ 4 \\ 6 \end{bmatrix}^T = \begin{bmatrix} 2 \times 1 & 4 \times 1 & 6 \times 1 \\ 2 \times 3 & 4 \times 3 & 6 \times 3 \\ 2 \times 5 & 4 \times 5 & 6 \times 5 \end{bmatrix} = \begin{bmatrix} 2 & 4 & 6 \\ 6 & 12 & 18 \\ 10 & 20 & 30 \end{bmatrix}$$

$$\mathbf{c} = \mathbf{Ma} = \begin{bmatrix} 1 & 0 & 2 \\ 0 & 1 & 0 \\ 2 & 0 & 1 \end{bmatrix} \begin{bmatrix} 1 \\ 3 \\ 5 \end{bmatrix} = \begin{bmatrix} 1 \times 1 & + & 0 \times 3 & + & 2 \times 5 \\ 0 \times 1 & + & 1 \times 3 & + & 0 \times 5 \\ 2 \times 1 & + & 0 \times 3 & + & 1 \times 5 \end{bmatrix} = \begin{bmatrix} 11 \\ 3 \\ 7 \end{bmatrix}$$

$$\mathbf{P} = \mathbf{MN} = \begin{bmatrix} 1 & 0 & 2 \\ 0 & 1 & 0 \\ 2 & 0 & 1 \end{bmatrix} \begin{bmatrix} 1 & 0 & -1 \\ 0 & 2 & 0 \\ -1 & 0 & 3 \end{bmatrix} = \begin{bmatrix} -1 & 0 & 5 \\ 0 & 2 & 0 \\ 1 & 0 & 1 \end{bmatrix}$$

corresponds to

```
s = a'*b;
T = a*b';
c = M*a;
P = M*N;
```
                     (*MatLab* eda01_07)

In *MatLab*, standard vector and matrix multiplication is performed just by using the normal multiplications sign, * (the asterisk). There are cases, however, where one needs to violate these rules and multiply the quantities element-wise (for example, create a vector, **d**, with elements $d_i = a_i b_i$). *MatLab* provides a special element-wise version of the multiplication sign, denoted .* (a period followed by an asterisk):

```
d = a.*b;
```
                     (*MatLab* eda01_07)

1.8 Element access

Individual elements of vectors and matrices can be accessed by specifying the relevant row and column indices in parentheses; for example, a(2) is the second element of the column vector **a** and M(2,3) is the second row, third column element of the matrix, **M**. Ranges of rows and columns can be specified using the : operator; for example, M(:,2) is the second column of matrix, **M**, M(2,:) is the second row of matrix, **M**, and M(2:3,2:3) is the 2×2 submatrix in the lower right-hand corner of the 3×3 matrix, **M** (the expression, M(2:end,2:end), would work as well). These operations are further illustrated below:

$$\mathbf{a} = \begin{bmatrix} 1 \\ 2 \\ 3 \end{bmatrix} \quad \text{and} \quad \mathbf{M} = \begin{bmatrix} 1 & 2 & 3 \\ 4 & 5 & 6 \\ 7 & 8 & 9 \end{bmatrix}$$

$$s = a_2 = 2 \quad \text{and} \quad t = M_{23} = 6 \quad \text{and} \quad \mathbf{b} = \begin{bmatrix} M_{12} \\ M_{22} \\ M_{32} \end{bmatrix} = \begin{bmatrix} 2 \\ 5 \\ 8 \end{bmatrix}$$

$$\mathbf{c} = \begin{bmatrix} M_{21} & M_{22} & M_{23} \end{bmatrix}^T = \begin{bmatrix} 4 \\ 5 \\ 6 \end{bmatrix} \quad \text{and} \quad \mathbf{T} = \begin{bmatrix} M_{22} & M_{23} \\ M_{32} & M_{33} \end{bmatrix} = \begin{bmatrix} 5 & 6 \\ 8 & 9 \end{bmatrix}$$

correspond to:

```
s = a(2);
t = M(2,3);
b = M(:,2);
c = M(2,:)';
T = M(2:3,2:3);                                            (MatLab eda01_08)
```

The colon notion can be used in other contexts as well. For instance, [1:4] is the row vector [1, 2, 3, 4]. The syntax, 1:4, which omits the square brackets, works fine in *MatLab*. However, we usually use square brackets, as they draw attention to the presence of a vector. Finally, we note that two colons can be used in sequence to indicate the spacing of elements in the resulting vector. For example, the expression [1:2:9] is the row vector [1, 3, 5, 7, 9] and the expression [10:-1:1] is a row vector whose elements are in the reverse order from [10:1].

1.9   Representing functions

A vector can be used to represent an arbitrary function $x(t)$. First one prepares a time vector **t**, which consists of N values of time, running from some minimum value t_{min} to some maximum value t_{max}, with even spacing $\Delta t = (t_{max} - t_{min})/(N - 1)$. Next, one prepares a vector **x** that gives the value of the function at these times (Figure 1.2). The *MatLab* code below is for the exemplary function $x(t) = \sin(\pi t)$:

```
% independent variable t
N=21;
tmin=0;
tmax=1;
Dt=(tmax-tmin)/(N-1);
t=tmin+Dt*[0:N-1]';

% exemplary function
x=sin(pi*t);                                              (MatLab eda01_09)
```

Many operations on functions are especially easy to peform when they are represented in this fashion, which is called a *time series*. For instance, the slope (first derivative)

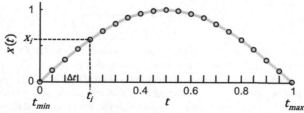

Figure 1.2 Time series representation of a function $x(t)$. The time axis is divided into N intervals of length Δt between t_{min} and t_{max}. The time vector **t** has elements $t_i = (i - 1)\Delta t$ and the corresponding time series vector **x** corresponds to the values of the function at these times. It has elements $x_i = x(t_i)$. *MatLab* script eda01_09.

$s(t) = dx/dt$ can be approximated by the slope of the line segments connecting adjacent points in the time series (Figure 1.3 A):

$$s(t_0) = \frac{dx}{dt}\bigg|_{x=x_0} \approx \frac{x(t_0 + \Delta t) - x(t_0)}{\Delta t}$$

A time series **s** of slopes (Figure 1.3 B) is calculated as:

```
sapprox=(x(2:N)-x(1:N-1))/Dt;                   (MatLab eda01_09)
```

Note that the vector **s** is of length $N-1$, since no value of the function is available for time $t_{max} + \Delta t$.

The area (integral) $a(t_0) = \int_0^{t_0} x(t)\,dt$ can be computed using the Riemann Summation approximation of the integral:

$$a(t_0) = \int_0^{t_0} x(t)\,dt \approx \Delta t \sum_{i=1}^{K} x(t_i) \quad \text{with} \quad K = t_0/\Delta t$$

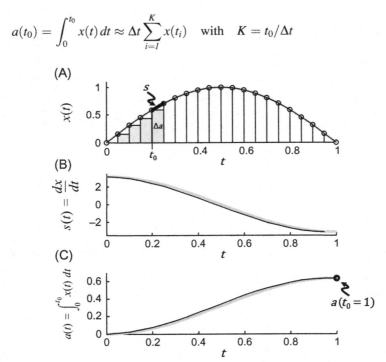

Figure 1.3 Approximations for the derivative and integral of a function $x(t)$. (A) The smooth function $x(t)$ (gray curve) is represented as a time series; that is, by its values (circles) at a sequence of equally-spaced values of t (vertical bars). The derivative $s(t_0)$ is approximated as the slope of a line segment (bold) connecting two adjacent values, the leftmost of which is at t_0. The integral $a(t_0)$ is approximated as the sum of the areas Δa of all the rectangles subtending the curve up to the position t_0 (shaded). (B) The resulting approximation for the derivative (black curve) closely approximates $s(t)$ (grey curve). (C) The resulting approximation for the integral (black curve) closely approximates $A(t)$ (grey curve). The value of the integral $a(t_0 = 1)$ at the right hand end of the interval is shown (circle). *MatLab* script eda01_09.

This approximation corresponds to adding up the areas of rectangles of width Δt that extend below the function, from the leftmost rectangle to the one at position t_0 (Figure 1.3 A). A time series **a** of areas (Figure 1.3 B) is calculated as:

```
aapprox = Dt*cumsum(x);
```
(MatLab eda01_09)

The *MatLab* function cumsum(x) returns the cumulative sum (running sum) of the elements of the vector x. The last element of **a** corresponds to the total area under the curve. It can be calculated more simiply as (Figure 1.3 C):

```
atotal = Dt*sum(x);
```
(MatLab eda01_09)

The *MatLab* function sum(x) returns the sum of elements of the vector x.

1.10   To loop or not to loop

MatLab provides a looping mechanism, the for command, which can be useful when the need arises to sequentially access the elements of vectors and matrices. Thus, for example,

```
M = [ [1, 4, 7]', [2, 5, 8]', [3, 6, 9]' ];
for i = [1:3]
    a(i) = M(i,i);
end
```
(MatLab eda01_10)

executes the a(i) = M(i,i) formula three times, each time with a different value of i (in this case, i = 1, i = 2, and i = 3). The net effect is to copy the diagonal elements of the matrix **M** to the vector, **a**, that is, $a_i = M_{ii}$. Note that the end statement indicates the position of the bottom of the loop. Subsequent commands are not part of the loop and are executed only once.

Loops can be nested; that is, one loop can be inside another. Such an arrangement is necessary for accessing all the elements of a matrix in sequence. For example,

```
M = [ [1, 4, 7]', [2, 5, 8]', [3, 6, 9]' ];
for i = [1:3]
for j = [1:3]
    N(i,4-j) = M(i,j);
end
end
```
(MatLab eda01_11)

copies the elements of the matrix, **M**, to the matrix, **N**, but reverses the order of the elements in each row; that is, $N_{i,j} = M_{i,4-j}$. Loops are especially useful in conjunction with *conditional* commands. For example

```
a = [ 1, 2, 1, 4, 3, 2, 6, 4, 9, 2, 1, 4 ]';
for i = [1:12]
    if ( a(i) >= 6 )
        b(i) = 6;
    else
```

```
        b(i) = a(i);
    end
end
```
(MatLab eda01_12)

sets $b_i = a_i$ if $a_i < 6$ and sets $b_i = 6$ otherwise (a process called *clipping* a vector, for it lops off parts of the vector that are larger than 6).

A purist might point out that *MatLab* syntax is so flexible that for loops are almost never really necessary. In fact, all three examples, above, can be computed with one-line formulas that omit for loops:

```
a = diag(M);
N = fliplr(M);
b = a.*(a<6)+6.*(a>=6);
```
(MatLab eda01_13)

The first two formulas are quite simple, but rely on the *MatLab* functions diag() and fliplr()whose existence we have not heretofore mentioned. One of the problems of a script-based environment is that learning the complete syntax of the scripting language can be pretty daunting. Writing a long script, such as one containing a for loop, will often be faster than searching through *MatLab* help files for a predefined function that implements the desired functionality in a single line of the script. The third formula points out a different problem: *MatLab* syntax is often pretty inscrutable. In this case, the expression (a<6) creates a column-vector of ones and zeros, depending on whether a given element of **a** is less-than or greater-than-or-equal-to 6. Element-wise multiplication is then used to create a vector a.*(a<6) whose elements are either a_i or 0. Similarly, 6.*(a>=6) is a vector whose elements are either 0 or 6. Their sum is a vector whose elements are either a_i or 6, depending on whether a_i is less-than or greater-than-or-equal-to 6. That's pretty complicated!

Because *MatLab's* syntax is so powerful, the same functionality can often be achieved in several different ways. Thus, for example, the commands

```
b = a;
b(find(a>6)) = 6;
```
(MatLab eda01_13)

will also clip the elements of the vector. The find() function returns a column-vector of the *indices* of the vector, a, that match the condition, and then that list is used to reset just those elements of b to 6, leaving the other elements unchanged.

When deciding between alternative ways of implementing a given functionality, you should always choose the one which *you* find clearest. Scripts that are terse or even computationally efficient are not necessarily a virtue, especially if they are difficult to debug. You should avoid creating formulas that are so inscrutable that you are not sure whether they will function correctly. Of course, the degree of inscrutability of any given formula will depend on your level of familiarity with *MatLab*. Your repertoire of techniques will grow as you become more practiced.

1.11 The matrix inverse

Recall that the matrix inverse is defined only for square matrices, and that it has the following properties:

$$\mathbf{A}^{-1}\mathbf{A} = \mathbf{A}\mathbf{A}^{-1} = \mathbf{I} \tag{1.1}$$

Here, **I** is the identity matrix, that is, a matrix with ones on its main diagonal and zeros elsewhere. In *MatLab*, the matrix inverse is computed as

```
B = inv(A);
```
(MatLab eda01_14)

In many of the formulas of data analysis, the matrix inverse either premultiplies or postmultiplies other quantities; for instance,

$$\mathbf{c} = \mathbf{A}^{-1}\mathbf{b} \quad \text{and} \quad \mathbf{D} = \mathbf{B}\mathbf{A}^{-1}$$

These cases do not actually require the explicit calculation of $\mathbf{A}^{-1}$; just the combinations $\mathbf{A}^{-1}\mathbf{b}$ and $\mathbf{B}\mathbf{A}^{-1}$, which are computationally simpler are sufficient. *MatLab* provides generalizations of the division operator that implements these two cases:

```
c = A\b;
D = B/A;
```
(MatLab eda01_15)

1.12 Loading data from a file

MatLab can read and write files with a variety of formats, but we start here with the simplest and most common one, the text file.

As an example, we load a hydrological dataset of stream flow from the Neuse River near Goldsboro NC. Our recommendation is that you always keep a file of notes about any dataset that you work with, and that these notes include information on where you obtained the dataset and any modifications that you subsequently made to it. Bill Menke provides the following notes for this one (Figure 1.4):

> *I downloaded stream flow data from the US Geological Survey's National Water Informatiuon Center for the Neuse River near Goldboro NC for the time period, 01/01/1974-12/31/1985. These data are in the file,* neuse.txt. *It contains two columns of data, time (in days starting on January 1, 1974) and discharge (in cubic feet per second, cfs). The data set contains 4383 rows of data. I also saved information about the data in the* file neuse_header.txt.

We reproduce the first few lines of neuse.txt, here:

```
1            1450
2            2490
3            3250
....         ....
```

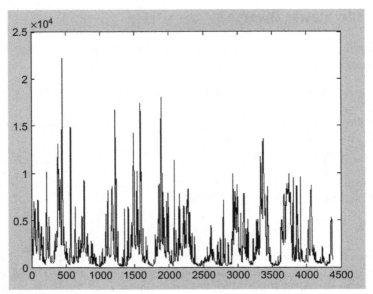

Figure 1.4 Preliminary plot of the Neuse River discharge dataset. *MatLab* script eda01_18.

The data is read into *MatLab* as follows:

```
D = load('neuse.txt');
t = D(:,1);
d = D(:,2);                                              (MatLab eda01_16)
```

The load() function reads the data into a 4383 × 2 array, **D**. Note that the filename, neuse.txt, needs to be surrounded by single quotes to indicate that it is a *character string* and not a variable name. The subsequent two lines break out **D** into two separate column-vectors, **t**, of time and **d**, of discharge. Strictly speaking, this step is not necessary, but our opinion is that fewer mistakes will be made if each of the different variables in the dataset has its own name.

1.13 Plotting data

One of the best things to do after loading a new dataset is to make a quick plot of it, just to get a sense of what it looks like. Such plots are very easily created in *MatLab*:

```
plot(t,d);
```

The resulting plot is quite functional, but lacks some graphical niceties such as labeled axes and a title. These deficiencies are easy to correct:

```
set(gca,'LineWidth',2);
plot(t,d,'k-','LineWidth',2);
```

```
title('Neuse River Hydrograph');
xlabel('time in days');
ylabel('discharge in cfs');                          (MatLab eda01_17)
```

The set command resets the line width of the axes, to make them easier to see. Several new arguments have been added to the plot() function. The 'k-' changes the plot color from its default value (a blue line, at least on our computer) to a black line. The 'LineWidth', 2 makes the line thicker (which is important if you print the plot to paper). A quick review of the plot indicates that the Neuse River discharge has some interesting properties, such as pattern of highs and lows that repeat every few hundred days. We will discuss it more extensively in Chapter 2.

Data can span a very large range of values and in some cases the difference bewteen the sizes of the small values is every bit as important as the difference between the sizes of the larger values. The simple, *linear* plot that was descibed above is ineffective in such a case, because small values are squeezed into one tiny corner of the graph. A *logarithmic* plot is preferred, because its axes give the same space to each order of magnitude of values. The $10 - 100$ decade, for example, is given the same prominence on the graph as the $100 - 1,000$ decade (Figure 1.5).

As an example, we consider an earthquake dataset in which the strength of each earthquake is characterized by its seismic energy:

> *I downloaded earthquake data from the US Geological Survey's earthquake database for the time period 01/01/2000-12/31/2010 and for the 5–10 magnitude range. The resulting dataset contains 13,258 earthquakes, each of which is described by 15 parameters, including its seismic magnitude. I then created a separate file containing just the list of the magnitudes, ordered chronologically. I converted these magnitudes to energy using the Gutenberg-Richter Energy-Magnitude Formula (Gutenberg, B. and C.F. Richter, Magnitude and energy of earthquakes, Ann. Geofis., 9, 1-15, 1956).*

The earthquakes span a large range of energies, from about 2×10^5 to 6×10^{12} joules. The number of earrthquakes in a given energy range varies widely, too. Several hundred of the least energetic earthquakes occur per yea, in contrast to only a few of the most energetic. The logarithmic plot (Figure 1.6) reveals an interesting pattern: the rate of occurrence r of earthquakes decrease systematically with energy, making approximately a straight line on the logarithmic plot. This pattern implies the power law $r = aE^{-b}$ where a and b are constants, since then $\log_{10}r = \log_{10}a + b\log_{10}E$ is a linear function of $\log_{10}E$.

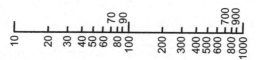

Figure 1.5 Exemplary horizontal axis on a log-log plot. Note that the each decade is the same length and that the distance between the minor tic marks is variable.

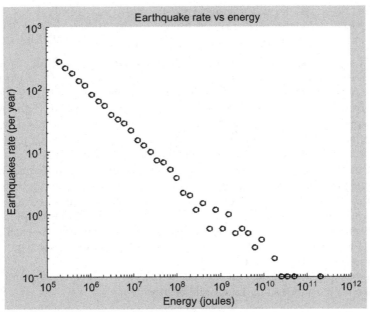

Figure 1.6 Earthquake rate, in number of events per year, plotted against their enegy, in joules. *MatLab* script eda01_19.

A log-log plot is easy to create in *MatLab*:

```
set(gca,'Xscale','log','Yscale','log');
hold on;
plot( E, r, 'ko', 'LineWidth', 2);
xlabel('energy (joules)');
ylabel('earthquakes rate (per year)');
title('Earthquake rate vs energy');        (MatLab eda01_18)
```

The set command specifies that both the *x* and *y* axes are logarithmic. The hold on command then prevents subsequent commands from overriding these settings. The plot command works the same way as described previously, except that we have set the the plot symbol to black circles with the 'ko' parameter and the boldness of the lines to 2 points with the 'LineWidth' parameter.

1.14   Saving data to a file

Data is saved to a to text file in a process that is more or less the reverse of the one used to read it. Suppose, for instance, that we want a version of neuse.txt that contains discharge in the metric units of m^3/s. After looking up the conversion factor, $f = 35.3146$, between cubic feet and cubic meters, we perform the conversion and write the data to a new file:

```
f = 35.3146;
dm = d/f;
Dm(:,1) = t;
Dm(:,2) = dm;
dlmwrite('neuse_metric.txt',Dm,'\t');      (MatLab eda01_19)
```

The function `dlmwrite()` (for "delimited write") writes the matrix, Dm, to the file `neuse_metric.txt`, putting a tab character (which in *MatLab* is represented with the symbol, `\t`) between the columns as a *delimiter*. Note that the filename and the delimiter are quoted; they are character strings.

1.15   Some advice on writing scripts

Practices that reduce the likelihood of scripting mistakes ("bugs") are almost always worthwhile, even though they may seem to slow you down a bit. They save time in the long run, as you will spend much less time debugging your scripts.

1.15.1   Think before you type

Think about what you want to do before starting to type in a script. Block out the necessary steps on a piece of scratch paper. Without some forethought, you can type for an hour and then realize that what you have been doing makes no sense at all.

1.15.2   Name variables consistently

MatLab automatically creates a new variable whenever you type a new variable name. That is convenient, but it means that a misspelled variable becomes a new variable. For instance, if you begin calling a quantity `xmin` but accidentally switch to `minx` halfway through, you will unknowingly have two variables in your script, and it will not function correctly. Do not tempt fate by creating two variables, such as `xmin` and `miny`, with an inconsistent naming pattern.

1.15.3   Save old scripts

Cannibalize an old script to make a new one, but keep a copy of the old one too, and make sure that the names are sufficiently different so that you will not confuse them with each other.

1.15.4   *Cut and paste sparingly*

Cutting and pasting segments of code from one script to another, tempting though it may be, is especially prone to error, particularly when variable names need to be changed. Read through the cut-and-pasted material carefully, to make sure that all necessary changes have been made.

1.15.5   *Start small*

Build scripts in small sections. Test each section thoroughly before going into the next. Check intermediate results, either by displaying variables to the Command Window, examining them with the spreadsheet tool in the Workspace Window, or by plotting them, to ensure that they *look right*.

1.15.6   *Test your scripts*

Build test datasets with known properties to test whether or not your scripts give the right answers. Test a script on a small, simple dataset before running it on large complicated datasets.

1.15.7   *Comment your scripts*

Use comments to communicate the big picture, not the minutia. Consider the two scripts in Figure 1.7. Which of the two styles of commenting code do you suppose will make a script easier to understand 2 years down the line?

1.15.8   *Don't be too clever*

An inscrutable script is *very* prone to error.

Case A

```
% Evaluate Normal distribution
% with mean dbar and covariance, C
CI=inv(C);
norm=1/(2*pi*sqrt(det(C)));
Pp=zeros(L,L);
for i = [1:L]
for j = [1:L]
dd = [d1(i)-d1bar, d2(j)-d2bar]';
Pp(i,j)=norm*exp( -0.5 * dd' * CI* dd );
end
end
```

Case B

```
CI=inv(C); % take inverse of C
norm=1/(2*pi*sqrt(det(C))); % compute norm
Pp=zeros(L,L);
% loop over i
for i = [1:L]
% loop over j
for j = [1:L]
% dd is a 2-vector
dd = [d1(i)-d1bar, d2(j)-d2bar]';
% compute exponential
Pp(i,j)=norm*exp( -0.5 * dd' * CI* dd );
end
end
```

Figure 1.7 The same script, commented in two different ways.

Problems

1.1 Write *MatLab* scripts to evaluate the following equations:

(A) $y = ax^2 + bx + c$ with $a = 2, b = 4, c = 8, x = 3.5$

(B) $p = p_0 \exp(-cx)$ with $p_0 = 1.6, c = 4, x = 3.5$

(C) $z = h \sin\theta$ with $h = 4, \theta = 31°$

(D) $v = \pi h r^2$ with $h = 6.9, r = 3.7$

1.2 Write a *MatLab* script that defines a column vector, **a**, of length $N = 12$ whose elements are the number of days in the 12 months of the year, for a nonleap year. Create a similar column vector, **b**, for a leap year. Then merge **a** and **b** together into an $N \times M = 12 \times 2$ matrix, **C**.

1.3 Write a *MatLab* script that solves the following linear equation, $\mathbf{y} = \mathbf{M}\,\mathbf{x}$, for **x**:

$$\mathbf{M} = \begin{bmatrix} 1 & -1 & 0 & 0 \\ 0 & 1 & -1 & 0 \\ 0 & 0 & 1 & -1 \\ 0 & 0 & 0 & 1 \end{bmatrix} \quad \text{and} \quad \mathbf{y} = \begin{bmatrix} 1 \\ 2 \\ 3 \\ 5 \end{bmatrix}$$

You may find useful the *MatLab* function `zeros(N,N)`, which creates a $N \times N$ matrix of zeros. Be sure to check that your **x** solves the original equation.

1.4 Create a 50×50 version of the **M**, above. One possibility is to use a `for` loop. Another is to use the *MatLab* function `toeplitz()`, as **M** has the form of a *Toeplitz* matrix, that is, a matrix with constant diagonals. Type `help toeplitz` in the Command Window for details on how this function is called.

1.5 Rivers always flow downstream. Write a *MatLab* script to check that none of the Neuse River discharge data is negative.

2 A first look at data

2.1   Look at your data!

When presented with a new dataset, the most important action that a practitioner of data analysis can take is to look closely and critically at it. This examination has three important objectives:

Objective 1: Understanding the general character of the dataset.
Objective 2: Understanding the general behavior of individual parameters.
Objective 3: Detecting obvious problems with the data.

These objectives are best understood through examples, so we look at a sample dataset of temperature observations from the Black Rock Forest weather station (Cornwall, NY) that is in the file `brf_temp.txt`. It conxtains two columns of data. The first is time in days after January 1, 1997, and the second is temperature in degree Celsius.

We would expect that a weather station would record more parameters than just temperature, so a reasonable assumption is that this file is not the complete Black Rock Forest dataset, but rather some portion extracted from it. If you asked the person who provided the file—Bill Menke, in this case—he would perhaps say something like this:

> *I downloaded the weather station data from the International Research Institute (IRI) for Climate and Society at Lamont-Doherty Earth Observatory, which is the data center used by the Black Rock Forest Consortium for its environmental data. About 20 parameters were available, but I downloaded only hourly averages of temperature. My original file,* `brf_raw.txt` *has time in a format that I thought would be hard to work with, so I wrote a MatLab script,* `brf_convert.m`, *that converted it into time in days, and wrote the results into the file that I gave you (see Notes 2.1 and 2.2).*

So our dataset is neither complete nor original. The issue of originality is important, because mistakes can creep into a dataset every time it is copied, and especially when it is reformatted. A purist might go back to the data center and download his or her own unadulterated copy—not a bad idea, but one that would require him or her to deal with the time format problem. In many instances, however, a practitioner of data analysis has no choice but to work with the data as it is supplied, regardless of its pedigree.

Any information about how one's particular copy of the dataset came about can be extremely useful, especially when diagnosing problems with the data. One should always keep a file of notes that includes a description of how the data was obtained and any modifications that were made to it. Unaltered data file(s) should also be kept, in case one needs to check that a format conversion was correctly made.

Developing some expectations about the data before actually looking at it has value. We know that the Black Rock Forest data are sampled every hour, so the time index, which is in days, should increment by unity every 24 points. As New York's climate is moderate, we expect that the temperatures will range from perhaps -20 °C (on a cold winter night) to around $+40$ °C (on a hot summer day). The actual temperatures may, of course, stray outside of this range during cold snaps and heat waves, but probably not by much. We would also expect the temperatures to vary with the diurnal cycle (24 h) and with the annual cycle (8760 h), and be hottest in the daytime and the summer months of those cycles, respectively.

As the data is stored in a tabular form in a text file, we can make use of the load() function to read it into *MatLab*:

```
D=load('brf_temp.txt');
t=D(:,1);
d=D(:,2);
Ns=size(D);
N=Ns(1);
M=Ns(2);
N
M                                          (MatLab eda02_01)
```

The load() function reads the data into the matrix, D. We then copy time into the column vector t, and temperature into the column vector d. Knowing how much data was actually read is useful, so we query the size of D with the size() function. It returns a vector of the number of rows and columns, which we break out into the variables L and M and display. *MatLab* informs us that we read in a table of L = 110430 rows and M = 2 columns. That is about 4600 days or 12.6 years of data, at one observation per hour. A display of the first few data points, produced with the command, D(1:5,:), yields the following:

0	17.2700
0.0417	17.8500
0.0833	18.4200
0.1250	18.9400
0.1667	19.2900

The first column, time, does indeed increment by $1/24 = 0.0417$ of a day. The temperature data seems to have been recorded with the precision of hundredths of a °C.

We are now ready to plot the data:

```
clf;
set(gca,'LineWidth',2);
hold on;
plot(t,d,'k-','LineWidth',2);                    (MatLab eda02_02)
```

The resulting graph is shown in Figures 2.1 and 2.2. Most of the data range from about -20 to $+35\,°C$, as was expected. The data are oscillatory and about 12 major cycles—annual cycles, presumably — are visible. The scale of the plot is too small for diurnal cycles to be detectable but they presumably contribute to the fuzziness of the curve. The graph contains several unexpected features: Two brief periods of cold temperatures, or cold spikes, occur at around 400 and 750 days. In each case, the temperature dips below $-50\,°C$. Even though they occur during the winter parts of cycles, such cold temperatures are implausible for New York, which suggests some sort of error in the data. A hot spike, with a temperature of about $+40\,°C$ occurs around the time of the second cold spike. While not impossible, it too is suspicious. Finally, two periods of constant—and zero — temperature occur, one in the 1400–1500 day range and the other in the 4600–4700 day range. These are some sort of data drop-outs, time periods where the weather station was either not recording data at all or not properly receiving input from the thermometer and substituting a zero, instead.

Reality checks such as these should be performed on all datasets very early in the data analysis process. They will focus one's mind on what the data *mean* and help reveal misconceptions that one might have about the content of the data set as well as errors in the data themselves.

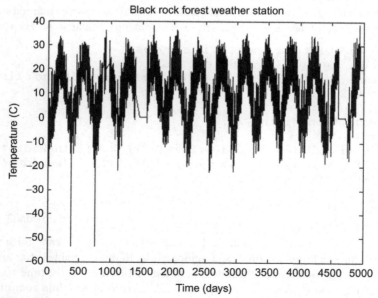

Figure 2.1 Preliminary plot of temperature against time. *MatLab* script eda02_02.

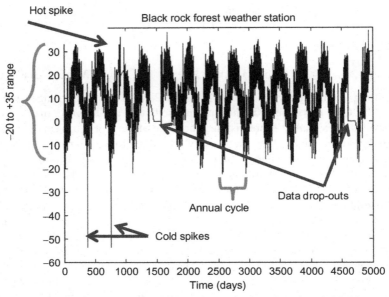

Figure 2.2 Annotated plot of temperature against time. *MatLab* script eda02_02.

The next step is to plot portions of the data on a finer scale. *MatLab*'s Figure Window has a tool for zooming in on portions of the plot. However, rather than to use it, we illustrate here a different technique, and create a new plot with a different scale, that enlarges a specified section of data, with the section specified by a mouse click. The advantage of the method is that scale of several successive enlarged sections can be made exactly the same, which helps when making comparisons. The script is:

```
w=1000; % width of new plot in samples
[tc, dc] = ginput(1); % detect mouse click
i=find((t>=tc),1); % find index i corresponding to click
figure(2);
clf;
set(gca,'LineWidth',2);
hold on;
plot(t(i-w/2:i+w/2), d(i-w/2:i+w/2), 'k-','LineWidth',2);
plot(t(i-w/2:i+w/2), d(i-w/2:i+w/2), 'k.','LineWidth',2);
title('Black Rock Forest Temp');
xlabel('time in days after Jan 1, 1997');
ylabel('temperature');
figure(1);                              (MatLab eda02_03)
```

This script needs careful explanation. The function `ginput()` waits for a mouse click and then returns its time and temperature in the variables `tc` and `dc`. The `find()` function returns the index, `i`, of the first element of the time vector, `t`, that is greater than or equal to `tc`, that is, an element near the time coordinate of the click. We now plot segments of the data, from `i-w/2` to `i+w/2`, where

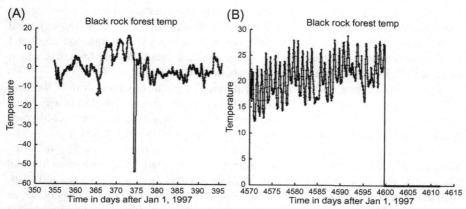

Figure 2.3 Enlargements of two segments of the temperature versus time data. (A) Segment with a cold spike. (B) Section with a drop-out. *MatLab* script eda02_03.

w=1000, to a new figure window. The figure() function opens this new window, and the clf command (for 'clear figure') clears any previous contents. The rest of the plotting is pretty standard, except that we plot the data twice, once with black lines (the 'k-' argument) and then again with black dots (the 'k.' argument). We need to place a hold on command before the two plot functions, so that the second does not erase the plot made by the first. Finally, at the end, we call figure() again to switch back to Figure 1 (so that when the script is rerun, it will again put the cursor on Figure 1). The results are shown in Figure 2.3.

The purpose behind plotting the data with both lines and symbols is to allow us to see the actual data points. Note that the cold spike in Figure 2.3A consists of two anomalously cold data points. The drop-out in Figure 2.3B consists of a sequence of zero-valued data, although examination of other portions of the dataset uncovers instances of missing data as well. The diurnal oscillations, each with a couple of dozens of data points, are best developed in the left part of Figure 2.3B. A more elaborate version of this script is given in eda02_04.

A complementary technique for examining the data is through its histogram, a plot of the *frequency* at which different values of temperature occur. The overall temperature range is divided into a modest number, say L_h, of *bins*, and the number of observations in each bin is counted up. In *MatLab*, a histogram is computed as follows:

```
Lh = 100;
dmin = min(d);
dmax = max(d);
bins = dmin + (dmax-dmin)*[0:Lh-1]'/(Lh-1);
dhist = hist(d, bins)';                          (MatLab eda02_05)
```

Here we use the min() and max() functions to determine the overall range of the data. The formula dmin+(dmax-dmin)*[0:Lh-1]'/(Lh-1) creates a column vector of length Lh of temperature values that are equally spaced between these two extremes. The histogram function hist() does the actual counting, returning a column-vector `dhist` whose elements are *counts*, that is, the number of observations in each of the bins. The value of `Lh` needs to be chosen with some care: too large and the bin size will be very small and the resulting histogram very rough; too small and the bins will be very wide and the histogram will lack detail. We use `Lh=100`, which divides the temperature range into bins about 1 °C wide.

The results (Figure 2.4) confirm our previous conclusions that most of the data fall in the range −20 to +35 °C and that the near-zero temperature bin is *way* over-represented in the dataset. The histogram does not clearly display the two cold-spike outliers (although two tiny peaks are visible in the −60 to −40 °C range).

The histogram can also be displayed as a grey-shaded column vector (Figure 2.5B). Note that the coordinate axes in this figure have been rotated with respect to those in Figure 2.4. The origin is in the upper-left and the positive directions are *down* and *right*. The intensity (darkness) of the grey-shade is proportional to the number of counts, as is shown by the *color bar* at the right of the figure. This display technique is most useful when only the pattern of variability, and not the numerical value, is of interest. Reading numerical values off a grey-scale plot is much less accurate than reading them off a standard graph! In many subsequent cases, we will omit the color bar, as only the pattern of variation, and not the numerical values, will be of interest. The *MatLab* commands needed to create this figure are described in the next section.

An important variant of the histogram is the moving-window histogram. The idea is to divide the overall dataset into smaller segments, and compute the histogram of each segment. The resulting histograms can then be plotted side-by-side using the grey-shaded column-vector technique, with time increasing from left to right

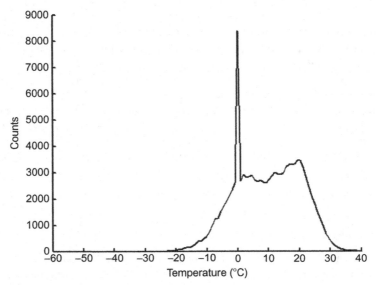

Figure 2.4 Histogram of the Black Rock Forest temperature data. Note the peak at a temperature of 0 °C. *MatLab* script eda02_05.

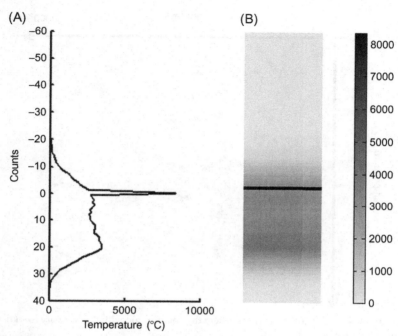

Figure 2.5 Alternate ways to display a histogram. (A) A graph. (B) A grey-shaded column vector. *MatLab* script eda02_06.

(Figure 2.6). The advantage is that the changes in the overall shape of the distribution with time are then easy to spot (as in the case of drop-outs). The *MatLab* code is as follows:

```
offset=1000;
Lw=floor(N/offset)-1;
Dhist = zeros(Lh, Lw);
for i = [1:Lw];
    j=1+(i-1)*offset;
    k=j+offset-1;
    Dhist(:,i) = hist(d(j:k), bins)';
end
```
(*MatLab* eda02_07)

Each segment of 1000 observations is offset by 1000 samples from the next (the variable, offset=1000). The number of segments, Lw, is the total length, N, of the dataset divided by the offset. However, as the result may be fractional, we round off to the nearest integer using the floor() function. Thus, we compute Lw histograms, each of length Lh, and store them in the columns of the matrix Dh. We first create an $L_w \times L_h$ matrix of zeros with the zeros() function. We then loop Lw times, each time creating one histogram from one segment, and copying the results into the proper columns of Dh. The integers j and k are the beginning and ending indices, respectively, of segment i, that is, d(j:k) is the i-th segment of data.

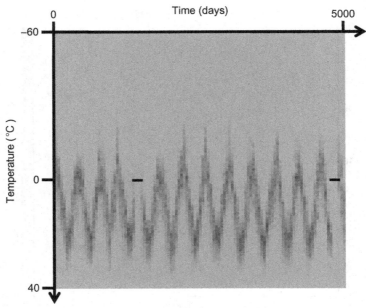

Figure 2.6 Moving-window histogram, where the counts scale with the intensity (darkness) of the grey. *MatLab* script eda02_07.

2.2 More on *MatLab* graphics

MatLab graphics are very powerful, but accessing that power requires learning what might, at first, seem to be a bewildering plethora of functions. Rather than attempting to review them in any detail, we provide some further examples. First consider

```
% create sample data, d1 and d2
N=51;
Dt = 1.0;
t = [0:N-1]';
tmax=t(N);
d1 = sin(pi*t/tmax);
d2 = sin(2*pi*t/tmax);

% plot the sample data
figure(7);
clf;
set(gca,'LineWidth',2);
hold on;
axis xy;
axis([0, tmax, -1.1, 1.1]);
plot(t,d1,'k-','LineWidth',2);
plot(t,d2,'k:','LineWidth',2);
```

```
title('data consisting of sine waves');
xlabel('time');
ylabel('data');                                    (MatLab eda02_08)
```

The resulting plot is shown in Figure 2.7. Note that the first section of the code creates a time variable, t, and two data variables, d1 and d1, the sine waves $\sin(\pi t/L)$ and $\sin(2\pi t/L)$, with L a constant. *MatLab* can create as many figure windows as needed. We plot these data in a new figure window, numbered 7, created using the figure () function. We first clear its contents with the clf command. We then use a hold on, which informs *MatLab* that we intend to overlay plots; so the second plot should not erase the first. The axis xy command indicates that the axis of the coordinate system is in the lower-left of the plot. Heretofore, we have been letting *MatLab* auto-scale plots, but now we explicitly set the limits with the axis() function. We then plot the two sine waves against time, with the first a solid black line (set with the 'k-') and the second a dotted black line (set with the 'k:'). Finally, we label the plot and axes.

MatLab can draw two side-by-side plots in the same Figure Window, as is illustrated below:

```
figure(8);
clf;
subplot(1,2,1);
set(gca,'LineWidth',2);
hold on;
axis([-1.1, 1.1, 0, tmax]);
axis ij;
plot(d1,t,'k-');
title('d1');
ylabel('time');
xlabel('data');
subplot(1,2,2);
set(gca,'LineWidth',2);
hold on;
axis ij;
axis([-1.1, 1.1, 0, tmax]);
plot(d2,t,'k-');
title('d2');
ylabel('time');
xlabel('data');                                    (MatLab eda02_09)
```

Here the subplot(1,2,1) function splits the Figure Window into 1 column and 2 rows of subwindows and directs *MatLab* to plot into the first of them. We plot data into this subwindow in the normal way. After finishing with the first dataset, the subplot (1,2,2) directs *MatLab* to plot the second dataset into the second subwindow. Note that we have used an axis ij command, which sets the origin of the plots to the upper-left (in contrast to axis xy, which sets it to the lower-left). The resulting plot is shown in Figure 2.8.

MatLab plots grey-scale and color images through the use of a *color-map*, that is, a 3-column table that converts a data value into the intensities of the red, green, and

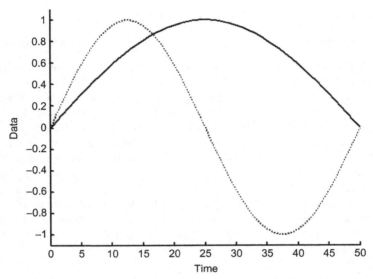

Figure 2.7 Plot of the functions $\sin(\pi t/L)$ (solid curve) and $\sin(2\pi t/L)$ (dashed curve), where $L = 50$. *MatLab* script eda02_08.

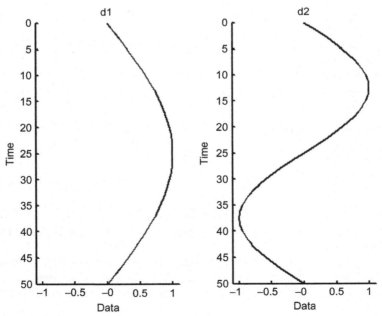

Figure 2.8 Plot of the functions $\sin(\pi t/L)$ (left plot) and $\sin(2\pi t/L)$ (right plot), where $L = 50$. *MatLab* script eda02_09.

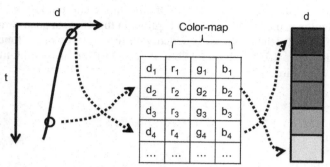

Figure 2.9 The data values are converted into color values through the color map.

blue colors of a pixel on a computer screen. If the data range from d_{min} to d_{max}, then the top row of the table gives the red, green, and blue values associated with d_{max} and the bottom row gives the red, green, and blue values associated with d_{min}, each of which range from 0 to 1 (Figure 2.9). The number of rows in the table corresponds to the smoothness of the color variation within the d_{min} to d_{max} range, with a larger number of rows corresponding to a smoother color variation. We generally use tables of 256 rows, as most computer screens can only display 256 distinct intensities of color.

While *MatLab* is capable of displaying a complete spectrum of colors, we use only black-and-white color maps in this book. A black-and-white color map has equal red, green, and blue intensities for any given data value. We normally use

```
% create grey-scale color map
bw=0.9*(256-[0:255]')/256;
colormap([bw,bw,bw]);                                    (MatLab eda02_06)
```

which colors the minimum data value, d_{min}, a light gray and the maximum data value, d_{max}, black. In this example, the column vector bw is of length 256 and ranges from 0.9 (light grey) to 0 (black). A vector, dhist (as in Figure 2.5B), can then be plotted as a grey-shade image using the following script:

```
axis([0, 1, 0, 1]);
hold on;
axis ij;
axis off;
imagesc( [0.4, 0.6], [0, 1], dhist);
text(0.66,-0.2,'dhist');
colorbar('vert');                                        (MatLab eda02_06)
```

Here, we set the axis to a simple 0 to 1 range using the axis() function, place the origin in the upper left with the axis ij command, and turn off the plotting of the axis and tick marks with the axis off command. The function,

```
imagesc( [0.4, 0.6], [0, 1], dhist);
```

plots the image. The quantities [0.4, 0.6] and [0, 1] are vectors **x** and **y**, respectively, which together indicate where dhist is to be plotted. They specify the positions of opposite corners of a rectangular area in the figure. The first element of dhist is plotted at the (x_1, y_1) corner of the rectangle and the last at (x_2, y_2). The text() function is used to place text (a caption, in this case) at an arbitrary position in the figure. Finally, the color bar is added with the colorbar() function.

A grey-shaded matrix, such as Dhist (Figure 2.6) can also be plotted with the imagesc() function:

```
figure(1);
clf;
axis([-Lw/8, 9*Lw/8, -Lh/8, 9*Lh/8]);
hold on;
axis ij;
axis equal;
axis off;
imagesc( [0, Lw-1], [0, Lh-1], Dhist);
text(6*Lw/16,17*Lw/16,'Dhist');            (MatLab eda02_06)
```

Here, we make the axes a little bigger than the matrix, which is Lw×Lh in size. Note the axis equal command, which ensures that the x and y axes have the same length on the computer screen. Note also that the two vectors in the imagesc function have been chosen so that the matrix plots in a square region of the window, as contrasted to the narrow and high rectangular area that was used in the previous case of a vector.

2.3   Rate information

We return now to the Neuse River Hydrograph (Figure 1.4). This dataset exhibits an annual cycle, with the river level being lowest in autumn. The data are quite spiky. An enlargement of a portion of the data (Figure 2.10A) indicates that the dataset contains many short periods of high discharge, each ∼5 days long and presumably corresponding to a rain storm. Most of these *storm events* seem to have an asymmetric shape, with a rapid rise followed by a slower decline. The asymmetry is a consequence of the river rising rapidly after the start of the rain, but falling slowly after its end, as water slowly drains from the land.

This qualitative assessment can be made more quantitative by estimating the time rate of change of the discharge—the discharge *rate* − using its finite-difference approximation:

$$\frac{dd}{dt} \approx \frac{\Delta d}{\Delta t} = \frac{d(t + \Delta t) - d(t)}{\Delta t} \quad \text{or} \quad \left[\frac{dd}{dt}\right]_i \approx \frac{d_{i+1} - d_i}{t_{i+1} - t_i} \tag{2.1}$$

The corresponding *MatLab* script is

```
dddt=(d(2:L)-d(1:L-1)) ./ (t(2:L)-t(1:L-1));
                                           (MatLab eda02_10)
```

Note that while the discharge data is of length L, its rate is of length L−1. The rate curve (Figure 2.10B) also contains numerous short events, although these are *two-sided* in shape. If, indeed, the typical storm event consists of a rapid rise followed

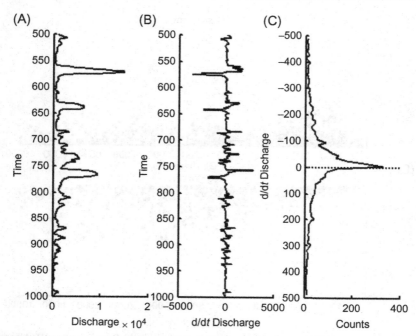

Figure 2.10 (A) Portion of the Neuse River Hydrograph. (B) Corresponding rate of change of discharge with time. (C) Histogram of rates for entire hydrograph. *MatLab* script eda01_10.

by a long decline, we would expect that the discharge rate would be negative more often than positive. This hypothesis can be tested by computing the histogram of discharge rate, and examining whether or not it is centered about a rate of zero. The histogram (Figure 2.10C) peaks at negative rates, lending support to the hypothesis that the typical storm event is asymmetric.

We can also use rate information to examine whether the river rises (or falls) faster at high water (large discharges) than at low water (small discharges). We segregate the data into two sets of (discharge, discharge rate) pairs, depending on whether the discharge rate is positive or negative, and then make scatter plots of the two sets (Figure 2.11). The *MatLab* code is as follows:

```
pos = find(dddt>0);
neg = find(dddt<0);
- - -
plot(d(pos),dddt(pos),'k.');
- - -
plot(d(neg),dddt(neg),'k.');                    (MatLab eda02_11)
```

Here, the "– – –" means that we have omitted lines of the script (standard plot setup commands, in this case). The find() function returns a column vector of *indices* of dddt that match the given test condition. For example, pos contains the indices of dddt for which dddt>0. Note that the quantities d(pos) and dddt(pos)

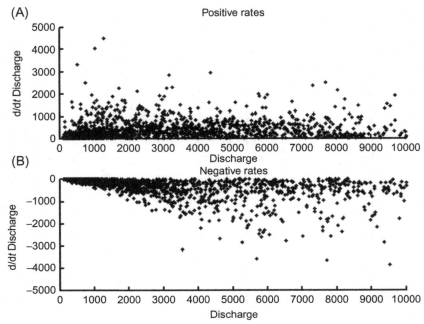

Figure 2.11 Scatter plot of discharge rate against discharge. (A) Positive rates. (B) Negative rates. *MatLab* script eda01_11.

are arrays of just the elements of d and dddt whose indices are contained in pos. Results are shown in Figure 2.11. Only the negative rates appear to correlate with discharge, that is, the river falls faster at high water than at low water. This pattern is related to the fact that a river tends to run faster when it's deeper, and can carry away water added by a rain storm quicker. The positive rates, which show no obvious correlation, are more influenced by meteorological conditions (e.g., the intensity and duration of the storms) than river conditions.

2.4 Scatter plots and their limitations

Both the Black Rock Forest temperature and the Neuse River discharge datasets are *time series*, that is, data that are organized sequentially in time. Many datasets lack this type or organization. An example is the Atlantic Rock Sample dataset, provided in the file, rocks.txt. Here are notes provided by Bill Menke, who created the file:

> *I downloaded rock chemistry data from PetDB's website at www.petdb.org. Their database contains chemical information about ocean floor igneous and metamorphic rocks. I extracted all samples from the Atlantic Ocean that had the following chemical species: SiO_2, TiO_2, Al_2O_3, FeO_{total}, MgO, CaO, Na_2O, and K_2O. My original file,* rocks_raw.txt *included a description of the rock samples, their geographic*

location, and other textual information. However, I deleted everything except the chemical data from the file, rocks.txt, so it would be easy to read into MatLab. The order of the columns is as given above and the units are weight percent.

Note that this Atlantic Rock dataset is just a fragment of the total data in the PetDB database. After loading the file, we determine that it contains $N = 6356$ chemical analyses.

Scatter plots (Figure 2.12) are a reasonably effective means to quickly review the data. In this case, the number, M, of columns of data is small enough that we can exhaustively review all of the $M^2/2$ combinations of chemical species. A *MatLab* script that runs through every combination uses a pair of nested for loops:

```
D = load('rocks.txt');
Ns = size(D);
N = Ns(1);
M = Ns(2);
for i = [1:M-1]
for j = [i+1:M]
    clf;
    axis xy;
    hold on;
    plot( D(:,i), D(:,j), 'k.' );
    xlabel(sprintf('element %d',i));
    ylabel(sprintf('element %d',j));
    [x, y]=ginput(1);
end
end
```
                                                          (*MatLab* eda02_12)

This nested for loop plots all combinations of species i with species j. Note that we can limit ourselves to the j>i, as the j=i case corresponds to plotting a species against itself, and the j<i plots are redundant. Note that the outer for loop variable, i, ranges from 1 to M−1 and the inner for loop variable, j, ranges over the interval from i+1 to M. The pause between successive plots is implemented with the ginput() command; clicking on the figure signals that it is time for the next graph to be displayed.

Note, also, the use of the sprintf() function (for "string print formatted"). It creates a character string that includes both text and the value of a variable. This is a useful, although fairly inscrutable function, and we refer readers to the *MatLab* help pages for a detailed description. Briefly, the function uses *placeholders* that start with the character % to indicate where in the character string the value of the variable should be placed. Thus,

```
i=2;
sprintf('element %d',i);
```

returns the character string 'element 2'. The %d is the *placeholder for an integer*. It is replaced with '2', the value of i. Several placeholders can be used in the same *format string*; for example, the script

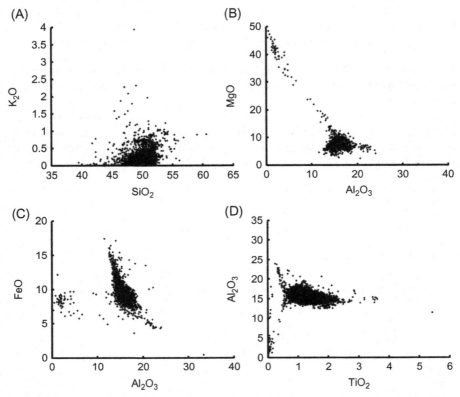

Figure 2.12 Scatter plot of four combinations of chemical components of the Atlantic Ocean rock sample dataset. *MatLab* script eda01_12.

```
i=2;
j=4;
sprintf('row %d column %d', i, j);
```

returns the character string `'row 2 column 4'`. If the variable is fractional, as contrasted to integer, the *floating-point placeholder*, %f, is used instead. For example,

```
a=4.71;
sprintf('a=%f', a);
```

returns the character string `'a=4.71'`. The `sprintf()` command can be used in any function that expects a character string, including `title()`, `xlabel()`, `ylabel()`, `load()`, `dlmwrite()`, and `disp()`. In the special case of `disp()`, an alternative is also available, with the command

```
fprintf('a=%f\n', a);
```

being equivalent to:

```
disp(sprintf('a=%f', a));
```

The \n (for *newline*) at the end of the format string ′a=%f\n′ indicates that subsequent characters should be written to a new line in the command window, rather than being appended to the end of the current line. In this book, our preference is to use disp(sprintf()), as it preserves regularity of usage.

Four of the resulting 32 plots are shown in Figure 2.12. Note their effectiveness in identifying both overall patterns in the data and data outliers that depart from the pattern. Figure 2.12A and Figure 2.12D both have a single outlier, in K_2O and TiO_2, respectively. We do not know whether they represent an unusual rock composition or an error in the data, but this issue could possibly be resolved with further information about the data. Note that two distinct groupings, or *populations*, of data are present in Figure 2.12C, whereas only one is evident in Figure 2.12A. Figure 2.12B has a well-defined linear variation of MgO with Al_2O_3, but Figure 2.12D has a more complicated Y-shaped relationship of Al_2O_3 and TiO_2.

On the one hand, this preliminary inspection has yielded interesting patterns that would be worth pursuing in a more detailed analysis. On the other hand, it has revealed one of the limitations of scatter plots when they are applied to multivariate data: *the plots all look different*! Some pairs of parameters (chemical species, in this case) seem uncorrelated, while others have strong correlations. Some have a single population, others two or even more. The problem becomes even worse when one considers that plots can also be made of combinations of parameters (e.g., a plot of MgO + FeO against NaO + K_2O). The problem is that the patterns within the dataset are inherently multidimensional, but a scatter plot reduces that pattern to just two dimensions.

This problem points to the need for more advanced data analysis tools that can get at the underlying multidimensional patterns—tools that we will discuss later in this book (e.g., the factor analysis discussed in Chapter 8).

Crib Sheet 2.1 When first examining a dataset

Archive your data

Keep an unaltered copy of your data, together with notes recording its source and any other useful information that you come across.

Questions to ask yourself

What was the purpose for which the data was collected?
What is the physical significance of each data type?
How were they measured?
What are the units of measurement?
How are missing data represented?

Look at the numerical values

How many significant digits are recorded?

Make preliminary plots

time series plots
symbols or lines or both?

Continued

Crib Sheet 2.1—cont'd

scatter plots
linear or logarithmic axes?
histograms
color images
plot error bars whenever possible!

Reality Checks
Does the data make sense in the context of what you already know?
sign
magnitude
minimum and maximum values
scale of variation with time and distance
daily, annual and other periodicities

Problems

2.1. Plot the Black Rock Forest temperature data on a graph whose time units are years. Check whether the prominent cycles are really *annual*.

2.2. What is the largest hourly change in temperature in the Black Rock Forest dataset? Ignore the changes that occur at the temperature spikes and rop-outs.

2.3. Examine the diurnal cycles in the Black Rock Forest dataset. Qualitatively, does their pattern vary with time of year?

2.4. Adapt the eda02_03 script to plot segments of the Neuse River Hydrograph dataset.

2.5. Create histograms for the eight chemical species in the Atlantic Rock dataset.

3 Probability and what it has to do with data analysis

3.1 Random variables

Every practitioner of data analysis needs a working knowledge of probability for one simple reason: *error*, an unavoidable aspect of measurement, is best understood using the ideas of probability.

The key concept that we draw upon is the *random variable*. If d is a random variable, then it has no fixed value until it is *realized*. Think of d as being in a box. As long as it is in the box, its value is fuzzy or indeterminate; but when taken out of the box and examined, d takes on a specific value. It has been realized. Put it back in the box, and its value becomes indeterminate again. Take d out again and it will have a different value, as it is now a different realization. This behavior is analogous to measurement in the presence of noise, so random variables are ideal for representing noisy data.

Even when the random variable, d, is in the box, we may know something about it. It may have a tendency to take on certain values more often than others. For example, suppose that d represents the number of H (hydrogen) atoms in a CH_4 (methane) molecule that are of the heavy variety called *deuterium*. Then d can take on only the discrete values 0 through 4, with $d = 0$ representing the no deuterium state and $d = 4$ representing the all deuterium state. The tendency of d to take on one of these five

Environmental Data Analysis with MATLAB®. http://dx.doi.org/10.1016/B978-0-12-804488-9.00003-3

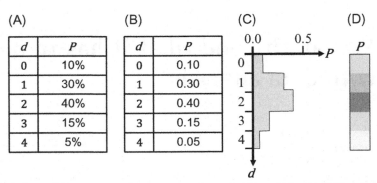

Figure 3.1 Four different ways of representing the probability, $P(d)$. (A) Table of percents; (B) Table of fractions; (C) Histogram, and (D) Shaded column vector.

values is represented by its probability, $P(d)$, which can be depicted by a table (Figure 3.1A).

Note that the probability is given in percent, that is, the percent of the realizations in which d takes on the given value. The probability necessarily sums to 100%, as d *must* take on one of the five possible values. We can write $\sum_{i=0}^{4} P(d_i) = 100\%$. An alternative way of quantifying probability is with the numbers 0-1, with 0 meaning 0% and 1 meaning 100% (Figure 3.1B). In this case, $\sum_{i=0}^{4} P(d_i) = 1$. The probability can also be represented graphically with a histogram (Figure 3.1C) or a shaded column vector (Figure 3.1D).

Not all random variables are discrete. Some may vary continuously between two extremes. Thus, for example, the depth, d, of a fish observed swimming in a 5-m-deep pond can take on any value, even fractional ones, between 0 and 5. In this continuous case, we quantify the probability that the fish is *near* depth, d, with the *probability density function*, $p(d)$. The probability, P, that the fish is observed between any two depths, say d_1 and d_2, is defined as the area under the curve $p(d)$ between d_1 and d_2 (Figure 3.2). This is equivalent to the integral

$$P(d_1, d_2) = \int_{d_1}^{d_2} p(d)\, \mathrm{d}d \qquad (3.1)$$

(Our choice of the variable name "d" for "data" makes the differential $\mathrm{d}d$ look a bit funny, but we will just have to live with it!) Note the distinction between upper-case and lower-case letters. Upper-case P, which quantifies probability, is a *number* between 0 and 1. Lower-case p is a *function* whose values are not easily interpretable, except to the extent that the larger the p, the more likely that a realization will have a value near d. One must calculate the area, which is to say, perform the integral, to determine how likely any given range of d is. Just as in the discrete case, d must take on *some* value between its minimum and maximum (in this case, $d_{\min} = 0$ and $d_{\max} = 5$). Thus

$$P(d_{\min}, d_{\max}) = \int_{d_{\min}}^{d_{\max}} p(d)\, \mathrm{d}d = 1 \qquad (3.2)$$

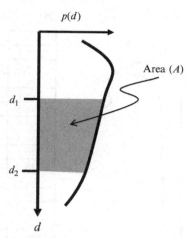

Figure 3.2 The probability, P, that the random variable, d, is between d_1 and d_2 is proportional to the area, A (shaded), under the probability density function, $p(d)$, from d_1 to d_2.

The function, $P(d_{\min}, d)$ (or $P(d)$, for short), which gives the total amount of probability less than d, is called the *probability distribution* (or, sometimes, the *cumulative probability distribution*) of the random variable, d.

Because all measurements contain noise, we view *every* measurement, d, as a random variable. Several repetitions of the same measurement will not necessarily yield the same value because of measurement error. On the other hand, repeated measurements usually have some sort of systematic behavior, such as scattering around a typical value. This systematic behavior will be represented by the probability density function, $p(d)$. Thus, $p(d)$ embodies both the "true" value of the quantity (if such a thing can be said to exist) and a description of the measurement noise.

Practitioners of data analysis very typically compute *derived quantities* from their data that are more relevant to the objective of their study. For example, temperature measurements made at different times might be differenced (subtracted) in order to determine a rate of warming. As we will discuss further below, functions of random variables are themselves random variables, because any quantity derived from noisy data itself contains error. The algebra of random variables will allow us to understand how measurement noise affects inferences made from the data.

3.2   Mean, median, and mode

The probability density function of a measurement, $p(d)$, is a function—possibly one with a complicated shape. As an aid to understanding it, one might try to derive from it two simple numbers, one that describes the typical measurement (i.e., the typical d) and the other which describes the variability of measurements (i.e., the amount of scatter around the typical measurement). Of course, the two numbers cannot completely capture the information in $p(d)$, but they can provide some insight into its behavior.

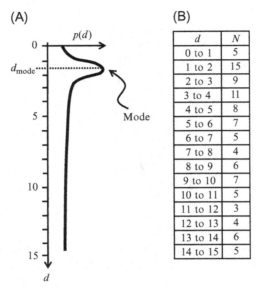

Figure 3.3 (A) The probability density function, $p(d)$, has its maximum value at the mode, $d = d_{\text{mode}} \approx 1.5$. (B) A binned table of 100 realizations of d. Note that while the bin with the largest number of measurements is at the mode, the majority of measurements are larger than the mode.

Several different approaches to calculating a typical value of d are in use. The simplest is the d at which $p(d)$ takes on its maximum value, which is called the *maximum likelihood point* or *mode*. The mode is useful because, in a list of repeated measurements, a particular value is often seen to occur more frequently than any other value. Modes can be deceptive, however, because while more measurements will be in the vicinity of the mode than in the vicinity of any other value of d, the majority of measurements are not necessarily in the vicinity of the mode. This effect is illustrated in Figure 3.3, which depicts a very skewed probability density function, $p(d)$ and a corresponding table of binned data. More data—15—are observed in the $1 < d < 2$ bin, which encloses the mode, than in any other bin. Nevertheless, full 80% of the measurements are larger than the mode, and 50% are quite far from it.

This effect is often encountered in real-world situations. Thus, for example, while ironwood (*Memecylon umbellatum*) is by far the most common of the ~50 species of trees in the evergreen forests of India's Eastern Ghats, a randomly chosen tree there would most likely not be ironwood, as this species accounts for only 21% of the individual trees (Chittibabu and Parthasarathy, 2000).

In *MatLab*, suppose that the column-vector, d, is d sampled at evenly spaced points, and that the column-vector, p, contains the corresponding values of $p(d)$. Then the mode is calculated as follows:

```
[pmax, i] = max(p);
themode = d(i);                                          (MatLab eda03_01)
```

Note that the function, `max(p)`, returns both the maximum value, `pmax`, of the column-vector, `p`, and the row index, `i`, at which the maximum value occurs.

Another way of defining the typical measurement is to pick the value below which 50% of probability falls and above which lies the other 50%. This quantity is called the *median* (Figure 3.4).

In *MatLab*, suppose that the column vector d contains the data d sampled at evenly spaced points, with spacing Dd, and that the column vector p contains the corresponding values of $p(d)$. Then, the median is calculated as follows:

```
pc = Dd*cumsum(p);
for i=[1:length(p)]
    if(pc(i) > 0.5 )
        themedian = d(i);
        break;
    end
end                                    (MatLab eda03_02)
```

The function `cumsum()` computes the cumulative sum (running sum) of p. The quantity, `Dd*cumsum(p)`, is thus an approximation for the indefinite integral, $\int_{d_{min}}^{d} p(d)\mathrm{d}d$, which is to say the area beneath $p(d)$. The `for` loop then searches for the first occurrence of the area that is greater than 0.5, terminating ("breaking") when this condition is satisfied.

Yet another common way to define a typical value is a generalization of the *mean* value of a set of measurements. The well-known formula for the sample mean

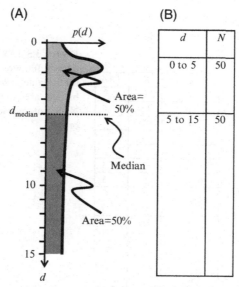

Figure 3.4 (A) Fifty percent of the probability lies on either side of the median, $d = d_{median} \approx 5$. (B) A binned table of 100 realizations of d has about 50 measurements on either side of the median.

is $\bar{d} = 1/N\sum_{i=0}^{N} d_i$. Let's approximate this formula with a histogram. First, divide the
d-axis into M small bins, each one centered at $d^{(s)}$. Now count up the number, N_s, of
data in each bin. Then, $\bar{d} \approx 1/N\sum_{s=0}^{M} d^{(s)}N_s$. Note that the quantity N_s/N is the
frequency of d_i; that is, the fraction of times that d_i is observed to fall in bin s. As this
frequency is, approximately, the probability, P_s, that the data falls in bin s,
$\bar{d} \approx \sum_{s=0}^{M} d^{(s)}P_s$. This relationship suggests that the mean of the probability density
function, $p(d)$, can be defined as

$$\bar{d} = \int_{d_{min}}^{d_{max}} d\, p(d)\, \mathrm{d}d \qquad (3.3)$$

Because of random variation, the mean of a set of measurements (the *sample* mean)
will not be the same as the mean of the probability density function from which the
measurements were drawn. However, as we will discuss later in this book, the sample
mean will usually be close to—will scatter around—the mean of the probability
density function (Figure 3.5).

In *MatLab*, suppose that the column-vector, d, contains the d sampled at evenly
spaced points, with spacing Dd, and that the column-vector, p, contains the corresponding values of $p(d)$. The definite integral $\int_{d_{min}}^{d} dp(d)\mathrm{d}d$ is approximated as
$\sum_i d_i p(d_i)\Delta d$ as follows:

        themean = Dd*sum(d.*p);   (*MatLab* eda03_03)

Note that the sum(v) function returns the sum of the elements of the column-
vector, v.

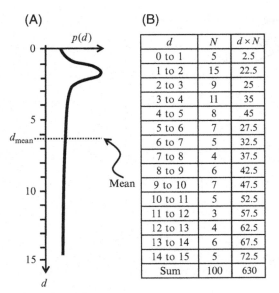

Figure 3.5 (A) The probability density function, $p(d)$, has its mean value at $d_{mean} \approx 6.3$. (B) A binned table of 100 realizations of d. Note that the mean is 630/100, or 6.3.

d	N	$d \times N$
0 to 1	5	2.5
1 to 2	15	22.5
2 to 3	9	25
3 to 4	11	35
4 to 5	8	45
5 to 6	7	27.5
6 to 7	5	32.5
7 to 8	4	37.5
8 to 9	6	42.5
9 to 10	7	47.5
10 to 11	5	52.5
11 to 12	3	57.5
12 to 13	4	62.5
13 to 14	6	67.5
14 to 15	5	72.5
Sum	100	630

3.3 Variance

The second part of the agenda that we put forward in Section 3.2 is to devise a number that describes the amount of scatter of the data around its typical value. This number should be large for a wide probability density function—one that corresponds to noisy measurements—and small for a narrow one. A very intuitive choice for a measure of the width of a probability density function, $p(d)$, is the length, d_{50}, of the d-axis that encloses 50% of the total probability and is centered around the typical value, d_{typical}. (This quantity is called the *interquartile*). Then, 50% of measurements would scatter between $d_{\text{typical}} - d_{50}/2$ and $d_{\text{typical}} + d_{50}/2$ (Figure 3.6). Probability density functions with a large d_{50} correspond to a high-noise measurement scenario and probability density functions with a small d_{50} correspond to a low-noise one. Unfortunately, this definition is only rarely used in the literature.

A much more commonly encountered—but much less intuitive—quantity is the *variance*. It is based on a different approach to quantifying width, one not directly related to probability. Consider the quadratic function $q(d) = (d - d_{\text{typical}})^2$. It is small near d_{typical} and large far from it. The product, $q(d)\,p(d)$, will be small everywhere if the probability density function is narrow, as near d_{typical}, large values of $p(d)$ will be offset by small values of $q(d)$ and far from d_{typical}, large values of $q(d)$ will be offset by small values of $p(d)$. The area under the product, $q(d)\,p(d)$, will be small in this case. Conversely, the area under the product, $q(d)\,p(d)$, will be large if the probability density function is wide. Thus, the area under $q(p)\,p(d)$ has the desired property of being small for narrow probability density functions and large for wide ones (Figure 3.7). With the special choice, $d_{\text{typical}} = \bar{d}$, it is called the *variance* and is given the symbol, σ_d^2:

$$\sigma_d^2 = \int_{d_{\min}}^{d_{\max}} (d - \bar{d})^2 p(d)\, dd \tag{3.4}$$

Variance has units of d^2, so the square root of variance, σ, is a measure of the width of the probability density function. A disadvantage of the variance is that the relationship

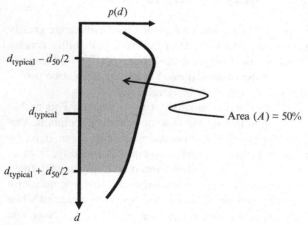

Figure 3.6 The shaded area of the probability density function, $p(d)$, encloses 50% of the probability and is centered on the typical value.

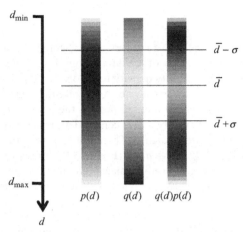

Figure 3.7 Calculation of the variance. The probability density function, $p(d)$, is multiplied by the quadratic function, $q(d)$. The area under the product $q(d)\, p(d)$ is then calculated via integration. See text for further discussion. *MatLab* script eda03_05.

between it and the probability that $\bar{d} \pm \sigma$ encloses is not immediately known. It depends on the functional form of the probability density function.

In *MatLab*, suppose that the column-vector, d, contains the data, d, sampled at evenly spaced points, with spacing Dd and that the column-vector, p, contains the corresponding values of $p(d)$. Then, the variance, σ^2, is calculated as follows:

```
q = (d-dbar).^2;
sigma2 = Dd*sum(q.*p);
sigma = sqrt(sigma2);                              (MatLab eda03_04)
```

3.4   Two important probability density functions

As both natural phenomena and the techniques that we use to observe them are greatly varied, it should come as no surprise that hundreds of different probability density functions, each with its own mathematical formula, have been put forward as good ways to model particular classes of noisy observations. Yet among these, two particular probability density functions stand out in their usefulness.

The first is the *uniform* probability density function, $p(d) =$ constant. This probability density function could be applied in the case of a measurement technique that can detect a fish in a lake, but which provides no information about its depth, d. As far as can be determined, the fish could be at any depth with equal probability, from its surface, $d = d_{min}$, to its bottom, $d = d_{max}$. Thus, the uniform probability density function is a good idealization of the limiting case of a measurement providing no useful information. The uniform probability density function is properly normalized when the constant is $1/(d_{max} - d_{min})$, where the data range from d_{min} to d_{max}. Note that the uniform probability density function can be defined only when the range is finite. It is not possible for data to *be anything* in the range from $-\infty$ to $+\infty$ with equal probability.

The second is the *Normal probability density function*:

$$p(d) = \frac{1}{\sqrt{2\pi}\,\sigma} \exp\left\{-\frac{(d-\bar{d})^2}{2\sigma^2}\right\} \tag{3.5}$$

The constants have been chosen so that the probability density function, when integrated over the range $-\infty < d < +\infty$, has unit area and so that its mean is $\bar{d}$ and its variance is σ^2. Not only is the Normal curve centered at the mean, but it is peaked at the mean and symmetric about the mean (Figure 3.8). Thus, both its mode and median are equal to its mean, $\bar{d}$. The probability, P, enclosed by the interval $\bar{d} \pm n\sigma$ (where n is an integer) is given by the following table:

n	$P, \%$
1	68.27
2	95.45
3	99.73

$$\tag{3.6}$$

It is easy to see why the Normal probability density function is seen as an attractive one with which to model noisy observations. The typical observation will be near its mean, $\bar{d}$, which is equal to the mode and median. Most of the probability (99.73%) is concentrated within $\pm 3\sigma$ of the mean and only very little probability (0.27%) lies outside that range. Because of the symmetry, the behavior of measurements less than the mean is the same as the behavior greater than the mean. Many measurement scenarios behave just in this way.

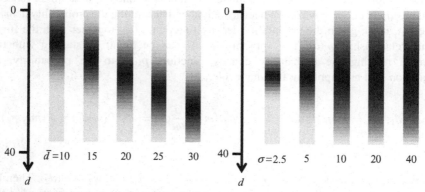

Figure 3.8 Examples of the Normal probability density functions. (Left) Normal probability density functions with the same variance ($\sigma^2 = 5^2$) but different means. (Right) Normal probability density functions with the same mean (20) but different variances. *MatLab* scripts eda03_06 and eda03_07. (See Note 3.2).

On the other hand, the Normal probability density function does have limitations. One limitation is that it is defined on the unbounded interval $-\infty < d < +\infty$, while in many instances data are bounded. The fish in the pond example, discussed above, is one of these. Any Normal probability density function, regardless of mean and variance, predicts *some* probability that the fish will be observed either in the air or buried beneath the bottom of the pond, which is unrealistic. Another limitation arises from the rapid falloff of probability away from the mean—behavior touted as good in the previous paragraph. However, some measurement scenarios are plagued by *outliers*, occasional measurements that are far from the mean. Normal probability density functions tend to under-predict the frequency of outliers.

While not all noise processes are Normally distributed, the Normal probability density function occurs in a wide range of situations. Its ubiquity is understood to be a consequence a mathematical result of probability theory called the *Central Limit Theorem*: under a fairly broad range of conditions, the sum of a sufficiently large number of random variables will be approximately Normally distributed, regardless of the probability density functions of the individual variables. Measurement error often arises from a several different sources of noise, which sum together to produce the overall error. The Central Limit Theorem predicts that, in this case, the overall error will be Normally distributed even when the component sources are not.

3.5   Functions of a random variable

An important aspect of data analysis is making inferences from data. The making of measurements is not an end unto itself, but rather the observations are used to make specific, quantitative predictions about the world. This process often consists of combining the data into a smaller number of more meaningful *model parameters*. These derived parameters are therefore functions of the data.

The simplest case is a model parameter, m, that is a function of a single random variable, d; that is, $m = m(d)$. We need a method of determining the probability density function, $p(m)$, given the probability density function, $p(d)$, together with the functional relationship, $m(d)$. The appropriate rule can be found by starting with the formula relating the probability density function, $p(d)$, to the probability, P, (Equation 3.1) and applying the chain rule.

$$P(d_1, d_2) = \int_{d_1}^{d_2} p(d)\, \mathrm{d}d = \int_{d_1(m_1)}^{d_2(m_2)} p[d(m)]\, \frac{\partial d}{\partial m}\, \mathrm{d}m = \int_{m_1}^{m_2} P(m)\, dm = P(m_1, m_2)$$

$$(3.7)$$

where (m_1, m_2) corresponds to (d_1, d_2); that is $m_1 = m(d_1)$ and $m_2 = m(d_2)$. Then by inspection, $p(m) = p[d(m)]\partial d/\partial m$. In some cases, such as the function $m(d) = 1/d$, the derivative $\partial d/\partial m$ is negative, corresponding to the situation where $m_1 > m_2$ and the direction of integration along the m-axis is reversed. To account for this case, we take the absolute value of $\partial d/\partial m$:

$$p(m) = p[d(m)] \, |\frac{\partial d}{\partial m}| \tag{3.8}$$

with the understanding that the integration is always performed from the smaller of (m_1, m_2) to the larger.

The significance of the $\partial d / \partial m$ factor can be understood by considering the uniform probability density function $P(d) = 1$ on the interval $0 \le d \le 1$ together with the function $m = 2d$. The interval $0 \le d \le 1$ corresponds to the interval $0 \le m \le 2$ and $\partial d / \partial m = \frac{1}{2}$. Equation (3.8) gives $p(m) = 1 \times \frac{1}{2} = \frac{1}{2}$, that is $p(m)$ is also a uniform probability density function, but with a different normalization than $p(d)$. The total area, A, beneath both $p(d)$ and $p(m)$ is the same, $A = 1 \times 1 = 2 \times \frac{1}{2} = 1$. Thus, $\partial d / \partial m$ acts as a scale factor that accounts for the way that the stretching or squeezing of the m-axis relative to the d-axis affects the calculation of area.

As linear functions, such as $m = cd$, where c is a constant, are common in data analysis, we mention one of their important properties here. Suppose that $p(d)$ has mean, $\bar{d}$, and variance, σ_d^2. Then, the mean, $\bar{m}$, and variance, σ_m^2, of $p(m)$ are as follows:

$$\bar{m} = \int m \, p(m) \, \mathrm{d}m = \int cd \, p[d(m)] \frac{\partial d}{\partial m} \frac{\partial m}{\partial d} \mathrm{d}d = c \int d \, p(d) \, \mathrm{d}d = c\bar{d} \tag{3.9}$$

$$\sigma_m^2 = \int (m - \bar{m})^2 \, p(m) \, \mathrm{d}m = \int (cd - c\bar{d})^2 \, p[d(m)] \frac{\partial d}{\partial m} \frac{\partial m}{\partial d} \mathrm{d}d$$
$$= c^2 \int (d - \bar{d})^2 \, p(d) \mathrm{d}d = c^2 \sigma_d^2 \tag{3.10}$$

Thus, in the special case of the linear function, $m = cd$, the formulas

$$\bar{m} = c\bar{d} \quad \text{and} \quad \sigma_m^2 = c^2 \sigma_d^2 \tag{3.11}$$

do not depend on the functional form of the probability density function, $p(d)$.

In another example of computing the probability density function of a function of a random variable, consider the probability density function $p(d) = 1$ on the interval $0 \le d \le 1$ and the function, $m = d^2$. The corresponding interval of m is $0 \le d \le 1$, $d = m^{1/2}$, and $\partial d / \partial m = \frac{1}{2} m^{-1/2}$. The probability density function, $p(m)$, is given as

$$p(m) = p[d(m)] \frac{\partial d}{\partial m} = 1 \times \frac{1}{2} m^{-1/2} = \frac{1}{2} m^{-1/2} \tag{3.12}$$

on the interval $0 \le m \le 1$. Unlike $p(d)$, the probability density function $p(m)$ is not uniform but rather has a peak (actually a singularity, but an integrable one) at $m = 0$ (Figure 3.9).

In this section, we have described how the probability density function of a function of a single observation, d, can be computed. For this technique to be truly useful,

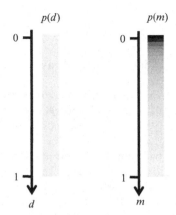

Figure 3.9 (Left) The probability density function, $p(d) = 1$, on the interval $0 \leq d \leq 1$. (Right) The probability density function, $p(m)$, which is derived from $p(d)$ together with the functional relationship, $m = d^2$. *MatLab* Script eda03_08.

we must generalize it so that we can compute the probability density function of a function of a set of *many* observations. However, before tackling this problem, we will need to discuss how to describe the probability density function of a set of observations.

3.6 Joint probabilities

Consider the following scenario: A certain island is inhabited by two species of birds, gulls and pigeons. Either species can be either tan or white in color. A census determines that 100 birds live on the island, 30 tan pigeons, 20 white pigeons, 10 tan gulls and 40 white gulls. Now suppose that we visit the island. The probability of sighting a bird of species, s, and color, c, can be summarized by a 2×2 table (Figure 3.10), which is called the *joint probability* of species, s, and color, c, and is denoted by $P(s, c)$. Note that the elements of the table must sum to 100%: $\sum_{i=1}^{2}\sum_{j=1}^{2}P(s_i, c_j) = 100\%$. $P(s, c)$ completely describes the situation, and other probabilities can be calculated from it. If we sum the elements of each row, we obtain the probability that the bird is a given

$P(s,c)$	Color (c)	
	Tan (t)	White (w)
Pigeon (p)	30%	20%
Gull (g)	10%	40%

Species (s)

Figure 3.10 Table of $P(s, c)$, the joint probability of color and species.

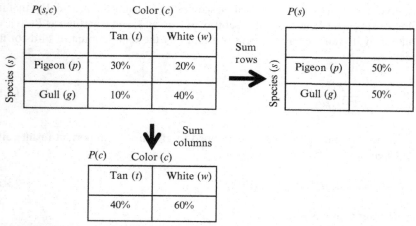

Figure 3.11 Computing $P(s)$ and $P(c)$ from $P(s, c)$.

species, irrespective of its color: $P(s) = \sum_{j=1}^{2} P(s, c_j)$. Likewise, if we sum the elements of each column, we obtain the probability that the bird is a given color, irrespective of its species: $P(c) = \sum_{i=1}^{2} P(s_i, c)$ (Figure 3.11).

Suppose that we observe the color of a bird but are not able to identify its species. Given that its color is c, what is the probability that it is of species s? This is called the *conditional probability* of s given c and is written as $P(s|c)$. We compute it by dividing every element of $P(s, c)$ by the total number of birds of that color (Figure 3.12):

$$P(s|c) = \frac{P(s,c)}{\sum_{i=1}^{2} P(s_i, c)} = \frac{P(s,c)}{P(c)} \tag{3.13}$$

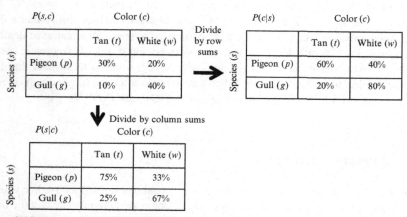

Figure 3.12 Computing $P(s|c)$ and $P(c|s)$ from $P(s, c)$.

Alternatively, we could ask, given that its species is s, what is the probability that it is of color c? This is the conditional probability of c given s, and is written as $P(c|s)$. We compute it by dividing every element of $P(s, c)$ by the total amount of birds of that species:

$$P(c|s) = \frac{P(s,c)}{\sum_{j=1}^{2}P(s,c_j)} = \frac{P(s,c)}{P(s)} \tag{3.14}$$

Equations (3.13) and (3.14) can be combined to give a very important result called *Bayes Theorem*:

$$P(s,c) = P(s|c)P(c) = P(c|s)P(s) \tag{3.15}$$

Bayes Theorem can also be written as

$$P(s|c) = \frac{P(c|s)P(s)}{P(c)} = \frac{P(c|s)P(s)}{\sum_i P(s_i,c)} = \frac{P(c|s)P(s)}{\sum_i P(c|s_i)P(s_i)}$$

$$P(c|s) = \frac{P(s|c)P(c)}{P(s)} = \frac{P(s|c)P(c)}{\sum_j P(s,c_j)} = \frac{P(s|c)P(c)}{\sum_j P(s|c_j)P(c_j)} \tag{3.16}$$

Note that we have used the following relations:

$$P(c) = \sum_i P(s_i,c) = \sum_i P(c|s_i)P(s_i)$$

$$P(s) = \sum_j P(s,c_j) = \sum_j P(s|c_j)P(c_j) \tag{3.17}$$

Note that the two conditional probabilities are *not* equal; that is, $P(s|c) \neq P(c|s)$. Confusion between the two is a major source of error in both scientific and popular circles! For example, the probability that a person who contracts pancreatic cancer "C" will die "D" from it is very high, $P(D|C) \approx 90\%$. In contrast, the probability that a dead person succumbed to pancreatic cancer, as contrasted to some other cause of death, is much lower, $P(C|D) \approx 1.4\%$. Yet, the news of a person dying of pancreatic cancer usually provokes more fear among people who have no reason to suspect that they have the disease than this low probability warrants (as after the tragic death of actor Patrick Swayze in 2009). They are confusing $P(C|D)$ with $P(D|C)$.

3.7 Bayesian inference

Suppose that we are told that an observer on the island has sighted a bird. We want to know whether it is a pigeon. Before being told its color, we can only say that the probability of its being a pigeon is 50%, because pigeons comprise 50% of the birds on the island. Now suppose that the observer tells us the bird is tan. We can use Bayes Theorem (Equation 3.15) to *update* our probability estimate:

$$P(s=p|c=t) = \frac{P(c=t|s=p)P(s=p)}{P(c=t|s=p)P(s=p) + P(c=t|s=g)P(s=g)}$$

$$= \frac{0.60 \times 0.5}{0.60 \times 0.5 + 0.20 \times 0.5} = \frac{0.30}{0.40} = 75\%$$

(3.18)

The probability that it is a pigeon improves from 50% to 75%. Note that the numerator in Equation (3.18) is the percentage of tan pigeons, while the denominator is the percentage of tan birds. As we will see later, Bayesian inference is very widely applied as a way to assess how new measurements improve our state of knowledge.

3.8 Joint probability density functions

All of the formalism developed in the previous section for discrete probabilities carries over to the case where the observations are continuously changing variables. With just two observations, d_1 and d_2, the probability that the observations are near (d_1, d_2) is described by a two-dimensional probability density function, $p(d_1, d_2)$. Then, the probability, P, that d_1 is between d_1^L and d_1^R, and d_2 is between d_2^L and d_2^R is given as follows:

$$P(d_1^L, d_1^R, d_2^L, d_2^R) = \int_{d_1^L}^{d_1^R} \int_{d_2^L}^{d_2^R} p(d_1, d_2) \, \mathrm{d}d_1 \mathrm{d}d_2$$

(3.19)

The probability density function for one datum, irrespective of the value of the other, can be obtained by integration (Figure 3.13):

$$p(d_1) = \int_{d_2^{\min}}^{d_2^{\max}} p(d_1, d_2) \mathrm{d}d_2 \quad \text{and} \quad p(d_2) = \int_{d_1^{\min}}^{d_1^{\max}} p(d_1, d_2) \, \mathrm{d}d_1$$

(3.20)

Here, $d_1^{\min}$, $d_2^{\max}$ is the overall range of d_1 and $d_2^{\min}$, $d_2^{\max}$ is the overall range of d_2. Note that the joint probability density function is normalized so that the total probability is unity:

$$P(d_1^{\min}, d_1^{\max}, d_2^{\min}, d_2^{\max}) = \int_{d_1^{\min}}^{d_1^{\max}} \int_{d_2^{\min}}^{d_2^{\max}} p(d_1, d_2) \, \mathrm{d}d_1 \mathrm{d}d_2 = 1$$

(3.21)

The mean and variance are computed in a way exactly analogous to a univariate probability density function:

$$\bar{d}_1 = \int\int d_1 \, p(d_1, d_2) \, \mathrm{d}d_1 \, \mathrm{d}d_2 \quad \text{and} \quad \bar{d}_2 = \int\int d_2 \, p(d_1, d_2) \, \mathrm{d}d_1 \, \mathrm{d}d_2$$

$$\sigma_1^2 = \int\int (d_1 - \bar{d}_1)^2 \, p(d_1, d_2) \, \mathrm{d}d_1 \, \mathrm{d}d_2 \quad \text{and} \quad \sigma_2^2 = \int\int (d_2 - \bar{d}_2)^2 \, p(d_1, d_2) \, \mathrm{d}d_1 \, \mathrm{d}d_2$$

(3.22)

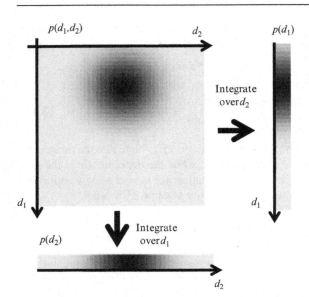

Figure 3.13 Computing the univariate probability density functions, $p(d_1)$ and $p(d_2)$, from the joint probability density function, $p(d_1, d_2)$. *MatLab* script eda03_09

Note, however, that these formulas can be simplified to just the one-dimensional formulas (Equations 3.3 and 3.4), as the factors multiplying $p(d_1, d_2)$ can be moved outside of one integral. The interior integral then reduces $p(d_1, d_2)$ to $p(d_1)$:

$$\bar{d}_1 = \int\int d_1 p(d_1, d_2)\, \mathrm{d}d_1\, \mathrm{d}d_2 = \int d_1 \int p(d_1, d_2)\, \mathrm{d}d_2\, \mathrm{d}d_1 = \int d_1 p(d_1)\, \mathrm{d}d_1$$

$$\sigma_1^2 = \int\int (d_1 - \bar{d}_1)^2\, p(d_1, d_2)\, \mathrm{d}d_1\, \mathrm{d}d_2 = \int (d_1 - \bar{d}_1)^2 \int p(d_1, d_2)\, \mathrm{d}d_2\, \mathrm{d}d_1$$

$$= \int (d_1 - \bar{d}_1)^2\, p(d_1)\, \mathrm{d}d_1 \tag{3.23}$$

and similarly for $\bar{d}_2$ and σ_2^2. In *MatLab*, we represent the joint probability density function, $p(d_1, d_2)$, as the matrix, P, where d_1 varies along the rows and d_2 along the columns. Normally, we would choose d_1 and d_2 to be evenly sampled with spacing Dd so that they can be represented by column vectors of length, L. A uniform probability density function is then computed as follows:

```
d1 = Dd*[0:L−1]';
d2 = Dd*[0:L−1]';
P = ones(L,L);
norm = (Dd^2)*sum(sum(P));
P = P/norm;                                      (MatLab eda03_10)
```

Note that the sum of all the elements of a matrix P is sum(sum(P)). The first sum() returns a row vector of column sums and the second sums up that vector and produces a scalar.

An example of a spatially variable probability density function is

$$p(d_1, d_2) = \frac{1}{2\pi\sigma_1\sigma_2} \exp\left\{ -\frac{(d_1 - \bar{d}_1)^2}{2\sigma_1^2} - \frac{(d_2 - \bar{d}_2)^2}{2\sigma_2^2} \right\} \tag{3.24}$$

where $\bar{d}_1$, $\bar{d}_2$, σ_1^2, and σ_2^2 are constants. This probability density function is a generalization of the Normal probability density functions, and is defined so that the $\bar{d}_1$ and $\bar{d}_2$ are means and the σ_1^2 and σ_2^2 are variances. We will discuss it in more detail, below. In *MatLab*, this probability density function is computed:

```
d1 = Dd* [0:L−1]';
d2 = Dd* [0:L−1]';
norm=1/(2*pi*s1*s2);
p1=exp(−((d1−d1bar).^2)/(2*s1*s1));
p2=exp(−((d2−d2bar).^2)/(2*s2*s2));
P=norm*p1*p2';                                         (MatLab eda03_11)
```

Here, d1bar, d2bar, s1, and s2 correspond to $\bar{d}_1$, $\bar{d}_1$, σ_1, and σ_2, respectively. Note that we have made use here of a vector product of the form, p1*p2', which creates a matrix, P, whose elements are $P_{ij} = p_i p_j$.

In *MatLab*, the joint probability density function is reduced to a univariate probability density function by using the sum() function to approximate an integral:

```
% sum along columns, which integrates P along d2 to get p1
p1 = Dd*sum(P,2);
% sum along rows, which integrates P along d1 to get p2
p2 = Dd*sum(P,1)';
```

The mean is then calculated as

```
d1mean = Dd*sum(d1 .* p1 );
d2mean = Dd*sum(d2 .* p2 );                            (MatLab eda03_12)
```

and the variance is computed as

```
sigma12 = Dd*sum( ((d1−d1mean).^2) .* p1 );
sigma22 = Dd*sum( ((d2−d2mean).^2) .* p2 ); (MatLab eda03_13)
```

Finally, we define the conditional probability density functions, $p(d_1|d_2)$ and $p(d_2|d_1)$ in a way that is analogous to the discrete case (Figure 3.14). Bayes Theorem then becomes as follows:

$$p(d_1|d_2) = \frac{p(d_2|d_1)\, p(d_1)}{p(d_2)} = \frac{p(d_2|d_1)\, p(d_1)}{\int p(d_1, d_2)\, dd_1} = \frac{p(d_2|d_1)\, p(d_1)}{\int p(d_2|d_1)\, p(d_1)\, dd_1}$$

$$p(d_2|d_1) = \frac{p(d_1|d_2)\, p(d_2)}{p(d_1)} = \frac{p(d_1|d_2)\, p(d_2)}{\int p(d_1, d_2)\, dd_2} = \frac{p(d_1|d_2)\, p(d_2)}{\int p(d_1|d_2)\, p(d_2)\, dd_2}$$

$$\tag{3.25}$$

Here, we have relied on the relations

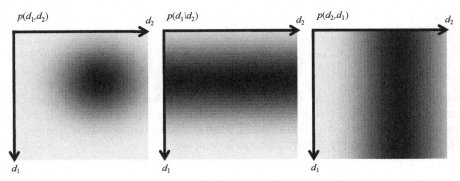

Figure 3.14 The joint probability density function, $p(d_1, d_2)$, and the two conditional probability density functions computed from it, $p(d_1|d_2)$, and $p(d_2|d_1)$. *MatLab* script eda03_14.

$$p(d_1, d_2) = p(d_1|d_2)p(d_2) \text{ and } p(d_1) = \int p(d_1, d_2)\, \mathrm{d}d_2 = \int p(d_1|d_2)\, p(d_2)\, \mathrm{d}d_2$$

$$p(d_1, d_2) = p(d_2|d_1)p(d_1) \text{ and } p(d_2) = \int p(d_1, d_2)\, \mathrm{d}d_1 = \int p(d_2|d_1)\, p(d_1)\, \mathrm{d}d_1$$

$$(3.26)$$

In *MatLab*, the conditional probability density functions are computed using Equation (3.25):

```
% sum along columns, which integrates P along d2 to get p1=p(d1)
p1 = Dd*sum(P,2);
% sum along rows, which integrates P along d1 to get p2=p(d2)
p2 = Dd*sum(P,1)';
% conditional distribution P1g2 = P(d1|d2) = P(d1,d2)/p2
P1g2 = P ./ (ones(L,1)*p2');
% conditional distribution P2g1 = P(d2|d1) = P(d1,d2)/p1
P2g1 = P ./ (p1*ones(L,1)');        (MatLab eda03_14)
```

Note that the *MatLab* sum() function, when operating on a matrix, returns a column vector of row sums or a row vector of column sums, depending on whether its second argument is 1 or 2, respectively.

3.9 Covariance

In addition to describing the behavior of d_1 and d_2 individually, the joint probability density function $p(d_1, d_2)$ also describes the degree to which they correlate. The sequence of *pairs* of measurements, (d_1, d_2), might contain a systematic pattern where unusually high values of d_1 occur along with unusually high values of d_2, and unusually low values of d_1 occur along with unusually low values of d_2. In this case, d_1 and d_2 are said to be positively correlated (Figure 3.15). Alternatively, high values of d_1 can occur along with unusually low values of d_2, and unusually low values of d_1 can occur along with unusually high values of d_2. This is a negative correlation.

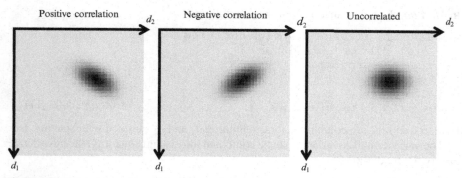

Figure 3.15 The observations d_1 and d_2 can have either a positive or a negative correlation or be uncorrelated, according to the shape of the joint probability density function, $p(d_1, d_2)$. *MatLab* scripts eda03_15 and eda03_16.

These cases correspond to joint probability density functions, $p(d_1, d_2)$, that have a slanting ridge of high probability. Probability density functions that have neither a positive nor a negative correlation are said to be *uncorrelated*.

Suppose that we divide the (d_1, d_2) plane into four quadrants of alternating sign, centered on a typical value such as the mean, $(\bar{d}_1, \bar{d}_2)$. A positively correlated probability density function will have most of its probability in the positive quadrants, and a negatively correlated probability density function will have most of its probability in the negative quadrants. This suggests a strategy for quantifying correlation: multiply $p(d_1, d_2)$ by a function, say $s(d_1, d_2)$, that has a four-quadrant alternating sign pattern and integrate (Figure 3.16). The resulting number quantifies the degree of correlation. When $s(d_1, d_2) = (d_1 - \bar{d}_1)(d_2 - \bar{d}_2)$, the result is called the covariance, $\sigma_{1,2}$:

$$\sigma_{1,2} = \int\int (d_1 - \bar{d}_1)(d_2 - \bar{d}_2)\, p(d_1, d_2)\, \mathrm{d}d_1\, \mathrm{d}d_2 \tag{3.27}$$

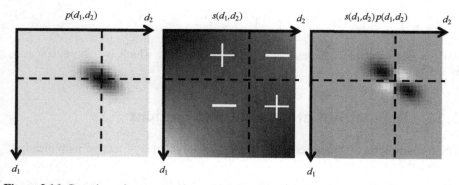

Figure 3.16 Covariance is computed by multiplying the probability density functions, $p(d_1, d_2)$, by the four-quadrant function, $s(d_1, d_2)$, and integrating to obtain a number called the covariance. In the positively correlated case shown, $p(d_1, d_2)$ has more area in the positive quadrants, so the covariance is positive. *MatLab* script eda03_17.

In *MatLab*, the covariance is computed as follows:

```
% make the alternating sign function
S = (d1-d1bar) * (d2-d2bar)';
% form the product
SP = S .* P;
% integrate
cov = (Dd^2)*sum(sum(SP));                    (MatLab eda03_17)
```

Here, d1 and d2 are column vectors containing d_1 and d_2 sampled with spacing, Dd.

The variance and covariance can be combined into a single quantity, the covariance matrix, $\mathbf{C}$, by defining

$$C_{ij} = \int \int (d_i - \bar{d}_i)(d_j - \bar{d}_j)p(d_1, d_2) \, \mathrm{d}d_1 \, \mathrm{d}d_2 \tag{3.28}$$

Its diagonal elements are the variances, $\sigma_1{}^2$ and $\sigma_2{}^2$, and its off-diagonal elements are the covariance, $\sigma_{1,2}$. Note that the matrix, $\mathbf{C}$, is symmetric.

3.10 Multivariate distributions

In the previous section, we have examined joint probability density functions of exactly two observations, d_1 and d_2. In practice, the number of observations can be arbitrarily large, $(d_1, \ldots d_N)$. The corresponding multivariate probability density function, $p(d_1, \ldots d_N)$, gives the probability that a set of observations will be in the vicinity of the point $(d_1, \ldots d_N)$. We will write this probability density function as p($\mathbf{d}$), where $\mathbf{d}$ is a column-vector of observations $\mathbf{d} = [d_1, \ldots d_N]^{\mathrm{T}}$. The mean, $\bar{\mathbf{d}}$, is a length-N column-vector whose components are the means of each observation. The covariance matrix, $\mathbf{C}$, is a $N \times N$ matrix whose i-th diagonal element is the variance of observation, d_i, and whose (i, j) off-diagonal element is the covariance of observations d_i and d_j.

$$\bar{d}_i = \int d_i \, p(\mathbf{d})\mathrm{d}^{\mathrm{N}}d \quad \text{and} \quad C_{ij} = \int (d_i - d_i)(d_j - d_j) \, p(\mathbf{d})\mathrm{d}^{\mathrm{N}}d \tag{3.29}$$

All the integrals are N-dimensional multiple integrals, which we abbreviate here as $\int \mathrm{d}^{\mathrm{N}}d$.

3.11 The multivariate Normal distributions

The formula for a N-dimensional Normal probability density function with mean, $\bar{\mathbf{d}}$, and covariance matrix, $\mathbf{C}$, is

$$p(\mathbf{d}) = \frac{1}{(2\pi)^{N/2}|\mathbf{C}|^{\frac{1}{2}}} \exp\{-\tfrac{1}{2}(\mathbf{d} - \bar{\mathbf{d}})^{\mathrm{T}}\mathbf{C}^{-1}(\mathbf{d} - \bar{\mathbf{d}})\} \tag{3.30}$$

where $|\mathbf{C}|$ is the determinant of the covariance matrix, $\mathbf{C}$, and $\mathbf{C}^{-1}$ is its matrix inverse. This probability density function bears a resemblance to the univariate Normal probability density function (Equation 3.5)—indeed it reduces to it in the $N = 1$ case. Equation (3.30) can be understood in the following way: The leading factor of $(2\pi)^{-N/2}|\mathbf{C}|^{-\frac{1}{2}}$ is just a normalization factor, chosen so that $\int p(\mathbf{d})\mathrm{d}^N d = 1$. The rest of the equation is the key part of the Normal curve, and contains, as expected, an exponential whose argument is quadratic in $\mathbf{d}$. The most general quadratic that can be formed from $\mathbf{d}$ is $(\mathbf{d} - \mathbf{d}')\mathbf{M}(\mathbf{d} - \mathbf{d}')^\mathrm{T}$, where $\mathbf{d}$' is an arbitrary vector and $\mathbf{M}$ is an arbitrary matrix. The specific choices $\mathbf{d}' = \bar{\mathbf{d}}$ and $\mathbf{M} = \mathbf{C}^{-1}$ are controlled by the desire to have the mean and covariance of the probability density function be exactly $\bar{\mathbf{d}}$ and $\mathbf{C}$, respectively. Therefore, we need only to convince ourselves that Equation (3.30) has the requisite total probability, mean, and covariance.

Surprisingly, given the complexity of Equation (3.30), these requirements can be easily checked by direct calculation. All that is necessary is to transform the probability density function to the new variable $\mathbf{y} = \mathbf{C}^{-\frac{1}{2}}(\mathbf{d} - \bar{\mathbf{d}})$ and perform the requisite integrations. However, in order to proceed, we need to recall that the rule for transforming a multidimensional integral (the analog to Equation 3.7) is

$$\int p(\mathbf{d})\mathrm{d}^N d = \int p(\mathbf{d}(\mathbf{y}))\left|\frac{\partial \mathbf{d}}{\partial \mathbf{y}}\right|\mathrm{d}^N y = \int p(\mathbf{d}(\mathbf{y}))\, J(\mathbf{y})\mathrm{d}^N y \qquad (3.31)$$

where $J(\mathbf{y}) = |\partial \mathbf{d}/\partial \mathbf{y}|$ is the *Jacobian determinant*; that is, the determinant of the matrix whose elements are $\partial d_i/\partial y_j$. (The rule, $\mathrm{d}^N d = |\partial \mathbf{d}/\partial \mathbf{y}|\mathrm{d}^N y$ is a multidimensional generalization of the ordinary chain rule, $\mathrm{d}d = |\mathrm{d}d/\mathrm{d}y|\mathrm{d}y$). In our case, $\mathbf{y} = \mathbf{C}^{-\frac{1}{2}}(\mathbf{d} - \bar{\mathbf{d}})$ and $|\partial \mathbf{d}/\partial \mathbf{y}| = |\mathbf{C}|^{\frac{1}{2}}$. Then, the area under $p(\boldsymbol{d})$ is

$$\int p(\boldsymbol{d})\mathrm{d}^N d = \frac{1}{(2\pi)^{N/2}|\mathbf{C}|^{\frac{1}{2}}}\int \exp\{-\tfrac{1}{2}\mathbf{y}^T\mathbf{y}\}|\mathbf{C}|^{\frac{1}{2}}\mathrm{d}^N y$$

$$= \frac{1}{(2\pi)^{N/2}}\int \exp\{-\tfrac{1}{2}\mathbf{y}^T\mathbf{y}\}\mathrm{d}^N y$$

$$= \prod_{i=1}^{N}\int \frac{1}{(2\pi)^{\frac{1}{2}}}\exp\{-\tfrac{1}{2}y_i^2\}\mathrm{d}y_i = \prod_{i=1}^{N}1 = 1 \qquad (3.32)$$

Note that the factor of $|\mathbf{C}|^{\frac{1}{2}}$ arising from the Jacobian determinant cancels the $|\mathbf{C}|^{-\frac{1}{2}}$ in the normalization factor. As the final integral is just a univariate Normal probability density function with zero mean and unit variance, its integral (total area) is unity. We omit here the integrals for the mean and covariance. They are algebraically more complex but are performed in an analogous fashion.

We run into a notational difficulty when computing a multivariate Normal probability density function in *MatLab*, of the sort that is frequently encountered when coding numerical algorithms. Heretofore, we have been using column vectors starting with lower-case "d" names, such as d1 and d2, to represent quantities sampled at

L evenly spaced increments (d_1 and d_2 in this case). Now, however, the equation for the Normal probability density function (Equation 3.30) requires us to group corresponding ds into an N-vector, $\mathbf{d} = [d_1, d_2, \ldots d_N]^T$, which appears in the matrix multiplication, $(\mathbf{d} - \bar{\mathbf{d}})^T \mathbf{C}^{-1} (\mathbf{d} - \bar{\mathbf{d}})$. Thus, we are tempted to define quantities that will also have lower-case "d" names, but will be N-vectors, which violates the naming convention. Furthermore, the *MatLab* syntax becomes more inscrutable, for the $N \times N$ matrix multiplication needs to be performed at each of L^2 combinations of (d_1, d_2) in order to evaluate $p(d_1, d_2)$ on a $L \times L$ grid. There is no really good way around these problems, but we put forward two different strategies for dealing with them.

The first is to use only L-vectors, and explicitly code the $N \times N$ matrix multiplication, instead of having *MatLab* perform it. The $N = 2$ case needed for the two-dimensional probability density function $p(d_1, d_2)$ is

$$(\mathbf{d} - \bar{\mathbf{d}})^T \mathbf{C}^{-1} (\mathbf{d} - \bar{\mathbf{d}})$$
$$= [\mathbf{C}^{-1}]_{11} (d_1 - \bar{d}_1)^2 + [\mathbf{C}^{-1}]_{22} (d_2 - \bar{d}_2)^2 + 2[\mathbf{C}^{-1}]_{12} (d_1 - \bar{d}_1)(d_2 - \bar{d}_2)$$
$$(3.33)$$

Note that we have made use of the fact that the covariance matrix, $\mathbf{C}$, is symmetric. The *MatLab* code for the Normal probability density function is then

```
CI=inv(C);
norm=1/(2*pi*sqrt(det(C)));
dd1=d1-d1bar;
dd2=d2-d2bar;
P=norm*exp(-0.5*CI(1,1)*(dd1.^2)*ones(N,1)'...
    -0.5*CI(2,2)*ones(N,1)*(dd2.^2)'...
    -CI(1,2)*dd1*dd2');                              (MatLab eda03_15)
```

Here, C is the covariance matrix, and d1 and d2 are d_1 and d_2 sampled at L evenly spaced increments.

The second strategy is to define both L-vectors and N-vectors and to use the N-vectors to compute $(\mathbf{d} - \bar{\mathbf{d}})^T \mathbf{C}^{-1} (\mathbf{d} - \bar{\mathbf{d}})$ using the normal *MatLab* syntax for matrix multiplication. In the $N = 2$ case, the 2-vectors must be formed explicitly for each of the $L \times L$ pairs of (d_1, d_2), which is best done inside a pair of for loops:

```
CI=inv(C);
norm=1/(2*pi*sqrt(det(C)));
P=zeros(L,L);
for i = [1:L]
for j = [1:L]
    dd = [d1(i)-d1bar, d2(j)-d2bar]';
    P(i,j)=norm*exp(-0.5 * dd' * CI * dd );
end
end                                                 (MatLab eda03_16)
```

The outer for loop corresponds to the L elements of d1 and the inner to the L elements of d2. The code for the second method is arguably a little more transparent than the code for the first, and is probably the better choice, especially for beginners.

3.12 Linear functions of multivariate data

Suppose that a column-vector of model parameters, $\mathbf{m}$, is derived from a column-vector of observations, $\mathbf{d}$, using the linear formula, $\mathbf{m} = \mathbf{Md}$, where $\mathbf{M}$ is some matrix. Suppose that the observations are random variables with a probability density function, $p(\mathbf{d})$, with mean, $\bar{\mathbf{d}}$, and covariance, $\mathbf{C}_d$. The model parameters are random variables too, with probability density function, $p(\mathbf{m})$. We would like to derive the functional form of the probability density function, $\mathbf{p}(\mathbf{m})$, as well as calculate its mean, $\bar{\mathbf{m}}$, and covariance, $\mathbf{C}_m$.

If $p(\mathbf{d})$ is a Normal probability density function, then $p(\mathbf{m})$ is also a Normal probability density function, as can be seen by transforming $p(\mathbf{d})$ to $p(\mathbf{m})$ using the rule (see Equation 3.31):

$$p(\mathbf{d}) = p(\mathbf{d}(\mathbf{m}))\left|\frac{\partial \mathbf{d}}{\partial \mathbf{m}}\right| = p(\mathbf{d}(\mathbf{m}))\, J(\mathbf{m}) \tag{3.34}$$

As $\mathbf{m} = \mathbf{Md}$, the Jacobian determinant is $J(\mathbf{m}) = |\partial \mathbf{d}/\partial \mathbf{m}| = |\mathbf{M}^{-1}| = |\mathbf{M}|^{-1}$. Then:

$$
\begin{aligned}
p(\mathbf{m}) &= p(\mathbf{d}(\mathbf{m}))\, J(\mathbf{m}) \\
&= \frac{1}{(2\pi)^{N/2}|\mathbf{C}_d|^{\frac{1}{2}}|\mathbf{M}|} \exp\{-\tfrac{1}{2}(\mathbf{M}^{-1}\mathbf{m} - \mathbf{M}^{-1}\mathbf{M}\bar{\mathbf{d}})^{\mathrm{T}}\mathbf{C}_d^{-1}(\mathbf{M}^{-1}\mathbf{m} - \mathbf{M}^{-1}\mathbf{M}\bar{\mathbf{d}})\} \\
&= \frac{1}{(2\pi)^{N/2}|\mathbf{M}\mathbf{C}_d\mathbf{M}^{\mathrm{T}}|^{\frac{1}{2}}} \exp\{-\tfrac{1}{2}(\mathbf{m} - \mathbf{M}\bar{\mathbf{d}})^{\mathrm{T}}[\mathbf{M}^{-1\mathrm{T}}\mathbf{C}_d^{-1}\mathbf{M}^{-1}](\mathbf{m} - \mathbf{M}\bar{\mathbf{d}})\} \\
&= \frac{1}{(2\pi)^{N/2}|\mathbf{C}_m|^{1/2}} \exp\{-\tfrac{1}{2}(\mathbf{m} - \bar{\mathbf{m}})^{\mathrm{T}}\mathbf{C}_m^{-1}(\mathbf{m} - \bar{\mathbf{m}})\}
\end{aligned}
$$

where $\bar{\mathbf{m}} = \mathbf{M}\bar{\mathbf{d}}$ and $\mathbf{C}_m^{-1} = \mathbf{M}^{-1\mathrm{T}}\mathbf{C}_d^{-1}\mathbf{M}^{-1}$ $\qquad\qquad$ (3.35)

Note that we have used the identities $(\mathbf{AB})^{\mathrm{T}} = \mathbf{B}^{\mathrm{T}}\mathbf{A}^{\mathrm{T}}$, $(\mathbf{AB})^{-1} = \mathbf{B}^{-1}\mathbf{A}^{-1}$, $|\mathbf{AB}| = |\mathbf{A}|\,|\mathbf{B}|$, $|\mathbf{C}^{\mathrm{T}}| = |\mathbf{C}|$, and $|\mathbf{C}^{-1}| = |\mathbf{C}|^{-1}$. Thus, the transformed mean and covariance matrix are given by the simple rule

$$\bar{\mathbf{m}} = \mathbf{M}\bar{\mathbf{d}} \quad \text{and} \quad \mathbf{C_m} = \mathbf{MC}_d\mathbf{M}^{\mathrm{T}} \tag{3.36}$$

Equation (3.36) is *very* important, for it shows how to calculate the mean and covariance matrix of the model parameters, given the mean and variance of the data. The covariance formula, $\mathbf{C}_m = \mathbf{MC}_d\mathbf{M}^{\mathrm{T}}$, can be thought of as a *rule for error propagation*. It links error in the data to error in the model parameters. The rule is shown to be true even when $\mathbf{M}$ is not square and when $\mathbf{M}^{-1}$ does not exist (see Note 3.1).

As an example, consider the case where we measure the masses of two objects A and B, by first putting A on a scale and then adding B, without first removing A. The observation, d_1, is the mass of object A and the observation d_2 is the combined mass of objects A and B. We assume that the measurements are a random process with probability density function $p(d_1, d_2)$, with means $\bar{d}_1$ and $\bar{d}_2$, variance, σ_d^2, and covariance, $\sigma_{1,2} = 0$. Note that both variances are the same, indicating that both measurements have the same amount of error, and that the covariance is zero, indicating that the two measurements are uncorrelated. Suppose that we want to compute the mass of object B and the difference in masses of objects B and A. Then, $B = (B + A) - A \rightarrow m_1 = d_2 - d_1$, and $B - A = (B + A) - 2A \rightarrow m_2 = d_2 - 2d_1$. The matrix, $\mathbf{M}$, is given by

$$\mathbf{M} = \begin{bmatrix} -1 & 1 \\ -2 & 1 \end{bmatrix} \tag{3.37}$$

The mean is

$$\bar{\mathbf{m}} = \mathbf{M}\bar{\mathbf{d}} = \begin{bmatrix} -1 & 1 \\ -2 & 1 \end{bmatrix} \begin{bmatrix} \bar{d}_1 \\ \bar{d}_2 \end{bmatrix} \quad \text{or} \quad \bar{m}_1 = \bar{d}_2 - \bar{d}_1 \quad \text{and} \quad \bar{m}_2 = \bar{d}_2 - 2\bar{d}_1 \tag{3.38}$$

The covariance matrix is

$$\mathbf{C}_m = \mathbf{M}\mathbf{C}_d\mathbf{M}^T = \begin{bmatrix} -1 & 1 \\ -2 & 1 \end{bmatrix} \begin{bmatrix} \sigma_d^2 & 0 \\ 0 & \sigma_d^2 \end{bmatrix} \begin{bmatrix} -1 & -2 \\ 1 & 1 \end{bmatrix} = \sigma_d^2 \begin{bmatrix} 2 & 3 \\ 3 & 5 \end{bmatrix}$$

or

$$\sigma_{m_1}^2 = 2\sigma_d^2 \quad \text{and} \quad \sigma_{m_2}^2 = 5\sigma_d^2 \quad \text{and} \quad \sigma_{m1,2} = 3\sigma_d^2 \tag{3.39}$$

Note that the *ms* have unequal variance even though the variances of the *ds* are equal, and that the *ms* have a non-zero covariance even though the covariance of the *ds* is zero (Figure 3.17). This process of forming model parameters from data often results in

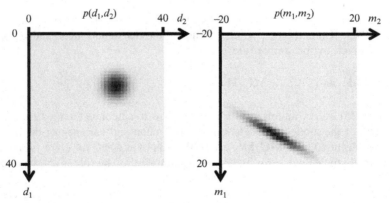

Figure 3.17 The joint probability density functions, $p(d_1, d_2)$ and $p(m_1, m_2)$ where $\mathbf{m} = \mathbf{M}\mathbf{d}$. See the text for the value of the matrix, $\mathbf{M}$. The data, (d_1, d_2), have mean (15, 25), variance (5, 5), and zero covariance. The model parameters, (m_1, m_2), have mean (10, −5), variance (10, 25), and covariance, 15. *MatLab* script eda03_18.

variances and degrees of correlation that are different from the underlying data. Depending on the situation, variance can be either reduced (good) or amplified (bad). While data are often observed through a measurement process that has uncorrelated errors (good), the model parameters formed from them usually exhibit strong correlations (not so good).

The rule for the mean and variance (Equation 3.36) is true even when the underlying probability density function, $p(\mathbf{m})$, is not Normal, as can be verified by a calculation analogous to that of Equations (3.9) and (3.10):

$$\bar{m}_i = \int m_i\, p(\mathbf{m})\, \mathrm{d}^N m$$

$$= \int \sum_j M_{ij} d_j\, p(\mathbf{d})\, \mathrm{d}^N d$$

$$= \sum_j M_{ij} \int d_j\, p(\mathbf{d}) d^N d = \sum_j M_{ij}\bar{d}_j \ \text{ or } \ \bar{\mathbf{m}} = \mathbf{M}\bar{\mathbf{d}} \tag{3.40}$$

$$[\mathbf{C_m}]_{ij} = \int (m_i - \bar{m}_i)(m_j - \bar{m}_j) p(\mathbf{m}) d^N m$$

$$= \int \sum_p M_{ip}(d_p - \bar{d}_p) \sum_q M_{jq}(d_q - \bar{d}_q) p(\mathbf{d}) d^N d$$

$$\tag{3.41}$$

$$= \sum_p M_{ip} \sum_q M_{jq} \int (d_p - \bar{d}_p)(d_q - \bar{d}_q) p(\mathbf{d}) d^N d$$

$$= \sum_p M_{ip} \sum_q M_{jq} [\mathbf{C_m}]_{pq} \quad \text{ or } \quad \mathbf{C}_m = \mathbf{M}\mathbf{C}_d\mathbf{M}^{\mathrm{T}}$$

Here, we have used the fact that

$$p(\mathbf{m}) d^N m = p(\mathbf{m}(\mathbf{d})) |\partial \mathbf{m}/\partial \mathbf{d}| |\partial \mathbf{d}/\partial \mathbf{m}| \mathrm{d}^N d = p(\mathbf{d}) \mathrm{d}^N d.$$

In the case of non-Normal probability density functions, these results need to be used cautiously. The relationship between variance and confidence intervals (e.g., the amount of probability falling between $m_1 - \sigma_{m1}$ and $m_1 + \sigma_{m1}$) varies from one probability density function to another.

Even in the case of Normal probability density functions, statements about confidence levels need to be made carefully, as is illustrated by the following scenario. Suppose that $p(d_1, d_2)$ represents the joint probability density function of two measurements, say the height and length of an organism, and suppose that these measurements are uncorrelated with equal variance, σ_d^2. As we might expect, the univariate probability density function $p(d_1) = \int p(d_1, d_2) \mathrm{d}d_2$, has variance, σ_d^2, and so the probability, P_1, that d_1 falls between $d_1 - \sigma_d$ and $d_1 + \sigma_d$ is 0.6827 or 68.27%. Likewise, the probability, P_2, that d_2 falls between $d_2 - \sigma_d$ and $d_2 + \sigma_d$ is also 0.6827 or 68.27%. But P_1 represents the probability of d_1 irrespective of the value of d_2 and P_2 represents the probability of d_2 irrespective of the value of d_1. The probability, P, that *both* d_1 and d_2 simultaneously fall within their respective one-sigma confidence intervals *is* $P = P_1 P_2 = (0.6827)^2 = 0.4660$ or 46.60%, which is significantly smaller than 68.27%.

Crib Sheet 3.1 Variance and error propagation

Variance $\sigma_{d_i}^2$ of Normally-distributed data d_i
68% of d_i's are within $\pm\sigma_{d_i}$ of the mean
95% of d_i's are within $\pm 2\sigma_{d_i}$ of the mean
a prior variance is based on knowledge of the measurement technique
a posterior variance is based on goodness of fit to a model

Covariance matrix of the data C_d
a matrix that expresses the statistical variability among data $\mathbf{d}$
diagonal elements express the variance of each datum d_i

$$[\mathbf{C}_d]_{ii} = \sigma_{d_i}^2$$

off-diagonal elements express correlation between pairs of data d_i and d_j
$[\mathbf{C}_d]_{ij} > 0$ when d_i and d_j are positively correlated
$[\mathbf{C}_d]_{ij} = 0$ when d_i and d_j are uncorrelated
$[\mathbf{C}_d]_{ij} < 0$ when d_i and d_j negatively correlated

Special cases
uncorrelated data with uniform variance σ_d^2

$$\mathbf{C}_d = \sigma_d^2\mathbf{I}$$

```
Cd = (sigmad^2)*eye(N,N);
```
uncorrelated data with nonuniform variance $\sigma_{d_i}^2$

$$\mathbf{C}_d = \mathrm{diag}\left(\sigma_{d_1}^2, \sigma_{d_2}^2, \ldots, \sigma_{d_N}^2\right)$$

Fundamental principal of error propagation
noisy data cause noisy results, because
any function of a random variable is itself a random variable

Linear function between model parameters m and data d
$\mathbf{m} = \mathbf{A}\mathbf{d}$ where $\mathbf{A}$ is a known matrix

Rule for error propagation
variability in model parameters $\mathbf{m}$ described by covariance matrix $\mathbf{C}_m$

$$\mathbf{C}_m = \mathbf{A}\mathbf{C}_d\mathbf{A}^{\mathrm{T}}$$

```
Cm = A*Cd*A';
```

95% confidence interval for model parameters

$$m_i^{est} - 2[\mathbf{C}_m]_{ii}^{1/2} < m_i^{true} < m_i^{est} + 2[\mathbf{C}_m]_{ii}^{1/2}$$

Problems

3.1 The univariate probability density function $p(d) = c(1 - d)$ is defined on the interval $0 \le d \le 1$. (A) What must the constant, c, be for the probability density function to be properly normalized? (B) Calculate the mode, median, and mean of this probability density function analytically.

3.2 The univariate *exponential probability density function* is $p(d) = \lambda \exp(-\lambda d)$ where d is defined on the interval $0 \le d < \infty$. The parameter, λ, is called the *rate parameter*. (A) Use *MatLab* to plot shaded column-vectors of this probability density function and to compute its mode, median, mean, and variance, for the cases $\lambda = 5$ and $\lambda = 10$. (B) Is it possible to control the mean and variance of this probability density function separately?

3.3 Suppose that $p(d)$ is a Normal probability density function with zero mean and unit variance. (A) Derive the probability density function of $m = |d|^{1/2}$, analytically. (B) Use *MatLab* to plot shaded column-vectors of this probability density function and to compute its mode, median, mean, and variance.

3.4 Suppose that a corpse is brought to a morgue and the coroner is asked to assess the probability that the cause of death was pancreatic cancer (as contrasted to some other cause of death). Before examining the corpse, the best estimate that the coroner can make is 1.4%, the death rate from pancreatic cancer in the general population. Now suppose the coroner performs a test for pancreatic cancer that is 99% accurate, both in the sense that if the cause of death was pancreatic cancer, the probability is 99% that the test results are positive, and if the cause of death was something else, the probability is 99% that the test results are negative. Let the cause of death be represented by the variable, D, which can take two discrete values, C, for pancreatic cancer and E, for something else. The test is represented by the variable, T, which can take two values, Y for positive and N for negative. (A) Write down the 2×2 table of the conditional probabilities $P(D|T)$. (B) Suppose the test results are negative. Use Bayesian Inference to assess the probability that the cause of death was pancreatic cancer. (C) How can the statement that the test is 99% accurate be used in a misleading way?

3.5 Suppose that two measurements, d_1 and d_2, are uncorrelated and with equal variance, σ_d^2. What is the variance and covariance of two model parameters, m_1 and m_2, that are the sum and difference of the ds?

3.6 Suppose that the vectors, $\mathbf{d}$, of N measurements are uncorrelated and with equal variance, σ_d^2. (A) What is the form of the covariance matrix, $\mathbf{C}_d$? (B) Suppose that $\mathbf{m} = \mathbf{Md}$. What property must $\mathbf{M}$ have to make the m's, as well as the d's, uncorrelated? (C) Does the matrix from Problem 3.5 have this property?

References

Chittibabu, C.V, Parthasarathy, N, 2000. Attenuated tree species diversity in human-impacted tropical evergreen forest sites at Kolli hills, Eastern Ghats, India. Biodiv. Conserv. 9, 1493–1519.

4 The power of linear models

4.1 Quantitative models, data, and model parameters

The purpose of any data analysis is to gain *knowledge* through a systematic examination of the data. While knowledge can take many forms, we assume here that it is primarily numerical in nature. We analyze data to infer, as best we can, the values of numerical quantities—*model parameters*, in the parlance of this book. The inference process is possible because the data and model parameters are linked through a *quantitative model*. In the very broadest sense, the model parameters and the data are linked though a functional relationship:

the data = a function of the model parameters

or

$$d_1 = g_1(m_1, m_2, \ldots, m_M)$$
$$d_2 = g_2(m_1, m_2, \ldots, m_M)$$
$$\vdots$$
$$d_N = g_N(m_1, m_2, \ldots, m_M)$$

or

$$d_i = g_i(m_1, m_2, \ldots, m_M)$$

or

$$\mathbf{d} = \mathbf{g}(\mathbf{m}) \tag{4.1}$$

Environmental Data Analysis with MATLAB®. http://dx.doi.org/10.1016/B978-0-12-804488-9.00004-5

The data are represented by a length-N vector, $\mathbf{d}$, and the model parameters by a length-M column vector, $\mathbf{m}$. The function, $\mathbf{g(m)}$, that relates them is called the *quantitative model*. We may find, however, that no prediction of the model, regardless of the value of $\mathbf{m}$ that is used, matches the observations, $\mathbf{d}^{\mathrm{obs}}$, because of observational noise. Then, Equation (4.1) must be understood in a more abstract sense: if we knew the true values, $\mathbf{m}^{\mathrm{true}}$, of the model parameters, then we could make predictions, $\mathbf{d}^{\mathrm{pre}}$, of data that would match those obtained through noise-free observations, were such observations possible. Alternatively, we could write Equation (4.1) as $\mathbf{d} = \mathbf{g(m)} + \mathbf{n}$, where the length-$N$ column vector, $\mathbf{n}$, represents measurement noise.

The function, $\mathbf{d} = \mathbf{g(m)}$, can be arbitrarily complicated. However, in *very* many important cases it is either linear or can be approximated as linear. In those cases, Equation (4.1) simplifies to

the data = a linear function of the model parameters

or

$$d_i = G_{i1}m_1 + G_{i2}m_2 + \cdots + G_{iM}m_M$$

or

$$d_1 = G_{11}m_1 + G_{12}m_2 + \cdots + G_{1M}m_M$$
$$d_2 = G_{21}m_1 + G_{22}m_2 + \cdots + G_{2M}m_M$$
$$\vdots$$
$$d_N = G_{N1}m_1 + G_{N2}m_2 + \cdots + G_{NM}m_M$$

or

$$
\begin{bmatrix} d_1 \\ d_2 \\ d_3 \\ \cdots \\ d_N \end{bmatrix}
=
\begin{bmatrix}
G_{11} & G_{12} & G_{13} & \cdots & G_{1M} \\
G_{21} & G_{22} & G_{23} & \cdots & G_{2M} \\
G_{31} & G_{32} & G_{33} & \cdots & G_{3M} \\
\cdots & \cdots & \cdots & \cdots & \cdots \\
G_{N1} & G_{N2} & G_{N3} & \cdots & G_{NM}
\end{bmatrix}
\begin{bmatrix} m_1 \\ m_2 \\ m_3 \\ \cdots \\ m_M \end{bmatrix}
$$

or

$$\mathbf{d} = \mathbf{Gm} \qquad\qquad (4.2)$$

Here, the matrix, $\mathbf{G}$, contains the coefficients of the linear relationship. It relates N data to M model parameters and so is $N \times M$. The matrix, $\mathbf{G}$, is often called the *data kernel*. In most typical cases, $N \neq M$, so $\mathbf{G}$ is *not* a square matrix (and, consequently, has no inverse).

Equation (4.2) can be used in several complementary ways. If the model parameters are known—let us call them $\mathbf{m}^{\mathrm{est}}$—then Equation (4.2) can be *evaluated* to provide a *prediction* of the data:

$$\mathbf{d}^{\mathrm{pre}} = \mathbf{Gm}^{\mathrm{est}} \qquad\qquad (4.3)$$

Alternatively, if the data are observed—we call them $\mathbf{d}^{\text{obs}}$—then Equation (4.2) can be *solved* to determine an *estimate* of the model parameters:

$$\text{find the } \mathbf{m}^{\text{est}} \text{ so that } \mathbf{d}^{\text{obs}} \approx \mathbf{Gm}^{\text{est}} \qquad (4.4)$$

Note that an estimate of the model parameters will not necessarily equal their true value, that is, $\mathbf{m}^{\text{est}} \neq \mathbf{m}^{\text{true}}$ because of observational noise.

4.2 The simplest of quantitative models

The simplest linear model is that in which the data are all equal to the same constant. This is the case of repeated observations, in which we make the same measurement N times. It corresponds to the equation

the data = a constant

or

$$\begin{bmatrix} d_1 \\ d_2 \\ d_3 \\ \vdots \\ d_N \end{bmatrix} = \begin{bmatrix} 1 \\ 1 \\ 1 \\ \vdots \\ 1 \end{bmatrix} [m_1]$$

or

$$\mathbf{d} = \mathbf{Gm} \qquad (4.5)$$

Here, the constant is given by the single model parameter, m_1 so that $M = 1$. The data kernel is the matrix, $\mathbf{G} = [1, 1, 1, \ldots, 1]^{\text{T}}$. In practice, N observations of the same thing actually would result in N different ds because of observational error. Thus, Equation (4.5) needs to be understood in the abstract sense: if we knew the value of the constant, m_1, and if the observations, $\mathbf{d}$, were noise-free, then they would all satisfy $d_i = m_1$. Alternatively, we could write Equation (4.5) as $\mathbf{d} = \mathbf{Gm} + \mathbf{n}$, where the length-$N$ column vector, $\mathbf{n}$, represents measurement noise.

Far from being trivial, this one-parameter model is arguably the most important of all the models in data analysis. It is equivalent to the idea that the data scatter around an average value. As we will see later on in this chapter, when the observational data are Normally distributed, a good estimate of the model parameter, m_1, is the sample mean, $\bar{d}$:

$$m_1^{\text{est}} = \bar{d} = \frac{1}{N} \sum_{i=1}^{N} d_i \qquad (4.6)$$

4.3   Curve fitting

Another simple but important model is the idea that the data fall on—or scatter around—a straight line. The data are assumed to satisfy the relationship

the data $=$ a linear function of x

or

$$d_i = m_1 + m_2 x_i$$

or

$$\begin{bmatrix} d_1 \\ d_2 \\ d_3 \\ \vdots \\ d_N \end{bmatrix} = \begin{bmatrix} 1 & x_1 \\ 1 & x_2 \\ 1 & x_3 \\ \vdots & \vdots \\ 1 & x_N \end{bmatrix} \begin{bmatrix} m_1 \\ m_2 \end{bmatrix}$$

or

$$\mathbf{d} = \mathbf{Gm} \tag{4.7}$$

Here, m_1 is the intercept and m_2 is the slope of the line. In order for this relationship to be linear in form, the data kernel, $\mathbf{G}$, must not contain any data or model parameters. Thus, we must assume that the x's are neither model parameters nor data, but rather *auxiliary* parameters whose values are exactly known. This may be an accurate, or nearly accurate, assumption in cases where the xs represent distance or time as, compared to most other types of data, time and distance can be determined so accurately as to have negligible error. In other cases, it can be a poor assumption.

A *MatLab* script that creates this data kernel is

```
M=2;
G=zeros(N,M);
G(:,1)=1;
G(:,2)=x;                                (MatLab eda04_01)
```

where x is a column vector of length M of x's. Note the call to zeros(N,M), which creates a matrix with N rows and M columns. Strictly speaking, this command is not necessary, but it helps in bug detection.

The formula for a straight line can easily be generalized to any order polynomial, by simply adding additional model parameters that represent the coefficients of higher powers of x's and by adding corresponding columns to the data kernel containing powers of the x's. For example, the quadratic case has $M = 3$ model parameters, $\mathbf{m} = [m_1, m_2, m_3]^{\mathrm{T}}$, and the equation becomes the data = a quadratic function of x's

or

$$d_i = m_1 + m_2 x_i + m_3 x_i^2$$

or

$$\begin{bmatrix} d_1 \\ d_2 \\ d_3 \\ \vdots \\ d_N \end{bmatrix} = \begin{bmatrix} 1 & x_1 & x_1^2 \\ 1 & x_2 & x_2^2 \\ 1 & x_3 & x_3^2 \\ \vdots & \vdots & \vdots \\ 1 & x_N & x_N^2 \end{bmatrix} \begin{bmatrix} m_1 \\ m_2 \\ m_3 \end{bmatrix}$$

or

$$\mathbf{d} = \mathbf{Gm} \tag{4.8}$$

A *MatLab* script that creates this data kernel is

```
M=3;
G=zeros(N,M);
G(:,1)=1;
G(:,2)=x;
G(:,3)=x.^2;
```
                                                    (*MatLab* eda04_02)

where x is a column vector of length M of x's. Note the use of the element-wise multiplication, x.^2, which creates a column vector with elements, x_i^2. The data kernel for the case of a polynomial of arbitrary degree is computed as

```
G=zeros(N,M);
G(:,1) = 1; % handle first column individually
for i = [2:M] % loop over remaining columns
    G(:,i) = x.^(i-1);
end
```
                                                    (*MatLab* eda04_03)

This method is not limited to polynomials; rather, it can be used to represent any curve of the form

the data = a sum of functions, f, of known form

or

$$d_i = m_1 f_1(x_i) + m_2 f_2(x_i) + \cdots + m_M f_M(x_i)$$

or

$$\begin{bmatrix} d_1 \\ d_2 \\ d_3 \\ \vdots \\ d_N \end{bmatrix} = \begin{bmatrix} f_1(x_1) & f_2(x_1) & \cdots & f_M(x_1) \\ f_1(x_2) & f_2(x_2) & \cdots & f_M(x_2) \\ f_1(x_3) & f_2(x_3) & \cdots & f_M(x_3) \\ \vdots & \vdots & \vdots & \vdots \\ f_1(x_N) & f_2(x_N) & \cdots & f_M(x_N) \end{bmatrix} \begin{bmatrix} m_1 \\ m_2 \\ \vdots \\ m_M \end{bmatrix} \tag{4.9}$$

Note that the model parameter, m_j, represents the amount of the function, f_j, in the representation of the data.

One important special case—called *Fourier analysis*—is the modeling of data with a sum of cosines and sines of different wavelength, λ_i:

the data = a sum of cosines and sines

or

$$d_i = m_1\cos\left(\frac{2\pi x_i}{\lambda_1}\right) + m_2\sin\left(\frac{2\pi x_i}{\lambda_1}\right) + \cdots + m_{M-1}\cos\left(\frac{2\pi x_i}{\lambda_{M/2}}\right) + m_M\sin\left(\frac{2\pi x_i}{\lambda_{M/2}}\right)$$

or $\mathbf{d} = \mathbf{Gm}$ with

$$\mathbf{G} = \begin{bmatrix} \cos\left(\frac{2\pi x_1}{\lambda_1}\right) & \sin\left(\frac{2\pi x_1}{\lambda_1}\right) & \cos\left(\frac{2\pi x_1}{\lambda_2}\right) & \sin\left(\frac{2\pi x_1}{\lambda_2}\right) & \cdots & \cos\left(\frac{2\pi x_1}{\lambda_{M/2}}\right) & \sin\left(\frac{2\pi x_1}{\lambda_{M/2}}\right) \\ \cos\left(\frac{2\pi x_2}{\lambda_1}\right) & \sin\left(\frac{2\pi x_2}{\lambda_1}\right) & \cos\left(\frac{2\pi x_2}{\lambda_2}\right) & \sin\left(\frac{2\pi x_2}{\lambda_2}\right) & \cdots & \cos\left(\frac{2\pi x_2}{\lambda_{M/2}}\right) & \sin\left(\frac{2\pi x_2}{\lambda_{M/2}}\right) \\ \cos\left(\frac{2\pi x_3}{\lambda_1}\right) & \sin\left(\frac{2\pi x_3}{\lambda_1}\right) & \cos\left(\frac{2\pi x_3}{\lambda_2}\right) & \sin\left(\frac{2\pi x_3}{\lambda_2}\right) & \cdots & \cos\left(\frac{2\pi x_3}{\lambda_{M/2}}\right) & \sin\left(\frac{2\pi x_3}{\lambda_{M/2}}\right) \\ \vdots & \vdots & \vdots & \vdots & \vdots & \vdots & \vdots \\ \cos\left(\frac{2\pi x_N}{\lambda_1}\right) & \sin\left(\frac{2\pi x_N}{\lambda_1}\right) & \cos\left(\frac{2\pi x_N}{\lambda_2}\right) & \sin\left(\frac{2\pi x_N}{\lambda_2}\right) & \cdots & \cos\left(\frac{2\pi x_N}{\lambda_{M/2}}\right) & \sin\left(\frac{2\pi x_N}{\lambda_{M/2}}\right) \end{bmatrix}$$

$$(4.10)$$

As we will discuss in more detail in Section 6.1, we normally choose pairs of sines and cosines of the same wavelength, λ_i. The total number of model parameters is M, which represents the amplitude coefficients of the $M/2$ sines and $M/2$ cosines.

A *MatLab* script that creates this data kernel is

```
G=zeros(N,M);
Mo2=M/2;
for i = [1:Mo2]
    ic = 2*i-1;
    is = 2*i;
    G(:,ic) = cos( 2*pi*x/lambda(i) );
    G(:,is) = sin( 2*pi*x/lambda(i) );
end                                         (MatLab eda04_04)
```

This example assumes that the column vector of wavelengths, lambda, omits the lambda=0 case, as it would cause a division-by-zero error. We use the variable *wavenumber*, $k = 2\pi/\lambda$, instead of wavelength, λ, in subsequent discussions of Fourier sums to avoid this problem.

Gray-shaded versions of the polynomial and Fourier data kernels are shown in Figure 4.1.

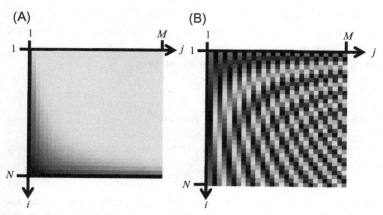

Figure 4.1 Grey-shaded plot of the data kernel, **G**, for the (A) polynomial and (B) Fourier cases. *MatLab* scripts eda04_03 and eda04_04.

4.4 Mixtures

Suppose that we view the data kernel as a concatenation of its columns, say $\mathbf{c}^{(j)}$ (Figure 4.2):

the data kernel = a concatenation of column-vectors

or

$$G = \begin{bmatrix} G_{11} & G_{12} & G_{13} \\ G_{21} & G_{22} & G_{23} \\ G_{31} & G_{32} & G_{33} \end{bmatrix} = \begin{bmatrix} \begin{bmatrix} G_{11} \\ G_{21} \\ G_{31} \end{bmatrix} & \begin{bmatrix} G_{12} \\ G_{22} \\ G_{32} \end{bmatrix} & \begin{bmatrix} G_{13} \\ G_{23} \\ G_{33} \end{bmatrix} \end{bmatrix} = \begin{bmatrix} \mathbf{c}^{(1)} & \mathbf{c}^{(2)} & \mathbf{c}^{(3)} \end{bmatrix} \qquad (4.11)$$

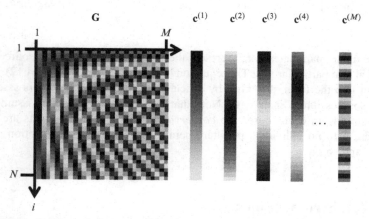

Figure 4.2 The data kernel, **G**, can be thought of as a concatenation of its M columns, $\mathbf{c}^{(j)}$, each of which is a column vector. *MatLab* script eda04_05.

Then the equation $\mathbf{d} = \mathbf{Gm}$ can be understood to mean that $\mathbf{d}$ is constructed by adding together the columns of $\mathbf{G}$ in proportions specified by the model parameters, m_j:

the data = a linear mixture of column-vectors

or

$$\mathbf{d} = m_1 \mathbf{c}^{(1)} + m_2 \mathbf{c}^{(2)} + m_3 \mathbf{c}^{(3)} + \ldots + m_M \mathbf{c}^{(M)}$$

or

$$\mathbf{d} = [\,\mathbf{c}^{(1)} \quad \mathbf{c}^{(2)} \quad \mathbf{c}^{(3)}\,]\mathbf{m} \tag{4.12}$$

This summation can be thought of as a *mixing* process. The data are a mixture of the columns of the data kernel. Each model parameter represents the amount of the corresponding column-vector in the mixture. Indeed, it can be used to represent literal mixing. For example, suppose that a city has M major sources of pollution, such as power plants, industrial facilities, vehicles (taken as a group), and so on. Each source emits into the atmosphere its unique combination of N different pollutants. An air sample taken from an arbitrary point within the city will then contain a mixture of pollutants from these sources:

pollutants in air = mixture of sources

or

$$\begin{bmatrix} \text{pollutant 1 in air} \\ \text{pollutant 2 in air} \\ \text{pollutant 3 in air} \\ \vdots \\ \text{pollutant } N \text{ in air} \end{bmatrix} = m_1 \begin{bmatrix} \text{pollutant 1 in source 1} \\ \text{pollutant 2 in source 1} \\ \text{pollutant 3 in source 1} \\ \vdots \\ \text{pollutant } N \text{ in source 1} \end{bmatrix} + \cdots + m_M \begin{bmatrix} \text{pollutant 1 in source } M \\ \text{pollutant 2 in source } M \\ \text{pollutant 3 in source } M \\ \vdots \\ \text{pollutant } N \text{ in source } M \end{bmatrix}$$

$$\tag{4.13}$$

where the model parameters, m_j, represent the contributions of the j-th source to the pollution at that particular site. This equation has the form of Equation (4.12), and so can be put into the form, $\mathbf{d} = \mathbf{Gm}$, by concatenating the column vectors associated with the sources into a matrix, $\mathbf{G}$. Note that this quantitative model assumes that the pollutants from each source are *conservative*, that is, the pollutants are mixed as a group, with no individual pollutant being lost because of degradation, slower transport, and so on.

4.5   Weighted averages

Suppose that we view the data kernel as a column vector of its rows, say $\mathbf{r}^{(i)}$:

the data kernel = a concatenation of row-vectors

or

$$\mathbf{G} = \begin{bmatrix} G_{11} & G_{12} & G_{13} \\ G_{21} & G_{22} & G_{23} \\ G_{31} & G_{32} & G_{33} \end{bmatrix} = \begin{bmatrix} [G_{11} & G_{12} & G_{13}] \\ [G_{21} & G_{22} & G_{23}] \\ [G_{31} & G_{32} & G_{33}] \end{bmatrix} = \begin{bmatrix} \mathbf{r}^{(1)} \\ \mathbf{r}^{(2)} \\ \mathbf{r}^{(3)} \end{bmatrix} \qquad (4.14)$$

Then the equation, $\mathbf{d} = \mathbf{Gm}$, can be understood to mean that the i-th datum, d_i, is constructed by taking the dot product, $\mathbf{r}^{(i)}\mathbf{m}$. For example, suppose that $M = 9$ and

$$\mathbf{r}^{(5)} = [0,0,0,\tfrac{1}{4},\tfrac{1}{2},\tfrac{1}{4},0,0,0] \text{ then } d_5 = \mathbf{r}^{(5)}\mathbf{m} = \tfrac{1}{4}m_4 + \tfrac{1}{2}m_5 + \tfrac{1}{4}m_6 \qquad (4.15)$$

Thus, the 5-th datum is a *weighted average* of the 4-th, 5-th, and 6-th model parameters. This scenario is especially useful when a set of observations are made along one spatial dimension—a profile. Then, the data correspond to a smooth version of the model parameters, with the amount of smoothing being described by the width of averaging. The three-point averaging of Equation (4.15) corresponds to a data kernel, $\mathbf{G}$, of the form

$$G = \begin{bmatrix} \tfrac{1}{2} & \tfrac{1}{4} & 0 & 0 & 0 & 0 & 0 & 0 & 0 \\ \tfrac{1}{4} & \tfrac{1}{2} & \tfrac{1}{4} & 0 & 0 & 0 & 0 & 0 & 0 \\ 0 & \tfrac{1}{4} & \tfrac{1}{2} & \tfrac{1}{4} & 0 & 0 & 0 & 0 & 0 \\ 0 & 0 & \tfrac{1}{4} & \tfrac{1}{2} & \tfrac{1}{4} & 0 & 0 & 0 & 0 \\ 0 & 0 & 0 & \tfrac{1}{4} & \tfrac{1}{2} & \tfrac{1}{4} & 0 & 0 & 0 \\ 0 & 0 & 0 & 0 & \tfrac{1}{4} & \tfrac{1}{2} & \tfrac{1}{4} & 0 & 0 \\ 0 & 0 & 0 & 0 & 0 & \tfrac{1}{4} & \tfrac{1}{2} & \tfrac{1}{4} & 0 \\ 0 & 0 & 0 & 0 & 0 & 0 & \tfrac{1}{4} & \tfrac{1}{2} & \tfrac{1}{4} \\ 0 & 0 & 0 & 0 & 0 & 0 & 0 & \tfrac{1}{4} & \tfrac{1}{2} \end{bmatrix} \qquad (4.16)$$

Note that each row must sum to unity for the operation to represent a true weighted average; otherwise, the average of three identical data would be unequal to their common value. Thus, the top and bottom rows of the matrix pose a dilemma. A three-point weighted average is not possible for these rows, because no m_0 or m_{M+1} exists. This problem can be solved in two ways: just continue the pattern, in which case these rows do not correspond to true weighted averages (as in Equation 4.16), or use coefficients on those rows that make them true two-point weighted averages (e.g., $\tfrac{1}{4}$ and $\tfrac{3}{4}$ for the first row and $\tfrac{3}{4}$ and $\tfrac{1}{4}$ for the last).

In *MatLab*, a data kernel, $\mathbf{G}$, corresponding to a weighted average can be created with the following script:

```
w = [2, 1]';
Lw = length(w);
n = 2*sum(w)-w(1);
w = w/n;
r = zeros(M,1); c = zeros(N,1);
r(1:Lw)=w; c(1:Lw)=w;
G = toeplitz(c,r);
```

(*MatLab* eda04_06)

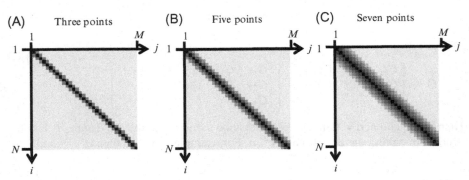

Figure 4.3 The data kernel, **G**, for weighted averages of different lengths. (A) Length of 3, (B) 5, (C) 7. *MatLab* script eda04_06.

We assume that the weighted average is symmetric around its central point, but allow it to be of any length, Lw. It is specified in the column vector, w, which contains only the central weight and the nonzero weights to the right of it. Only the relative size of the elements of w is important, as the weights are normalized through computation of and division by a normalization factor, n. Thus, w=[2, 1]' corresponds to the case given in Equation (4.16). The matrix, G, is Toeplitz, meaning that all its diagonals are constant, so it can be specified through its left column, c, and its top row, r, using the toeplitz() function. Grey-scale images of the **G**s for weighted averages of different lengths are shown in Figure 4.3.

An important class of weighted averages that are not symmetric around the central value is the *causal filter*, which appears in many problems that involve time. An instrument, such as an aqueous oxygen sensor (a device that measures the oxygen concentration in water), does not measure the present oxygen concentration, but rather a weighted average of concentrations over the last few instants of time. This behavior arises from a limitation in the sensor's design. Oxygen must diffuse though a membrane within the sensor before it can be measured. Hence, the sensor reading (observation, d_i) made at time t_i is a weighted average of oxygen concentrations (model parameters, m_j), at times $t_j \leq t_i$. The weights are called filter coefficients, **f**, with

the data = a weighted average of present and past values of m's

or

$$d_i = f_1 m_i + f_2 m_{i-1} + f_3 m_{i-2} + f_4 m_{i-3} + \cdots \qquad (4.17)$$

The corresponding data kernel, **G**, has the following form:

$$\mathbf{G} = \begin{bmatrix} f_1 & 0 & 0 & 0 & 0 & \cdots \\ f_2 & f_1 & 0 & 0 & 0 & \cdots \\ f_3 & f_2 & f_1 & 0 & 0 & \cdots \\ f_4 & f_3 & f_2 & f_1 & 0 & \cdots \\ f_5 & f_4 & f_3 & f_2 & f_1 & \cdots \\ \cdots & \cdots & \cdots & \cdots & \cdots & \cdots \end{bmatrix} \qquad (4.18)$$

The data kernel, **G**, is both Toeplitz and lower triangular. For filters of length L, the first $L - 1$ elements of **d** are inaccurately computed, because they require knowledge of unavailable model parameters, those corresponding to times earlier than t_1. We will discuss filters in more detail in Chapter 7.

4.6 Examining error

Suppose that we have somehow obtained an estimate of the model parameters, $\mathbf{m}^{\text{est}}$— for example, by guessing! One of the most important questions that can then be asked is

How do the predicted data, $\mathbf{d}^{\text{pre}} = \mathbf{Gm}^{\text{est}}$, *compare with the observed data,* $\mathbf{d}^{\text{obs}}$?

This question motivates us to define an error vector, **e**:

$$\mathbf{e} = \mathbf{d}^{\text{obs}} - \mathbf{d}^{\text{pre}} = \mathbf{d}^{\text{obs}} - \mathbf{Gm}^{\text{est}} \tag{4.19}$$

When the error, e_i, is small, the corresponding datum, d_i^{obs}, is well predicted and, conversely, when the error, e_i, is large, the corresponding datum, d_i^{obs}, is poorly predicted. A measure of *total* error, E, is as follows:

$$E = \mathbf{e}^{\text{T}}\mathbf{e} = [\mathbf{d}^{\text{obs}} - \mathbf{Gm}^{\text{est}}]^{\text{T}}[\mathbf{d}^{\text{obs}} - \mathbf{Gm}^{\text{est}}] \tag{4.20}$$

The total error, E, is the length of the error vector, **e**, which is to say, the sum of squares of the individual errors:

$$E = \sum_{i=1}^{N} e_i^2 \tag{4.21}$$

The error depends on the particular choice of model parameters, **m**, so we can write $E(\mathbf{m})$. One possible choice of a best estimate of the model parameters is the choice for which $E(\mathbf{m}^{\text{est}})$ is a minimum. This is known as the *principle of least squares*.

Plots of error are an extremely important tool for understanding whether a model has the overall ability to fit the data as well as whether the dataset contains anomalous points—*outliers*—that are unusually poorly fit. An example of the straight-line case is shown in Figure 4.4.

In *MatLab*, the error is calculated as follows:

```
dpre = G*mest;
e=dobs-dpre;
E = e'*e;                                    (MatLab eda04_07)
```

where mest, dobs, and dpre are the estimated model parameters, $\mathbf{m}^{\text{est}}$, observed data, $\mathbf{d}^{\text{obs}}$, and predicted data, $\mathbf{d}^{\text{pre}}$, respectively.

So far, we have not said anything useful regarding how one might arrive at a reasonable estimate, $\mathbf{m}^{\text{est}}$, of the model parameters. In cases, such as the straight line,

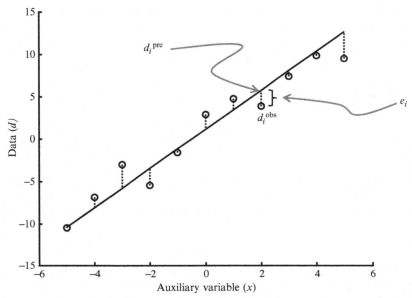

Figure 4.4 Observed data, $\mathbf{d}^{obs}$, predicted error, $\mathbf{d}^{pre}$, and error, $\mathbf{e}$, for the case of a straight line. *MatLab* script eda04_07.

where the number of model parameters is small, one might try a *grid search*. The idea is to systematically evaluate the total error, $E(\mathbf{m})$, for many different $\mathbf{m}$'s—a grid of $\mathbf{m}$'s—and choose $\mathbf{m}^{est}$ as the $\mathbf{m}$ for which $E(\mathbf{m})$ is the smallest (Figure 4.5). Note, however, that the point of minimum error, E^{min}, is surrounded by a region of almost-minimum error, that is, a region in which the error is only slightly larger than its value at the minimum. Any $\mathbf{m}$ chosen from this region is almost as good an estimate as is $\mathbf{m}^{est}$. As we will see in Section 4.9, this region defines confidence intervals for the model parameters.

For a grid search to be effective, one must have a general idea of the value of the solution, $\mathbf{m}^{est}$, so as to be able to choose the boundaries of a grid that contains it. We note that a plot of the logarithm of error, $ln[E(\mathbf{m})]$, is often visually more effective than a plot of $E(\mathbf{m})$, because it has less overall range.

In *MatLab*, a two-dimensional grid search is performed as follows:

```
% define grid
L1=100; L2=100;
m1min=0; m1max=4;
m2min=0; m2max=4;
m1=m1min+(m1max-m1min)*[0:L1-1]'/(L1-1);
m2=m2min+(m2max-m2min)*[0:L2-1]'/(L2-1);
% evaluate error at each grid point
E=zeros(L1,L2);
for i = [1:L1]
for j = [1:L2]
```

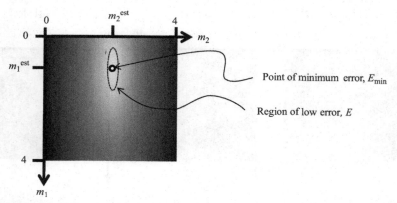

Figure 4.5 Grey-shaded plot of the logarithm of the total error, $E(m_1, m_2)$. The point of minimum error, E_{min}, is shown with a circle. The coordinates of this point are the least squares estimates of the model parameters, $m_1{}^{est}$ and $m_2{}^{est}$. The point of minimum error is surrounded by a region (dashed) of low error. *MatLab* script eda04_08.

```
mest = [ m1(i), m2(j) ]';
dpre = G*mest;
e = dobs-dpre;
E(i,j) = e'*e;
end
end
% search grid for minimum E
[Etemp, k] = min(E);
[Emin, j] = min(Etemp);
i=k(j);
m1est = m1(i);
m2est = m2(j);                                              (MatLab eda04_08)
```

The first section defines the grid of possible values of model parameters, m1 and m2. The second section, with the two nested for loops, computes the total error, E, for every point on the grid, that is, for every combination of m1 and m2. The third section searches the matrix, E, for its minimum value, Emin, and determines the corresponding best estimates of the model parameters, m1est and m2est. This search is a bit tricky. The function min(E) returns a row vector, Etemp, containing the minimum values in each column of E, as well as a row vector, k, of the row index at which the minimum occurs. The function min(Etemp) searches for the minimum of Etemp (which is also the minimum, Emin, of matrix, E) and also returns the column index, j, at which the minimum occurs. The minimum of E is therefore at row, i=k(j), and column, j, and the best-estimates of the model parameters are m1(i) and m2(j).

The overall amplitude of $E(\mathbf{m})$ depends on the amount of error in the observed data, $\mathbf{d}^{obs}$. The shape of the error surface, however, is mainly dependent on the *geometry* of observations. For example, merely shifting the *xs* to the left or right has a major impact on the overall shape of the error, E (Figure 4.6).

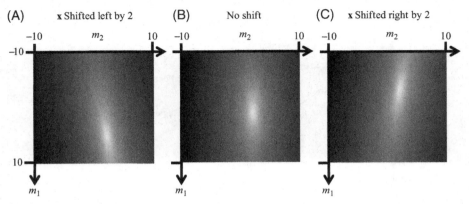

Figure 4.6 Grey-shaded plot of the logarithm of the total error, $E(m_1, m_2)$, for the straight-line case. (A) The values of **x** are shifted to the left by $\Delta x = 2$. (B) No shift. (C) The values of **x** are shifted to the right by $\Delta x = 2$. Note that the shape of the error function is different in each case. *MatLab* script eda04_09.

4.7 Least squares

In this section, we show that the least squares estimate of the model parameters can be determined directly, without recourse to a grid search. First, however, we return to some of the ideas of probability theory that we put forward in Chapter 3.

Suppose that the measurement error can be described by a Normal probability density function, and that observations, d_i, are uncorrelated and with equal variance, σ_d^2. Then, the probability density function, $p(d_i)$, for one observation, d_i, is as follows:

$$p(d_i) = \frac{1}{\sqrt{2\pi}\sigma_d} \exp\{-(d_i - \bar{d}_i)^2/(2\sigma_d^2)\} \tag{4.22}$$

where $\bar{d}$ is the mean. As the observations are uncorrelated, the joint probability density function, $p(\mathbf{d})$, is just the product of the individual probability density functions:

$$p(\mathbf{d}) = \frac{1}{(2\pi)^{N/2}\sigma^N} \exp\left\{ -\frac{1}{2\sigma_d^2} \sum_{i=1}^{N} (d_i - \bar{d}_i)^2 \right\}$$

$$= \frac{1}{(2\pi)^{N/2}\sigma^N} \exp\left\{ -\frac{1}{2\sigma_d^2} (\mathbf{d} - \bar{\mathbf{d}})^T (\mathbf{d} - \bar{\mathbf{d}}) \right\} \tag{4.23}$$

We now assume that the model predicts the *mean* of the probability density functions, that is, $\bar{\mathbf{d}} = \mathbf{Gm}$. The resulting probability density function is

$$p(\mathbf{d}) = \frac{1}{(2\pi)^{N/2}\sigma^N}\exp\left\{-\frac{1}{2\sigma_d^2}(\mathbf{d}-\mathbf{Gm})^{\mathrm{T}}(\mathbf{d}-\mathbf{Gm})\right\} = \frac{1}{(2\pi)^{N/2}\sigma^N}\exp\left\{-\frac{1}{2\sigma_d^2}E(\mathbf{m})\right\}$$

$$\text{with} \quad E(\mathbf{m}) = (\mathbf{d}-\mathbf{Gm})^{\mathrm{T}}(\mathbf{d}-\mathbf{Gm})$$

$$(4.24)$$

In Chapter 3, we noted that the mean and mode of a Normal probability density function occur at the same value of $\mathbf{m}$. Thus, the mean of this probability density function occurs at the point at which $p(\mathbf{d})$ is maximum (the mode), which is the same as the point where $E(\mathbf{m})$ is minimum. But this is just the principle of least squares. The $\mathbf{m}^{\mathrm{est}}$ that minimizes $E(\mathbf{m})$ is also the $\mathbf{m}$ such that $\mathbf{Gm}^{\mathrm{est}}$ is the mean of $p(\mathbf{d})$. The two are one and the same.

The actual value of $\mathbf{m}^{\mathrm{est}}$ is calculated by minimizing $E(\mathbf{m})$ with respect to a model parameter, m_k. Taking the derivative, $\partial E/\partial m_k$, and setting the result to zero yields

$$0 = \frac{\partial E}{\partial m_k} = \frac{\partial}{\partial m_k}\sum_{i=1}^{N}\left(d_i - \sum_{j=1}^{M}G_{ij}m_j\right)^2$$

We then apply the chain rule to obtain

$$0 = -2\sum_{i=1}^{N}\left(\sum_{j=1}^{M}G_{ij}\frac{\partial m_j}{\partial m_k}\right)\left(d_i - \sum_{j=1}^{M}G_{ij}m_j\right)$$

As m_j and m_k are independent variables, the derivative, $\partial m_j/\partial m_k$, is zero except when $j = k$, in which case it is unity (this relationship is sometimes written as $\partial m_j/\partial m_k = \delta_{jk}$, where δ_{jk}, called the *Kronecker delta symbol,* is an element of the identity matrix). Thus, we can perform the first summation trivially, that is, by replacing j with k and deleting the derivative and first summation sign:

$$0 = -2\sum_{j=1}^{M}G_{ik}\left(d_i - \sum_{j=1}^{M}G_{ij}m_j\right) \quad \text{or} \quad 0 = -\mathbf{G}^{\mathrm{T}}\mathbf{d} + \mathbf{G}^{\mathrm{T}}\mathbf{Gm} \quad \text{or} \quad [\mathbf{G}^{\mathrm{T}}\mathbf{G}]\mathbf{m} = \mathbf{G}^{\mathrm{T}}\mathbf{d}$$

As long as the inverse of the $M \times M$ matrix, $[\mathbf{G}^{\mathrm{T}}\mathbf{G}]$, exists, the least-squares solution is

$$\mathbf{m}^{\mathrm{est}} = [\mathbf{G}^{\mathrm{T}}\mathbf{G}]^{-1}\mathbf{G}^{\mathrm{T}}\mathbf{d}^{\mathrm{obs}} \qquad (4.25)$$

Note that the estimated model parameters, $\mathbf{m}^{\mathrm{est}}$, are related to the observed data, $\mathbf{d}^{\mathrm{obs}}$, by multiplication by a matrix, $\mathbf{M} = [\mathbf{G}^{\mathrm{T}}\mathbf{G}]^{-1}\mathbf{G}^{\mathrm{T}}$, that is, $\mathbf{m}^{\mathrm{est}} = \mathbf{Md}^{\mathrm{obs}}$. According to the rules of error propagation developed in Chapter 3, the covariance of the estimated model parameters, $\mathbf{C}_m$, is related to the covariance of the observed data, $\mathbf{C}_d$, by $\mathbf{C}_m = \mathbf{MC}_d\mathbf{M}^T$. In the present case, we have assumed that the data are uncorrelated with equal variance, σ_d^2, so $\mathbf{C}_d = \sigma_d^2\mathbf{I}$. The covariance of the estimated model parameters is, therefore

$$\mathbf{C}_m = [[\mathbf{G}^T\mathbf{G}]^{-1}\mathbf{G}^T]\sigma_d^2\mathbf{I}[[\mathbf{G}^T\mathbf{G}]^{-1}\mathbf{G}^T]^T = \sigma_d^2[\mathbf{G}^T\mathbf{G}]^{-1} \tag{4.26}$$

Here, we have used the rule $(\mathbf{AB})^T = \mathbf{B}^T\mathbf{A}^T$. Note that $\mathbf{G}^T\mathbf{G}$ is a symmetric matrix, and that the inverse of a symmetric matrix is symmetric.

The derivation above assumes that quantities are purely real, which is the most common case. See Note 4.1 for a discussion of least squares in the case where quantities are complex.

4.8 Examples

In Section 4.2, we put forward the simplest linear problem, where the data are constant, which has $M = 1$, $\mathbf{m} = [m_1]$ and $\mathbf{G} = [1, 1, \ldots, 1]^T$. Then,

$$\mathbf{G}^T\mathbf{G} = [1 \quad 1 \quad 1 \quad \ldots \quad 1]\begin{bmatrix} 1 \\ 1 \\ 1 \\ \vdots \\ 1 \end{bmatrix} = N \quad \text{and} \quad \mathbf{G}^T\mathbf{d} = [1 \quad 1 \quad 1 \quad \ldots \quad 1]\begin{bmatrix} d_1 \\ d_2 \\ d_3 \\ \ldots \\ d_N \end{bmatrix} = \sum_{i=1}^{N} d_i \tag{4.27}$$

Then,

$$\mathbf{m}^{est} = m_1^{est} = [\mathbf{G}^T\mathbf{G}]^{-1}\mathbf{G}^T\mathbf{d} = \frac{1}{N}\sum_{i=1}^{N} d_i \quad \text{and} \quad \mathbf{C}_m = \frac{\sigma_d^2}{N} \tag{4.28}$$

As stated in Section 4.2, $\mathbf{m}^{est}$ is the mean of the data—the *sample mean*. The result for the variance of the sample mean, $\mathbf{C}_m = \sigma_d^2/N$, is a very important one. The variance of the mean is less than the variance of the data by a factor of N^{-1}. Thus, the more the measurements, the greater is the precision of the mean. However, the confidence intervals of the mean, which depend on the square root of the variance, decline slowly with additional measurements: $\sigma_m = \sigma_d/\sqrt{N}$.

Similarly, for the straight line case, we have

$$
\mathbf{G}^T\mathbf{G} = \begin{bmatrix} 1 & 1 & 1 & 1 & 1 \\ x_1 & x_2 & x_3 & \cdots & x_N \end{bmatrix} \begin{bmatrix} 1 & x_1 \\ 1 & x_2 \\ 1 & x_3 \\ \vdots & \vdots \\ 1 & x_N \end{bmatrix} = \begin{bmatrix} N & \sum_{i=1}^{N} x_i \\ \sum_{i=1}^{N} x_i & \sum_{i=1}^{N} x_i^2 \end{bmatrix}
$$

$$
\mathbf{G}^T\mathbf{d} = \begin{bmatrix} 1 & 1 & 1 & 1 & 1 \\ x_1 & x_2 & x_3 & \cdots & x_N \end{bmatrix} \begin{bmatrix} d_1 \\ d_2 \\ d_3 \\ \vdots \\ d_N \end{bmatrix} = \begin{bmatrix} \sum_{i=1}^{N} d_i \\ \sum_{i=1}^{N} x_i d_i \end{bmatrix}
$$

$$
\mathbf{m}^{est} = [\mathbf{G}^T\mathbf{G}]^{-1}\mathbf{G}^T\mathbf{d} = \frac{1}{N\sum_{i=1}^{N}x_i^2 - \left[\sum_{i=1}^{N}x_i\right]^2} \begin{bmatrix} \sum_{i=1}^{N} x_i^2 & -\sum_{i=1}^{N} x_i \\ -\sum_{i=1}^{N} x_i & N \end{bmatrix} \begin{bmatrix} \sum_{i=1}^{N} d_i \\ \sum_{i=1}^{N} x_i d_i \end{bmatrix}
$$

$$
= \frac{1}{N\sum_{i=1}^{N}x_i^2 - \left[\sum_{i=1}^{N}x_i\right]^2} \begin{bmatrix} \sum_{i=1}^{N} x_i^2 \sum_{i=1}^{N} d_i - \sum_{i=1}^{N} x_i \sum_{i=1}^{N} x_i d_i \\ N\sum_{i=1}^{N} x_i d_i - \sum_{i=1}^{N} x_i \sum_{i=1}^{N} d_i \end{bmatrix}
$$

$$
\mathbf{C}_m = \sigma_d^2[\mathbf{G}^T\mathbf{G}]^{-1} = \frac{\sigma_d^2}{N\sum_{i=1}^{N}x_i^2 - \left[\sum_{i=1}^{N}x_i\right]^2} \begin{bmatrix} \sum_{i=1}^{N} x_i^2 & -\sum_{i=1}^{N} x_i \\ -\sum_{i=1}^{N} x_i & N \end{bmatrix}
$$

$$
(4.29)
$$

Here, we have used the fact that the inverse of a 2×2 matrix is

$$
\begin{bmatrix} a & b \\ c & d \end{bmatrix}^{-1} = \frac{1}{ad - bc} \begin{bmatrix} d & -b \\ -c & a \end{bmatrix}
\qquad (4.30)
$$

In *MatLab*, all these quantities are computed by first defining the data kernel, **G**, and then forming all the other quantities using linear algebra:

```
G=zeros(N,M);
G(:,1)=1;
G(:,2)=x;
mest = (G'*G)\(G'*dobs);
dpre = G*mest;
e=dobs-dpre;
E = e'*e;
sigmad2 = E/(N-M);
Cm = sigmad2*inv(G'*G);                          (MatLab eda04_10)
```

Note that we use the backslash operator,\, when evaluating the formula, $\mathbf{m}^{est} = [\mathbf{G}^T\mathbf{G}]^{-1}\mathbf{G}^T\mathbf{d}$. An extremely important issue is how the variance of the data, σ_d^2, is obtained. In some cases, determining the observational error might be possible in advance, based on some knowledge of the measurement system that is being used. If, for example, a ruler has 1 mm divisions, then one might assume that σ_d is about 1 mm. This is called a *prior* estimate of the variance. Another possibility is to use the total error, E, to estimate the variance in the data

$$\sigma_d^2 = \frac{E}{(N-M)} \qquad (4.31)$$

as is done in the eda04_10 script above. This is essentially approximating the variance by the mean squared error $E/N = (e_1^2 + e_1^2 + \ldots + e_N^2)/N$. The factor of M is added to account for the ability of an M-parameter model to predict M data exactly (e.g., a straight line fan fits any two points, exactly). The actual variance of the data is larger than E/N. This estimate is called a *posterior* estimate of the variance. The quantity, $N - M$, is called the *degrees of freedom* of the problem.

One problem with posterior estimates is that they are influenced by the quality of the quantitative model. If the model is not a good one, then it will not fit the data well and a posterior estimate of variance will be larger than the true variance of the observations.

We now return to the Black Rock Forest temperature data discussed in Chapter 2. One interesting question is whether we can observe a long-term increase or decrease in temperature over the 12 years of observations. The problem is detecting such a trend, which is likely to be just fractions of a degree against the very large annual cycle. One possibility is to model both:

$$d_i = m_1 + m_2 t_i + m_3 \cos\frac{2\pi t_i}{T} + m_4 \sin\frac{2\pi t_i}{T} \qquad (4.32)$$

where T is a period of 1 year or 365.25 days. While we solve for four model parameters, only m_2, which quantifies the rate of increase of temperature with time, is of interest. The other three model parameters are included to provide a better fit of the model to the data.

Before proceeding, we need to have some strategy to deal with the errors that we detect in the dataset. One strategy is to identify *bad* data points and throw them out. This is a dangerous strategy because the data that are thrown out might not all be bad, because the data that are included might not all be good, and especially because the reason why bad data are present in the data has never been determined. Nevertheless, it is the only viable strategy, in some cases.

The Black Rock Forest dataset has three types of bad data: cold spikes that fall below $-40°C$, warm spikes that with temperatures above $38°C$, and dropouts with a temperature of exactly $0 °C$. The following *MatLab* script eliminates them:

```
Draw=load('brf_temp.txt');
traw=Draw(:,1);
draw=Draw(:,2);
n = find((draw~=0) & (draw>-40) & (draw<38));
t=traw(n);
d=draw(n);                                          (MatLab eda04_11)
```

The find() function returns a column vector, n, of indices that satisfy a logical expression, in this case

$$(draw\sim =0) \ \& \ (draw>-40) \ \& \ (draw<38) \qquad (MatLab\ eda04\_11)$$

which means the elements of $\mathbf{d}^{raw}$ that satisfy $d_i^{raw} \neq 0$, $d_i^{raw} > -40$, and $d_i^{raw} < 38$. Note that, in *MatLab*, the tilde, $\sim$, means *not*, so that $\sim=$ means *not equal*. The vector, n, is then used in the statements t=traw(n) and d=draw(n), which form two new versions of data, $\mathbf{d}$, and time, $\mathbf{t}$, containing only good data.

The *MatLab* code that creates the data kernel, $\mathbf{G}$, is

```
Ty=365.25;
G=zeros(N,4);
G(:,1)=1;
G(:,2)=t;
G(:,3)=cos(2*pi*t/Ty);
G(:,4)=sin(2*pi*t/Ty);                               (MatLab eda04_11)
```

The results of the fit are shown in Figure 4.7. Note that the model does poorly in fitting the overall amplitude of the seasonal cycle, mainly because the annual oscillations, while having a period of 1 year, do not have a sinusoidal shape. The fit could be substantially improved by adding oscillations of half-year, third-year and quarter-year periods (see Problem 4.5). The estimated long-term slope is $m_2 = -0.03°C/yr$, corresponding to a slight cooling trend. The prior error of the slope, based on an estimate of $\sigma_d = 0.01°C$ (the resolution of the data, which is recorded to hundredths of a degree), is about $\sigma_{m2} = 10^{-5} °C/yr$. The error based on the posterior variance of the data, $\sigma_d = 5.6 °C$, is larger, $\sigma_{m2} = 0.0046 °C/yr$. In both cases, the slope is significantly different from zero to better than 95% confidence, in the sense that $m_2 + 2\sigma_{m2} < 0$, so we may be justified in claiming that this site has experienced a slight cooling trend. However, the poor overall fit of the model to the data should give any practitioner of data analysis pause. More effort should be put into improving the model.

Crib Sheet 4.1 Standard error of the mean

The model

the data d_i all have the same mean $m_1^{true} = \bar{d}^{true}$

Least squares estimate of the mean

the estimated mean is the sample mean

$$\bar{d}^{est} = m_1^{est} = \frac{1}{N} \sum_{i=1}^{N} d_i$$

```
dbarest = mean(d);
```

The variance of the data

the estimated (posterior) variance is the sample standard deviation

$$\left(\sigma_d^{pos}\right)^2 = \frac{1}{N-1} \sum_{i=1}^{N} \left(d_i - \bar{d}^{est}\right)^2$$

```
sigmadpos = std(d);
```

Standard error of the estimated mean

square root of N error reduction

$$\sigma_{\bar{d}}^{pos} = \frac{\sigma_d^{pos}}{\sqrt{N}}$$

```
sigmadbarpos = sigmadpos/sqrt(N);
```

95% confidence intervals for the true mean

two standard errors

$$\bar{d}^{true} = \bar{d}^{est} \pm 2\sigma_{\bar{d}}^{pos}$$

Crib Sheet 4.2 Steps in simple least squares

Step 1: State the problem in words

how are the data related to the model?

Step 2: Organize the problem in standard form

identify the data $\mathbf{d}$ (length N) the model parameters $\mathbf{m}$ (length M)
define the data kernel $\mathbf{G}$ so that $\mathbf{d}^{obs} = \mathbf{Gm}$

Step 4: Establish the accuracy of the data

state a prior variance σ_d^2 based on accuracy of the measurement technique

Step 5: Estimate model parameters $\mathbf{m}^{est}$ and their covariance $\mathbf{C}_m$

$$\mathbf{m}^{est} = \left[\mathbf{G}^T\mathbf{G}\right]^{-1}\mathbf{G}^T\mathbf{d}^{obs} \quad \text{and} \quad \mathbf{C}_m = \sigma_d^2\left[\mathbf{G}^T\mathbf{G}\right]^{-1}$$

```
mest = (G'*G)\(G'*dobs);
Cm = sigma2d * inv(G'*G);
```

Continued

Crib Sheet 4.2—cont'd

Step 6: Compute the variance of the estimated model parameters
$$\sigma^2_{m_i} = [\mathbf{C}_m]_{ii}$$

Step 7: State 95% confidence intervals for the model parameters
$$m_i^{true} = m_i^{est} \pm 2\sigma_{m_i}$$

Step 8: Compute and examine the error and the posterior variance
$$\mathbf{e} = \mathbf{d}^{obs} - \mathbf{Gm}^{est} \quad \text{and} \quad E = \mathbf{e}^\mathrm{T}\mathbf{e} \quad \text{and} \quad \left(\sigma_d^{pos}\right)^2 = \frac{E}{N-M}$$

```
e = dobs - G * mest;
E = e' * e;
sigma2dpos = E/(N-M);
```

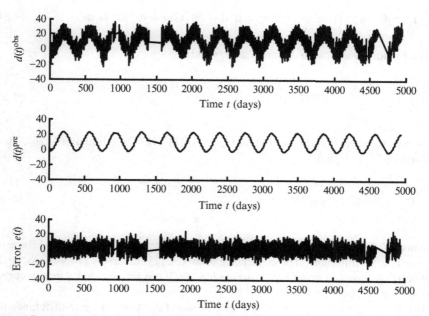

Figure 4.7 (Top) Clean version of the observed Black Rock Forest temperature data, $\mathbf{d}^{obs}$. (Middle) Predicted data, $\mathbf{d}^{pre}$, from $M = 4$ parameter model. (Bottom) Prediction error, $\mathbf{e}$. *MatLab* script eda04_11.

4.9   Covariance and the behavior of error

Back in Section 4.6, during our discussion of the grid search, we alluded to a relationship between the shape of the total error, $E(\mathbf{m})$ and the corresponding error in the estimated model parameters, $\mathbf{m}^{\text{est}}$. We defined an elliptical region of near-minimum errors, centered on the point, $\mathbf{m}^{\text{est}}$, of minimum error, and claimed that any $\mathbf{m}$ within this region was almost as good as $\mathbf{m}^{\text{est}}$. We asserted that the size of the elliptical region is related to the confidence intervals, and its orientation to correlations between the individual ms. This reasoning suggests that there is a relationship between the covariance matrix, $\mathbf{C}_m$, and the shape of $E(\mathbf{m})$ near its minimum.

This relationship can be understood by noting that, near its minimum, the total error, $E(\mathbf{m})$, can be approximated by the first two nonzero terms of its Taylor series (see Section 11.6) :

$$E(\mathbf{m}) \approx E(\mathbf{m}^{\text{est}}) + \sum_{i=1}^{M} \sum_{j=1}^{M} \frac{1}{2} [m_i - m_i^{\text{est}}][m_j - m_j^{\text{est}}] \frac{\partial E}{\partial m_i \partial m_j} \bigg|_{\mathbf{m} = \mathbf{m}^{\text{est}}} \tag{4.33}$$

Note that the linear term is missing, as $\partial E / \partial m_i$ is zero at the minimum of $E(\mathbf{m})$. The error is related to the data kernel, $\mathbf{G}$, via

$$E(\mathbf{m}) = [\mathbf{d} - \mathbf{Gm}]^{\text{T}}[\mathbf{d} - \mathbf{Gm}] = \mathbf{d}^{\text{T}}\mathbf{d} - 2\mathbf{d}^{\text{T}}\mathbf{Gm} + \mathbf{m}^{\text{T}}[\mathbf{G}^{\text{T}}\mathbf{G}]\mathbf{m} \tag{4.34}$$

This equation, when twice differentiated, yields

$$\frac{\partial E}{\partial m_i \partial m_j} \bigg|_{\mathbf{m} = \mathbf{m}^{\text{est}}} = 2[\mathbf{G}^{\text{T}}\mathbf{G}]_{ij} \tag{4.35}$$

However, we have already shown that $\mathbf{C}_m = \sigma_d{}^2[\mathbf{G}^{\text{T}}\mathbf{G}]^{-1}$ (see Equation 4.26). Thus, Equation (4.35) implies

$$\mathbf{C}_m = 2\sigma_d^2 \mathbf{D}^{-1} \quad \text{where} \quad D_{ij} = \frac{\partial E}{\partial m_i \partial m_j} \bigg|_{\mathbf{m} = \mathbf{m}^{\text{est}}} \tag{4.36}$$

The matrix, $\mathbf{D}$, of second derivatives of the error describes the curvature of the error surface near its minimum. The covariance matrix, $\mathbf{C}_m$, is inversely proportional to the curvature. A steeply curved error surface has small covariance, and a gently curved surface has large covariance.

Equation (4.36) is of practical use in grid searches, where a finite-difference approximation to the second derivative can be used to estimate the second-derivative matrix, $\mathbf{D}$, which can then be used to estimate the covariance matrix, $\mathbf{C}_m$.

Problems

4.1. Suppose that a person wants to determine the weight, m_j, of $M = 40$ objects by weighing the first, and then weighing the rest in pairs: the first plus the second, the second plus the third, the third plus the fourth, and so on. (A) What is the corresponding data kernel, $\mathbf{G}$? (B) Write a *MatLab* script that creates this data kernel and computes the covariance matrix, $\mathbf{C}_m$, assuming that the observations are uncorrelated and have a variance, $\sigma_d^2 = 1 \text{ kg}^2$. (C) Make a plot of σ_{mj} as a function of the object number, j, and comment on the results.

4.2. Consider the equation $d_i = m_1 \exp(-m_2 t_i)$. Why cannot this equation be arranged in the linear form, $\mathbf{d} = \mathbf{Gm}$? (A) Show that the equation can be *linearized* into the form, $\mathbf{d}' = \mathbf{G}'\mathbf{m}'$, where the primes represent new, transformed variables, by taking the logarithm of the equation. (B) What would have to be true about the measurement error in order to justify solving this linearized problem by least squares? (Notwithstanding your answer, this problem is often solved with least squares in a *let's-hope-for-the-best* mode).

4.3. (A) What is the relationship between the elements of the matrix, $\mathbf{G}^T\mathbf{G}$, and the columns, $\mathbf{c}^{(j)}$, of $\mathbf{G}$? (B) Under what circumstance is $\mathbf{G}^T\mathbf{G}$ a diagonal matrix? (C) What is the form of the covariance matrix in this case? (D) What is the form least-squares solution in this case? Is it harder or easier to compute than the case where $\mathbf{G}^T\mathbf{G}$ is not diagonal? (E) Examine the straight line case in this context.

4.4. The dataset shown in Figure 4.4 is in the file, `linedata01.txt`. Write *MatLab* scripts to least-squares fit polynomials of degree 2, 3, and 4 to the data. Make plots that show the observed and predicted data. Display the value of each coefficient and its 95% confidence limits. Comment on your results.

4.5. Modify the *MatLab* script, `eda04_11`, to try to achieve a better fit to the Black Rock Forest temperature dataset. (A) Add additional periods of $T_y/2$ and $T_y/3$, where T_y is the period of 1 year, in an attempt to better capture the shape of the annual variation. (B) In addition to the periods in part A, add additional periods of T_d, $T_d/2$, and $T_d/3$, where T_d is the period of 1 day. (C) How much does the total error change in the cases? If it goes up, your code has a bug! (D) How much do the slope, m_2, and its confidence intervals change?

5 Quantifying preconceptions

5.1   When least square fails

The least-squares solution fails when $[\mathbf{G}^T\mathbf{G}]$ has no inverse, or equivalently, when its determinant is zero. In the straight line case, the $[\mathbf{G}^T\mathbf{G}]$ is 2×2 and the determinant, D, can readily be computed from Equation (4.29):

$$D = N\sum_{i=1}^{N} x_i^2 - \left(\sum_{i=1}^{N} x_i\right)^2$$

Two different scenarios lead to the determinant being zero. If only one observation is available (i.e., $N = 1$), then

$$D = x_1^2 - (x_1)^2 = 0$$

This case corresponds to the problem of trying to fit a straight line to a single point. The determinant is also zero when $N > 1$, but the data are all measured at the same value of x_i (say $x_i = x^*$). Then,

$$D = NN(x^*)^2 - (Nx^*)^2 = 0$$

This case corresponds to the problem of trying to fit a straight line to many points, all with the same x. In both cases, the problem is that more than one choice of $\mathbf{m}$ has minimum error. In the first case, any line that passes through the point (x_1, d_1) has *zero* error, regardless of its slope (Figure 5.1A). In the second case, all lines that pass

Environmental Data Analysis with MATLAB®. http://dx.doi.org/10.1016/B978-0-12-804488-9.00005-7

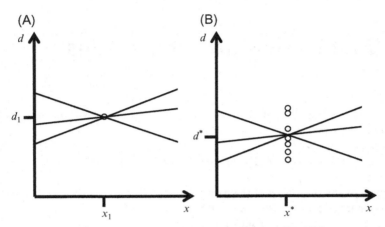

Figure 5.1 (A) All lines passing through (x_i, d_i) have zero error. (B) All lines passing through (x^*, d^*) have the same error.

through the point, (x^*, d^*), where d^* is an arbitrary value of d, will have the *same* error, regardless of the slope, and one of these will correspond to the minimum error ($d^* = \bar{d}$, actually) (Figure 5.1B).

In general, the method of least squares fails when the data do not *uniquely* determine the model parameters. The problem is associated with the data kernel, **G**, which describes the geometry or structure of the experiment, and not with the actual values of the data, **d**, themselves. Nor is the problem limited to the case where $[\mathbf{G}^T\mathbf{G}]^{-1}$ is exactly singular. Solutions when $[\mathbf{G}^T\mathbf{G}]^{-1}$ is almost singular are useless as well, because the covariance of the model parameters is proportional to $[\mathbf{G}^T\mathbf{G}]^{-1}$, and it has very large elements in this case. If almost no data constrains the value of a model parameter, then its value is very sensitive to measurement noise. In these cases, the matrix, $\mathbf{G}^T\mathbf{G}$, is said to be *ill-conditioned*.

Methods are available to spot deficiencies in **G** that lead to $\mathbf{G}^T\mathbf{G}$ being ill-conditioned. However, they are usually of little practical value, because while they can identify the problem, they offer no remedy for it. We take a different approach here, which is to assume that most experiments have deficiencies that lead to at least a few model parameters being poorly determined.

We will not concern ourselves too much with which model parameters are causing the problem. Instead, we will use a modified form of the least-squares methodology that leads to a solution in all cases. This methodology will, in effect, fill in *gaps in information*, but without providing much insight into the nature of those gaps.

5.2   Prior information

Usually, we know *something* about the model parameters, even before we perform any observations. Even before we measure the density of a soil sample, we know that its density will be around 1500 kg/m³, give or take 500 or so, and that negative densities are nonsensical. Even before we measure a topographic profile across a range of hills, we know that it can contain no impossibly high and narrow peaks. Even before we measure the chemical components of an organic substance, we know that they should sum to 100%. Further, even before we measure the concentration of a pollutant in an underground reservoir, we know that its dispersion is subject to the diffusion equation.

These are, of course, just preconceptions about the world, and as such, they are more than a little dangerous. Observations might prove them to be wrong. On the other hand, most are based on experience, years of observations that have shown that, at least on Earth, most physical parameters commonly behave in well-understood ways. Furthermore, we often have a good idea of just how good a preconception is. Experience has shown that the *range* of plausible densities for sea water, for example, is much more restricted than, say, that for crude oil.

These preconceptions embody *prior information* about the results of observations. They can be used to supplement observations. In particular, they can be used to *fill in the gaps* in the information content of a dataset that prevent least squares from working.

We will express prior information probabilistically, using the Normal probability density function.

This choice gives us the ability to represent both the information itself, through the *mean* of the probability density function, and our uncertainty about the information, through its *covariance matrix*. The simplest case is when we know that the model parameters, $\mathbf{m}$, are near the values, $\bar{\mathbf{m}}$, where the uncertainty of the nearness is quantified by a *prior* covariance matrix, $\mathbf{C}_m^p$. Then, the prior information can be represented as the probability density function:

$$
p_p(\mathbf{m}) = \frac{1}{(2\pi)^{M/2} |\mathbf{C}_m^p|^{1/2}} \exp\left\{ -\frac{1}{2}(\mathbf{m} - \bar{\mathbf{m}})^{\mathrm{T}} [\mathbf{C}_m^p]^{-1}(\mathbf{m} - \bar{\mathbf{m}}) \right\}
$$

$$
= \frac{\exp\left\{ -\frac{1}{2} E_p(\mathbf{m}) \right\}}{(2\pi)^{M/2} |\mathbf{C}_m^p|^{1/2}} \quad \text{with} \quad E_p(\mathbf{m}) = (\mathbf{m} - \bar{\mathbf{m}})^{\mathrm{T}} [\mathbf{C}_m^p]^{-1}(\mathbf{m} - \bar{\mathbf{m}})
\tag{5.1}
$$

Note that we interpret the argument of the exponential as depending on a function, $E_p(\mathbf{m})$, which quantifies the degree to which the prior information is satisfied. It can be thought of as a measure of the *error in the prior information* (compare with Equation 4.24).

In the soil density case above, we would choose, $\bar{\mathbf{m}} = 1500\ \mathrm{kg/m^3}$ and $\mathbf{C}_m^p = \sigma_m^2 \mathbf{I}$, with $\sigma_m = 500\ \mathrm{kg/m^3}$. In this case, we view the prior information as uncorrelated, so $\mathbf{C}_m^p \propto \mathbf{I}$.

Note that the prior covariance matrix, $\mathbf{C}_m^p$, is *not* the same as the covariance matrix of the estimated model parameters, $\mathbf{C}_m$ (which is called the *posterior* covariance matrix). The matrix, $\mathbf{C}_m^p$, expresses the uncertainty in the prior information about the model parameters, before we make any observations. The matrix, $\mathbf{C}_m$, expresses the uncertainty of the estimated model parameters, after we include the observations.

A more general case is when the prior information can be represented as a linear function of the model parameters:

a linear function of the model parameters = a known value

or

$$\mathbf{Hm} = \bar{\mathbf{h}} \tag{5.2}$$

where $\mathbf{H}$ is a $K \times M$ matrix, where K is the number of rows of prior information. This more general representation can be used in the chemical component case mentioned above, where the concentrations need to sum to 100% or unity. This is a single piece of prior information, so $K = 1$ and the equation for the prior information has the form

sum of model parameters = unity

or

$$[1 \quad 1 \quad 1 \quad \ldots \quad 1]\mathbf{m} = 1$$

or

$$\mathbf{Hm} = \bar{\mathbf{h}} \tag{5.3}$$

The prior probability density function of the prior information is then

$$p_p(\mathbf{h}) = \frac{1}{(2\pi)^{M/2}|\mathbf{C}_h|^{1/2}} \exp\left\{-\frac{1}{2}(\mathbf{Hm} - \bar{\mathbf{h}})^{\mathrm{T}}[\mathbf{C}_h]^{-1}(\mathbf{Hm} - \bar{\mathbf{h}})\right\} = \frac{\exp\left\{-\frac{1}{2}E_p(\mathbf{m})\right\}}{(2\pi)^{M/2}|\mathbf{C}_h|^{1/2}}$$

where $E_p(\mathbf{m}) = (\mathbf{Hm} - \bar{\mathbf{h}})^{\mathrm{T}}[\mathbf{C}_h]^{-1}(\mathbf{Hm} - \bar{\mathbf{h}})$

note that

$$p_p(\mathbf{m}) = p_p[\mathbf{h}(\mathbf{m})]J(\mathbf{m}) \propto p_p[\mathbf{h}(\mathbf{m})] \tag{5.4}$$

Here the covariance matrix, $\mathbf{C}_h$, expresses the uncertainty to which the model parameters obey the linear equation, $\mathbf{Hm} = \bar{\mathbf{h}}$. Note that the Normal probability density function contains the quantity, $E_p(\mathbf{m})$, which is zero when the prior information, $\mathbf{Hm} = \bar{\mathbf{h}}$, is satisfied exactly, and positive otherwise. $E_p(\mathbf{m})$ quantifies the error in the prior information. The probability density function for $\mathbf{m}$ is proportional to the probability density function for $\mathbf{h}$, as the Jacobian determinant, $J(\mathbf{m})$, is constant (see Note 5.1).

5.3 Bayesian inference

Our objective is to combine prior information with observations. Bayes theorem (Equation 3.25) provides the methodology through the equation

$$p(\mathbf{m}|\mathbf{d}) = \frac{p(\mathbf{d}|\mathbf{m})p(\mathbf{m})}{p(\mathbf{d})} \tag{5.5}$$

We can interpret this equation as a rule for *updating* our knowledge of the model parameters. Let us ignore the factor of $p(\mathbf{d})$ on the right hand side, for the moment. Then the equation reads as follows:

the probability of the model parameters, **m**, given the data, **d**

is proportional to

the probability that the data, **d**, was observed, given a particular

set of model parameters, **m** multiplied by

the prior probability of that set of model parameters, **m**   (5.6)

We identify $p(\mathbf{m})$ with $p_p(\mathbf{m})$, that is, our best estimate of the probability of the model parameters, *before* the observations are made.. The conditional probability density function, $p(\mathbf{d}|m)$, is the probability that data, **d**, are observed, given a particular choice for the model parameters, **m**. We assume, as we did in Equation (4.23), that this probability density function is Normal:

$$p(\mathbf{d}|\mathbf{m}) = \frac{1}{(2\pi)^{N/2}|\mathbf{C}_d|^{1/2}} \exp\left\{-\frac{1}{2}(\mathbf{Gm}-\mathbf{d})^{\mathrm{T}}[\mathbf{C}_d]^{-1}(\mathbf{Gm}-\mathbf{d})\right\} = \frac{\exp\left\{-\frac{1}{2}E(\mathbf{m})\right\}}{(2\pi)^{N/2}|\mathbf{C}_d|^{\frac{1}{2}}}$$

$$\text{where } E(\mathbf{m}) = (\mathbf{Gm}-\mathbf{d})^{\mathrm{T}}[\mathbf{C}_d]^{-1}(\mathbf{Gm}-\mathbf{d}) \qquad (5.7)$$

where $\mathbf{C}_d$ is the covariance matrix of the observations. (Previously, we assumed that $\mathbf{C}_d = \sigma_d^2\mathbf{I}$, but now we allow the general case). Note that the Normal probability density function contains the quantity, $E(\mathbf{m})$, which is zero when the data are exactly satisfied and positive when they are not. This quantity is the *total data error*, as defined in Equation (4.24), except that the factor $\mathbf{C}_d^{-1}$ acts to weight each of the component errors. Its significance will be discussed later in the section.

We now return to the factor of $p(\mathbf{d})$ on the right-hand side of Bayes theorem (Equation 5.5). It is not a function of the model parameters, and so acts only as a normalization factor. Hence, we can write

$$p(\mathbf{m}|\mathbf{d}) \propto p(\mathbf{d}|\mathbf{m})\,p(\mathbf{m})$$

$$\propto \exp\left\{-\frac{1}{2}(\mathbf{Gm}-\mathbf{d})^{\mathrm{T}}[\mathbf{C}_d]^{-1}(\mathbf{Gm}-\mathbf{d})-\frac{1}{2}(\mathbf{Hm}-\bar{\mathbf{h}})^{\mathrm{T}}[\mathbf{C}_h]^{-1}(\mathbf{Hm}-\bar{\mathbf{h}})\right\}$$

$$= \exp\left\{-\frac{1}{2}[E(\mathbf{m})+E_p(\mathbf{m})]\right\} = \exp\left\{-\frac{1}{2}E_T(\mathbf{m})\right\}$$

$$\text{with } E_T(\mathbf{m}) = E(\mathbf{m})+E_p(\mathbf{m}) \qquad (5.8)$$

Note that $p(\mathbf{m}|\mathbf{d})$ contains the quantity, $E_T(\mathbf{m})$, that is the sum of two errors: the error in fitting the data; and the error in satisfying the prior information. We call it the *generalized error*. We do not need the overall normalization factor, because the only operation that we will perform with this probability density function is the computation of its mode (point of maximum likelihood), which (as in Equation 4.24) we will identify as the best estimate, $\mathbf{m}^{\mathrm{est}}$, of the model parameters. An example for the very simple $N = 1$, $M = 2$ case is shown in Figure 5.2. However, before proceeding with more

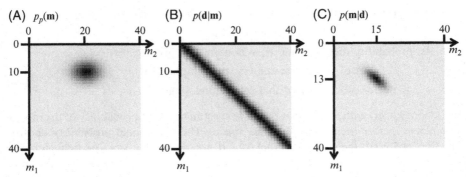

Figure 5.2 Example of the application of Bayes theorem to a $N = 1$, $M = 2$ problem. (A) The prior probability density function, $p_p(\mathbf{m})$, for the model parameters has its maximum at (20,10) and is uncorrelated with variance $(6^2, 10^2)$. (B) The conditional probability density function, $p(\mathbf{d}|\mathbf{m})$, is for one observation, $m_1 - m_2 = d_1 = 0$ with a variance of 3^2. Note that this observation, by itself, is not sufficient to uniquely determine two model parameters. The conditional probability density distribution, $p(\mathbf{m}|\mathbf{d}) \propto p(\mathbf{d}|\mathbf{m})p_p(\mathbf{m})$, has its maximum at $(m_1^{\text{est}}, m_2^{\text{est}}) = (13,15)$. The estimated model parameters do not exactly satisfy the observation, $m_1^{\text{est}} - m_2^{\text{est}} \neq d_1$, reflecting the observational error represented in the probability density function, $p(\mathbf{d}|\mathbf{m})$. They do not exactly satisfy the prior information, either, reflecting the uncertainty represented in $p_p(\mathbf{m})$. *MatLab* script eda05_01.

complex problems, we need to discuss an important issue associated with products of Normal probability density functions (as in Equation 5.8).

5.4   The product of Normal probability density distributions

The conditional probability density function, $p(\mathbf{m}|\mathbf{d})$, in Equation (5.8) is the product of two Normal probability density functions. One of the many useful properties of Normal probability density functions is that their products are themselves Normal (Figure 5.3). To verify that this is true, we start with three Normal probability density functions, $p_a(\mathbf{m})$, $p_b(\mathbf{m})$, and $p_c(\mathbf{m})$:

$$p_a(\mathbf{m}) \propto \exp\left\{ -\frac{1}{2}(\mathbf{m} - \bar{\mathbf{a}})^{\text{T}} \mathbf{C}_a^{-1}(\mathbf{m} - \bar{\mathbf{a}}) \right\}$$

$$p_b(\mathbf{m}) \propto \exp\left\{ -\frac{1}{2}(\mathbf{m} - \bar{\mathbf{b}})^{\text{T}} \mathbf{C}_b^{-1}(\mathbf{m} - \bar{\mathbf{b}}) \right\}$$

$$p_c(\mathbf{m}) \propto \exp\left\{ -\frac{1}{2}(\mathbf{m} - \bar{\mathbf{c}})^{\text{T}} \mathbf{C}_c^{-1}(\mathbf{m} - \bar{\mathbf{c}}) \right\}$$

$$= \exp\left\{ -\frac{1}{2}(\mathbf{m}^{\text{T}} \mathbf{C}_c^{-1}\mathbf{m} - 2\mathbf{m}^{\text{T}}\mathbf{C}_c^{-1}\bar{\mathbf{c}} + \bar{\mathbf{c}}^{\text{T}}\mathbf{C}_c^{-1}\bar{\mathbf{c}}) \right\} \qquad (5.9)$$

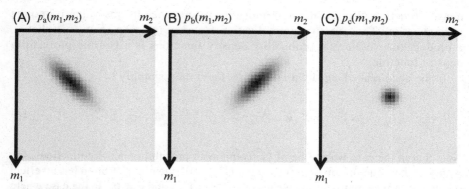

Figure 5.3 The product of two Normal distributions is itself a Normal distribution. (A) A Normal distribution, $p_a(m_1, m_2)$. (B) A Normal distribution, $p_b(m_1, m_2)$. (C) The product, $p_c(m_1, m_2) = p_a(m_1, m_2) \, p_b(m_1, m_2)$. *MatLab* script eda05_02.

Note that the second version of $p_c(\mathbf{m})$ is just the first with the expression within the braces expanded out. We now compute the product of the first two:

$$p_a(\mathbf{m})p_b(\mathbf{m}) \propto \exp\left\{ -\frac{1}{2}(\mathbf{m} - \bar{\mathbf{a}})^{\mathrm{T}} \mathbf{C}_a^{-1}(\mathbf{m} - \bar{\mathbf{a}}) - \frac{1}{2}(\mathbf{m} - \bar{\mathbf{b}})^{\mathrm{T}} \mathbf{C}_b^{-1}(\mathbf{m} - \bar{\mathbf{b}}) \right\}$$

$$= \exp\left\{ -\frac{1}{2}(\mathbf{m}^{\mathrm{T}}[\mathbf{C}_a^{-1} + \mathbf{C}_b^{-1}]\mathbf{m} - 2\mathbf{m}^{\mathrm{T}}[\mathbf{C}_a^{-1}\bar{\mathbf{a}} + \mathbf{C}_b^{-1}\bar{\mathbf{b}}] + [\mathbf{a}^{\mathrm{T}}\mathbf{C}_a^{-1}\mathbf{a} + \mathbf{b}^{\mathrm{T}}\mathbf{C}_b^{-1}\mathbf{b}]) \right\}$$

$$(5.10)$$

We now try to choose $\bar{c}$ and $\mathbf{C}_c$ in Equation (5.9) so that $p_c(\mathbf{m})$ in Equation (5.9) matches $p_a(\mathbf{m}) \, p_b(\mathbf{m})$ in Equation (5.10). The choice

$$\mathbf{C}_c^{-1} = \mathbf{C}_a^{-1} + \mathbf{C}_b^{-1} \tag{5.11}$$

matches the first pair of terms (the ones quadratic in $\mathbf{m}$) and gives, for the second pair of terms (the ones linear in $\mathbf{m}$)

$$2\mathbf{m}^{\mathrm{T}}(\mathbf{C}_a^{-1} + \mathbf{C}_b^{-1})\bar{\mathbf{c}} = 2\mathbf{m}^{\mathrm{T}}(\mathbf{C}_a^{-1}\bar{\mathbf{a}} + \mathbf{C}_b^{-1}\bar{\mathbf{b}}) \tag{5.12}$$

Solving for $\bar{c}$, we find that these terms are equal when

$$\bar{\mathbf{c}} = (\mathbf{C}_a^{-1} + \mathbf{C}_b^{-1})^{-1}(\mathbf{C}_a^{-1}\bar{\mathbf{a}} + \mathbf{C}_b^{-1}\bar{\mathbf{b}}) \tag{5.13}$$

Superficially, these choices do make the third pair of terms (the ones that do not contain $\mathbf{m}$) equal. However, as these terms do not depend on $\mathbf{m}$, they just correspond to the multiplicative factors that affect the normalization of the probability density function. We can always remove the discrepancy by absorbing it into the normalization. Thus, up to a normalization factor, $p_c(\mathbf{m}) = p_a(\mathbf{m})p_b(\mathbf{m})$; that is, a product of two Normal probability density functions is a Normal probability density function.

In the uncorrelated, equal variance case, these rules simplify to

$$\sigma_c^{-2} = \sigma_a^{-2} + \sigma_b^{-2} \quad \text{and} \quad \bar{\mathbf{c}} = (\sigma_a^{-2} + \sigma_b^{-2})^{-1}(\sigma_a^{-2}\bar{\mathbf{a}} + \sigma_b^{-2}\bar{\mathbf{b}}) \qquad (5.14)$$

Note that in the case where one of the component probability density functions, say $p_a(\mathbf{m})$, contains no information (i.e., when $\mathbf{C}_a^{-1} \rightarrow 0$), the multiplication has no effect on the covariance matrix or the mean (i.e., $\mathbf{C}_c^{-1} = \mathbf{C}_b^{-1}$ and $\bar{\mathbf{c}} = \bar{\mathbf{b}}$). In the case where both $p_a(\mathbf{m})$ and $p_b(\mathbf{m})$ contain information, the covariance of the product will, in general, be *smaller* than the covariance of either probability density function (Equation 5.11), and the mean, $\bar{\mathbf{c}}$, will be somewhere on a line connecting $\bar{\mathbf{a}}$ and $\bar{\mathbf{b}}$ (Equation 5.13).

Thus, $p(\mathbf{m}|\mathbf{d})$ in Equation (5.8), being the product of two Normal probability density functions, is itself a Normal probability density function.

5.5   Generalized least squares

We now return to the matter of deriving an estimate of model parameters that combines both observations and prior information by finding the peak (mode) of the Normal distribution in Equation (5.8). This Normal distribution depends on the *generalized error, $E_T(\mathbf{m})$*:

$$p(\mathbf{m}|\mathbf{d}) \propto \exp \quad \left\{ -\frac{1}{2}E_T(\mathbf{m}) \right\} \quad \text{where}$$

$$E_T(\mathbf{m}) = (\mathbf{Hm} - \bar{\mathbf{h}})^T[\mathbf{C}_h]^{-1}(\mathbf{Hm} - \bar{\mathbf{h}}) + (\mathbf{Gm} - \mathbf{d}^{obs})^T[\mathbf{C}_d]^{-1}(\mathbf{Gm} - \mathbf{d}^{obs})$$
$$(5.15)$$

The expression for the generalized error can be simplified by defining a matrix $\mathbf{F}$ and a vector $\mathbf{f}$ such that:

$$\mathbf{F} = \begin{bmatrix} \mathbf{C}_d^{-\frac{1}{2}}\mathbf{G} \\ \mathbf{C}_h^{-\frac{1}{2}}\mathbf{H} \end{bmatrix} \quad \text{and} \quad \mathbf{f}^{obs} = \begin{bmatrix} \mathbf{C}_d^{-\frac{1}{2}}\mathbf{d}^{obs} \\ \mathbf{C}_h^{-\frac{1}{2}}\bar{\mathbf{h}} \end{bmatrix} \qquad (5.16)$$

Here $\mathbf{C}_d^{-\frac{1}{2}}$ is the square root of $\mathbf{C}_d^{-1}$ (which obeys $\mathbf{C}_d^{-1} = \mathbf{C}_d^{-\frac{1}{2}}\mathbf{C}_d^{-\frac{1}{2}}$) and $\mathbf{C}_h^{-\frac{1}{2}}$ is the square root of $\mathbf{C}_h^{-1}$. In the commonly-encountered case where $\mathbf{C}_d^{-1}$ and $\mathbf{C}_h^{-1}$ are diagonal matrices, the square root is computed simply by taking the square root of the diagonal elements). The generalized error is then:

$$E_T(\mathbf{m}) = \left[\mathbf{f}^{obs} - \mathbf{F}\mathbf{m}^{est}\right]^T \mathbf{C}_f^{-1} \left[\mathbf{f}^{obs} - \mathbf{F}\mathbf{m}^{est}\right] \quad \text{with} \quad \mathbf{C}_f^{-1} = \mathbf{I} \qquad (5.17)$$

This equivalence can be shown by substituting Equation (5.16) into Equation (5.17) and multiplying out the expression. The generalized error has been manipulated into a form identical to the ordinary least squares error, implying that the solution is the ordinary least squares solution:

$$\left[\mathbf{F}^T\mathbf{F}\right]\mathbf{m}^{est} = \mathbf{F}^T\mathbf{f}^{obs} \quad \text{and} \quad \mathbf{m}^{est} = \left[\mathbf{F}^T\mathbf{F}\right]^{-1}\mathbf{F}^T\mathbf{f}^{obs} \qquad (5.18)$$

The covariance matrix $\mathbf{C}_m$ is computed by the usual rules of error propagation:

$$\mathbf{C}_m = \left(\left[\mathbf{F}^T\mathbf{F}\right]^{-1}\mathbf{F}^T\right)\mathbf{C}_f\left(\left[\mathbf{F}^T\mathbf{F}\right]^{-1}\mathbf{F}^T\right)^T = \left[\mathbf{F}^T\mathbf{F}\right]^{-1} \quad \text{since} \quad \mathbf{C}_f = \mathbf{I} \qquad (5.19)$$

These results are due to Tarantola and Valette (1982) and are further discussed by Menke (1989). When we substitute Equation (5.16) into Equation (5.19), we obtain the expressions:

$$\mathbf{m}^{est} = \left[\mathbf{G}^T[\mathbf{C}_d]^{-1}\mathbf{G} + \mathbf{H}^T[\mathbf{C}_h]^{-1}\mathbf{H}\right]^{-1}\left[\mathbf{G}^T[\mathbf{C}_d]^{-1}\mathbf{d}^{obs} + \mathbf{H}^T[\mathbf{C}_h]^{-1}\bar{\mathbf{h}}\right]$$

$$\mathbf{C}_m = \left[\mathbf{G}^T[\mathbf{C}_d]^{-1}\mathbf{G} + \mathbf{H}^T[\mathbf{C}_h]^{-1}\mathbf{H}\right]^{-1} \qquad (5.20)$$

However, these expressions are cumbersome and almost never necessary. Instead, we construct the equation:

$$\begin{bmatrix} \mathbf{C}_d^{-\frac{1}{2}}\mathbf{G} \\ \mathbf{C}_h^{-\frac{1}{2}}\mathbf{H} \end{bmatrix}\mathbf{m}^{est} = \begin{bmatrix} \mathbf{C}_d^{-\frac{1}{2}}\mathbf{d}^{obs} \\ \mathbf{C}_h^{-\frac{1}{2}}\bar{\mathbf{h}} \end{bmatrix} \qquad (5.21)$$

directly and solve it by ordinary least squares. In this equation, the rows of the data equations, $\mathbf{G}\mathbf{m} = \mathbf{d}^{obs}$ and the rows of the prior information equation $\mathbf{H}\mathbf{m}^{est} = \bar{\mathbf{h}}$ are combined into a single matrix equation, $\mathbf{F}\mathbf{m}^{est} = \mathbf{f}$, with the N rows of $\mathbf{G}\mathbf{m}^{est} = \mathbf{d}$ weighted by the certainty of the data (that is, by the factor σ_d^{-1}), and the K rows of $\mathbf{H}\mathbf{m}^{est} = \bar{\mathbf{h}}$ weighted by the certainty of the prior information (that is, by the factor σ_h^{-1}). Observations and prior information play symmetrical roles in this *generalized least squares solution*. Provided that enough prior information is added to "fill in the gaps", the generalized least squares solution, $\mathbf{m}^{est} = \left[\mathbf{F}^T\mathbf{F}\right]^{-1}\mathbf{F}^T\mathbf{f}^{obs}$, will be well-behaved, even when the ordinary least squares solution, $\mathbf{m}^{est} = \left[\mathbf{G}^T\mathbf{G}\right]^{-1}\mathbf{F}\mathbf{G}^T\mathbf{d}^{obs}$, fails. The prior information *regularizes* the matrix, $[\mathbf{F}^T\mathbf{F}]$.

One type of prior information that always regularizes a generalized least squares problem is the model parameters being close to a constant, $\bar{\mathbf{m}}$. This is the case where

$K = M$, $\mathbf{H} = \mathbf{I}$ and $\bar{\mathbf{h}} = \bar{\mathbf{m}}$. The special case of $\bar{\mathbf{h}} = \bar{\mathbf{m}} = 0$ is called damped least squares, and corresponds to the solution:

$$\mathbf{m}^{\text{est}} = \left[\mathbf{G}^{\text{T}}\mathbf{G} + \epsilon^2\mathbf{I}\right]\mathbf{G}^T\mathbf{d} \quad \text{with} \quad \epsilon^2 = \sigma_d^2/\sigma_m^2 \tag{5.22}$$

The attractiveness of the damped least squares is the ease by which it can be used. One merely adds a small number, ϵ^2, to the main diagonal of $[\mathbf{G}^{\text{T}}\mathbf{G}]$. However, while easy, damped least squares is only warranted when there is good reason to believe that the model parameters are actually near-zero.

In the generalized least squares formulation, all model parameters are affected by the prior information, even those that are well-determined by the observations. Unfortunately, alternative methods that target prior information at only underdetermined or poorly-determined model parameters are much more cumbersome to implement and are, in general, computationally unsuited to problems with a large number of model parameters (e.g. $M > 10^3$ or so). On the other hand, by choosing the magnitude of the elements of $\mathbf{C}_h^{-1}$ to be sufficiently small, a similar result can be achieved, though often trial-and-error is required to determine how *small is small*.

As an aside, we mention an interesting interpretation of the equation for the generalized least squares solution, in the special case where $M = K$ and $\mathbf{H}^{-1}$ exists, so we can write $\bar{\mathbf{m}} = \mathbf{H}^{-1}\bar{\mathbf{h}}$. Then, if we subtract $\left[\mathbf{G}^{\text{T}}[\mathbf{C}_d]^{-1}\mathbf{G} + \mathbf{H}^{\text{T}}[\mathbf{C}_h]^{-1}\mathbf{H}\right]\bar{\mathbf{m}}$ from both sides of Equation (5.20), we obtain:

$$\left[\mathbf{G}^{\text{T}}[\mathbf{C}_d]^{-1}\mathbf{G} + \mathbf{H}^{\text{T}}[\mathbf{C}_h]^{-1}\mathbf{H}\right](\mathbf{m}^{est} - \bar{\mathbf{m}}) = \mathbf{G}^{\text{T}}[\mathbf{C}_d]^{-1}(\mathbf{d} - \mathbf{G}\bar{\mathbf{m}}) \tag{5.23}$$

which involves the *deviatoric* quantities $\Delta\mathbf{m} = \mathbf{m}^{est} - \bar{\mathbf{m}}$ and $\Delta\mathbf{d} = \mathbf{d} - \mathbf{G}\bar{\mathbf{m}}$. In this view, the generalized least squares solution determines the deviation $\Delta\mathbf{m}$ of the solution away from the prior model parameters, $\bar{\mathbf{m}}$, using the deviation $\Delta\mathbf{d}$, of the data away from the prediction $\mathbf{G}\bar{\mathbf{m}}$ of the prior model parameters.

5.6 The role of the covariance of the data

Generalized least squares (Equation 5.21) adds an important nuance to the estimation of model parameters, even in the absence of prior information, because it weights the contribution of an observation, $\mathbf{d}$, to the error, $E(\mathbf{m})$, according to its *certainty* (the inverse of its variance):

$$E(\mathbf{m}) = (\mathbf{Gm} - \mathbf{d})^{\text{T}}[\mathbf{C}_d]^{-1}(\mathbf{Gm} - \mathbf{d}) = \mathbf{e}^{\text{T}}[\mathbf{C}_d]^{-1}\mathbf{e} \tag{5.24}$$

This effect is more apparent in the special case where the data are uncorrelated with variance, σ^2_{di}. Then, $\mathbf{C}_d$ is a diagonal matrix and the error is

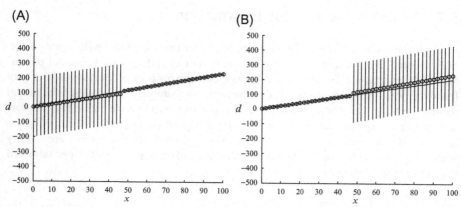

Figure 5.4 Example of least-squares fitting of a line to $N = 50$ data of unequal variance. The data values (circles) in (A) and (B) are the same, but their variance (depicted here by $2\sigma_d$ error bars) is different. (A) The variance of the first 25 data is much greater than that of the second 25 data. (B) The variance of the first 25 data is much less than that of the second 25 data. The best-fit straight line (solid line) is different in the two cases, and in each case more closely fits the half of the dataset with the smaller error. *MatLab* scripts eda05_03 and eda05_04.

$$E(\mathbf{m}) = \mathbf{e}^{\mathrm{T}} \begin{bmatrix} \sigma_{d1}^{-2} & 0 & \cdots & 0 \\ 0 & \sigma_{d2}^{-2} & \cdots & 0 \\ 0 & 0 & .. & 0 \\ \cdots & \cdots & \cdots & \cdots \\ 0 & 0 & 0 & \sigma_{dN}^{-2} \end{bmatrix} \mathbf{e} = \sum_{i=1}^{N} \frac{\mathbf{e}_i^2}{\sigma_{di}^2} \qquad (5.25)$$

Thus, poorly determined data contribute less to the total error than well-determined data and the resulting solution better fits the data with small variance (Figure 5.4).

The special case of generalized least squares that weights the data according to its certainty but includes no prior information is called *weighted least squares*. In *MatLab*, the solution is computed as

```
mest = (G'*Cdi*G)\(G'*Cdi*d);          (MatLab eda05_03)
```

where `Cdi` is the inverse of the covariance matrix of the data, $\mathbf{C}_d^{-1}$. In many cases, however, the covariance is diagonal, as in Equation (5.25). Then, defining a column vector, `sigmad`, with elements, σ_{di}, Equation (5.18) can be used, as follows:

```
for i=[1:N]
    F(i,:)=G(i,:)./sd(i);
end
f=d./sd;
mest = (F'*F)\(F*f);                    (MatLab eda05_04)
```

where `sd` is a column vector with elements, σ_{di}.

5.7 Smoothness as prior information

An important type of prior information is the belief that the model parameter vector, **m**, is smooth. This notion implies some sort of natural ordering of the model parameters in time or space, because smoothness characterizes how model parameters vary from one position or time to another one nearby. The simplest case is when the model parameters vary with one coordinate, such as position, x. They are then a discrete version of a function, $m(x)$, and their roughness (the opposite of smoothness) can be quantified by the second derivative, d^2m/dx^2. When the model parameters are evenly spaced in x, say with spacing Δx, the first and second derivative can be approximated with the finite differences (see Section 1.9):

$$\left. \frac{dm}{dx} \right|_{x_i} \approx \frac{m(x_i + \Delta x) - m(x_i)}{\Delta x} = \frac{1}{\Delta x}[m_{i+1} - m_i]$$

$$\left. \frac{d^2m}{dx^2} \right|_{x_i} \approx \frac{m(x_i + \Delta x) - 2m(x_i) + m(x_i - \Delta x)}{(\Delta x)^2} = \frac{1}{(\Delta x)^2}[m_{i+1} - 2m_i + m_{i-1}] \quad (5.26)$$

The smoothness condition implies that the roughness is small. We represent roughness with the equation, $\mathbf{Hm} = \bar{\mathbf{h}} = 0$, where each row of the equation corresponds to a second derivative centered at a different x-position. A typical row of **H** has elements proportional to

$$0 \quad \cdots \quad 0 \quad 1 \quad -2 \quad 1 \quad 0 \quad \cdots \quad 0$$

However, a problem arises with the first and last row, because the model parameters m_0 and m_{M+1} are unavailable. We can either omit these rows, in which case **H** will contain only $M - 2$ pieces of information, or use different prior information there. A natural choice is to require the slope (i.e., the first derivative, dm/dx) to be small at the ends (i.e., the ends are *flat*), which leads to

$$\mathbf{H} = \frac{1}{(\Delta x)^2} \begin{bmatrix} -\Delta x & \Delta x & 0 & 0 & 0 & \cdots & 0 \\ 1 & -2 & 1 & 0 & 0 & \cdots & 0 \\ 0 & 1 & -2 & 1 & 0 & \cdots & 0 \\ \cdots & \cdots & \cdots & \cdots & \cdots & \cdots & \cdots \\ 0 & \cdots & 0 & 1 & -2 & 1 & 0 \\ 0 & \cdots & 0 & 0 & 1 & -2 & 1 \\ 0 & \cdots & 0 & 0 & 0 & -\Delta x & \Delta x \end{bmatrix} \quad (5.27)$$

The vector, $\bar{h}$, is taken to be zero, as our intent is to make the roughness and steepness—the opposites of smoothness and flatness—as small as possible.

One simple application of smoothness information is the filling in of data gaps. The idea is to have the model parameters represent the values of a function on a grid, with the data representing the values on a subset of grid points whose values have been observed. The other grid points represent data gaps. The equation, $\mathbf{Gm} = \mathbf{d}$, reduces to $m_i = d_j$, which has a $\mathbf{G}$ as follows:

$$\mathbf{G} = \begin{bmatrix} 0 & 1 & 0 & 0 & 0 & 0 & 0 & \cdots & 0 \\ 0 & 0 & 0 & 1 & 0 & 0 & 0 & \cdots & 0 \\ \cdots & \cdots & \cdots & \cdots & \cdots & \cdots & 0 & \cdots & \cdots \\ 0 & 0 & 0 & 0 & 0 & 1 & 0 & \cdots & 0 \end{bmatrix} \quad (5.28)$$

Each row of $\mathbf{G}$ has $M - 1$ zeros and a single 1, positioned to match up an observation with the model parameter corresponding to at the same value of x. In *MatLab*, the matrix, $\mathbf{F}$, and vector, $\mathbf{f}$, are created in two stages. The first stage creates the top part of $\mathbf{F}$ and $\mathbf{f}$ (i.e., the part containing $\mathbf{G}$ and $\mathbf{d}$):

```
L=N+M;
F=zeros(L,M);
f=zeros(L,1);
for p = [1:N]
    F(p,rowindex(p)) = 1;
    f(p)=d(p);
end
```
(*MatLab* eda05_05)

Here, rowindex is a column vector that specifies the correspondence of observation, d_p, and its corresponding model parameter. For simplicity, we assume that the observations have unit variance, and so omit factors of σ_d^{-1}. The second stage creates the bottom part of $\mathbf{F}$ and $\mathbf{f}$ (i.e., the part containing $\mathbf{H}$ and $\mathbf{h}$)

```
shi = 1e-6;
for p = [1:M-2]
    F(p+N,p) = shi/Dx2;
    F(p+N,p+1) = -2*shi/Dx2;
    F(p+N,p+2) = shi/Dx2;
    f(p+N)=0.0;
end
F(L-1,1)=-shi*Dx;
F(L-1,2)=shi*Dx;
f(L-1)=0;
F(L,M-1)=-shi*Dx;
F(L,M)=shi*Dx;
f(L)=0;
```
(*MatLab* eda05_05)

Here, we assume that the prior information is uncorrelated and with equal variance, so we can use a single variable shi to represent σ_h^{-1}. We set it to a value much smaller than unity so that it will have much less weight in the solution than the data. This way, the solution will favor satisfying the data at points where data is available. A for loop is used to create this part of the matrix, $\mathbf{F}$, which corresponds to smoothness. Finally, the flatness information is put in the last two rows of $\mathbf{F}$. The estimated model parameters are then calculated by solving $\mathbf{Fm} = \mathbf{f}$ in the least squares sense:

```
mest = (F'*F)\(F'*f);
```
(*MatLab* eda05_05)

An example is shown in Figure 5.5.

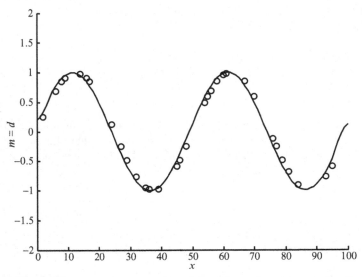

Figure 5.5 The model parameters, m_i, consist of the values of an unknown function, $m(x)$, evaluated at $M = 100$ equally spaced points, x_i. The data, d_i, consist of observations (circles) of the function at $N = 40$ of these points. Prior information, that the function is smooth, is used to fill in the gaps and produce an estimate (solid line) of the function at all points, x_i. *MatLab* script eda05_05.

Crib Sheet 5.1 Generalized least squares

Step 1: State the problem in words
How are the data related to the model

Step 2: Organize the problem in standard form
identify the data **d** (length N) the model parameters **m** (length M)
define the data kernel **G** so that $\mathbf{d}^{obs} = \mathbf{Gm}$

Step 3: Examine the data
make plots of the data

Step 4: Establish the accuracy of the data
state a prior variance σ_d^2 based on accuracy of the measurement technique

Step 5: State the prior information in words, for example:
the model parameters are close to a known values, $\mathbf{h}^{pri}$
the mean of the model parameters is close to a known value
the model parameters vary smoothly with space and time

Step 6: Organize the prior information in standard form:
$$\mathbf{h}^{pri} = \mathbf{Hm}$$

Continued

Crib Sheet 5.1—cont'd

Step 7: Establish the accuracy of the prior information
state a prior variance σ_h^2 based on the accuracy of the prior information

Step 8: Estimate model parameter $\mathbf{m}^{est}$ and their covariance $\mathbf{C}_m$

$$\mathbf{m}^{est} = \left[\mathbf{F}^T\mathbf{F}\right]^{-1}\mathbf{F}^T\mathbf{f}^{obs} \quad \text{and} \quad \mathbf{C}_m = \left[\mathbf{F}^T\mathbf{F}\right]^{-1}$$

$$\text{with} \quad \mathbf{F} = \begin{bmatrix} \sigma_d^{-1}\mathbf{G} \\ \sigma_h^{-1}\mathbf{H} \end{bmatrix} \quad \text{and} \quad \mathbf{f}^{obs} = \begin{bmatrix} \sigma_d^{-1}\mathbf{d}^{obs} \\ \sigma_h^{-1}\mathbf{h}^{pri} \end{bmatrix}$$

Step 9: State estimates and their 95% confidence intervals

$$m_i^{true} = m_i^{est} \pm 2\sigma_{mi}\,(95\%) \quad \text{with} \quad \sigma_{mi} = \sqrt{[\mathbf{C}_m]_{ii}}$$

Step 10: Examine the individual errors

$$\mathbf{d}^{pre} = \mathbf{G}\mathbf{m}^{est} \quad \text{and} \quad \mathbf{e} = \mathbf{d}^{obs} - \mathbf{d}^{pre}$$

$$\mathbf{h}^{pre} = \mathbf{H}\mathbf{m}^{est} \quad \text{and} \quad \mathbf{e}_p = \mathbf{h}^{pri} - \mathbf{h}^{pre}$$

plot $\mathbf{e_i}$ vs. i and plot $\mathbf{e}_{pi}$ vs. i
scatter plot of d_i^{pre} vs. d_i^{obs} and scatter plot of h_i^{pre} vs. h_i^{pri}
any unusually large errors?

Step 11: Examine the total error E_T

$$E_T = E + E_p \quad \text{with} \quad E = \sigma_d^{-2}\mathbf{e}^T\mathbf{e} \quad \text{and} \quad E_p = \sigma_h^{-2}\mathbf{e}_p^T\mathbf{e}_p$$

use a chi-squared test on E_T to assess the likelihood of the Null Hypothesis
that E_T is different than expected only because of random variation

Step 12: Two different models?
use an F-test on the E's of the two models to assess the likelihood
of the Null Hypothesis that the E's are
different from each other only because of random variation

5.8 Sparse matrices

Many perfectly reasonable problems have large number of model parameters—hundreds of thousands or even more. The gap-filling scenario discussed in the previous section is one such example. If, for instance, we were to use it to fill in the gaps in the Black Rock Forest dataset (see Chapter 2), we would have $N \approx M \approx 10^5$. The $L \times M$ matrix, $\mathbf{F}$, where $L = M + N$, would then have about $2N^2 \approx 2 \times 10^{10}$ elements—enough to tax the memory of a notebook computer, at least! On the other hand, only about $3N \approx 3 \times 10^5$ of these elements are nonzero. Such a matrix is said to be *sparse*.

A computer's memory is wasted storing the zero elements and its processing power is wasted multiplying other quantities by them (as the result is a foregone conclusion—zero). An obvious solution is to omit storing the zero elements of sparse matrices and to omit any multiplications involving them. However, such a solution requires special software support to properly organize the matrix's elements and to optimize arithmetic operations involving them.

In *MatLab*, a matrix needs to be defined as sparse, but once defined, *MatLab* more or less transparently handles all array-element access and arithmetic operations. The command

```
L=M+N;
F=spalloc(L,M,4*N);                      (MatLab eda05_06)
```

creates a $L \times M$ sparse matrix, **F**, capable of holding $4N$ nonzero elements. *MatLab* will properly process the command:

```
mest = (F'*F)\(F'*f);                    (MatLab eda05_05)
```

Nevertheless, we do not recommend solving for $\mathbf{m}^{\text{est}}$ this way, except when M is very small, because it does not utilize all the inherent efficiencies of the generalized least-squares equation, $\mathbf{F}^T\mathbf{Fm} = \mathbf{F}^T\mathbf{f}$. Our preferred technique is to use *MatLab's* `bicg()` function, which solves the matrix equation by the *biconjugate gradient* method. The simplest way to use this function is

```
mest=bicg(F'*F,F'*f,1e-10,3*L);          (MatLab eda05_06)
```

As you can see, two extra argument are present, in addition to the matrix, `F'*F`, and the vector, `F'*f`. They are a *tolerance* (set here to `1e-10`) and a *maximum number of iterations* (set here to `3*L`). The `bicg()` function works by iteratively improving an initial guess for the solution, with the tolerance specifying when the error is small enough for the solution to be considered done, and the maximum number of iterations specifying that the method should terminate after this limit is reached, regardless of whether or not the error is smaller than the tolerance. The actual choice of these two parameters needs to be adjusted by trial and error to suit a particular problem. Each time it is used, the `bicg()` function displays a line of information that can be useful in determining the accuracy of the solution.

This simple way of calling `bicg()` has one defect—it requires the computation of the quantity, $\mathbf{F}^T\mathbf{F}$. This is undesirable, for while $\mathbf{F}^T\mathbf{F}$ is sparse, it is typically not nearly as sparse as $\mathbf{F}$, itself. Fortunately, the biconjugate gradient method utilizes $\mathbf{F}^T\mathbf{F}$ in only one simple way: it multiplies various internally constructed vectors to form products such as $\mathbf{F}^T\mathbf{Fv}$. However, this product can be performed as $\mathbf{F}^T(\mathbf{Fv})$, that is, **v** is first premultiplied by **F** and the resulting *vector* is then premultiplied by $\mathbf{F}^T$ so that the matrix $\mathbf{F}^T\mathbf{F}$ is never actually calculated. *MatLab* provides a way to modify the `bicg()` function to perform the multiplication in this very efficient fashion. However, in order to use it, we must first write a *MatLab* function, stored in a separate file that performs the two multiplications (see Note 5.2). We call this function, `afun`, and the corresponding file, `afun.m`:

```
function y = afun(v,transp_flag)
global F;
temp = F*v;
y = F'*temp;
return                                    (MatLab afun.m)
```

We have not said anything about the *MatLab* `function` command so far, and will say little about it here (however, see Note 5.2). Briefly, *MatLab* provides a mechanism for a user to define functions of his or her own devising that act in analogous fashion to built-in functions such a `sin()` and `cos()`. However, as the `afun()` function will not need to be modified, the user is free to consider it a *black box*. In order to use this function, the two commands

```
clear F;
global F;                                 (MatLab eda05_06)
```

need to be placed at the top of the script that uses the `bicg()` function. They ensure that *MatLab* understands that the matrix, F, in the main script and in the function refers to the same variable. Then the `bicg()` function is called as follows:

```
mest=bicg(@afun,F'*f,1e-10,3*L);          (MatLab eda05_06)
```

Note that only the first argument is different than in the previous version, and that this argument is a reference (a *function handle*) to the `afun()` function, indicated with the syntax, `@afun`. Incidentally, we gave the function the name, `afun()`, to match the example in the *MatLab* help page for `bicg()` (which you should read). A more descriptive name might have been better.

5.9 Reorganizing grids of model parameters

Sometimes, the model parameters have a natural organization that is more complicated than can be represented naturally with a column vector, $\mathbf{m}$. For example, the model parameters may represent the values of a function, $m(x, y)$, on a two-dimensional (x, y) grid, in which case they are more naturally ordered into a matrix, $\mathbf{A}$, whose elements are $A_{ij} = m(x_i, y_j)$. Unfortunately, the model parameters must still be arranged into a column vector, $\mathbf{m}$, in order to use the formulas of least squares, at least as they have been developed in this book. One possible solution is to *unwrap* (reorganize) the matrix into a column vector as follows:

$$\mathbf{A} = \begin{bmatrix} A_{11} & A_{12} \\ A_{21} & A_{22} \end{bmatrix} \rightarrow \mathbf{m} = \begin{bmatrix} A_{11} \\ A_{12} \\ A_{21} \\ A_{22} \end{bmatrix} \quad \text{or} \quad m_k = A_{ij} \text{ with } k = (i - 1)J + j$$

$$(5.29)$$

Here, $\mathbf{A}$ is assumed to be an $I \times J$ matrix so that $\mathbf{m}$ is of length, $M = IJ$. In *MatLab*, the conversions from k to (i,j) and back to k are given by

```
k = (i-1)*J+j;

i = floor((k-1)/J)+1;
j = k-(i-1)*J;                                            (MatLab eda05_07)
```

The floor() function rounds down to the nearest integer. See Note 5.3 for a discussion of several advanced *MatLab* functions that can be used as alternatives to these formulas.

As an example, we consider a scenario in which spatial variations of pore pressure cause fluid flow in aquifer (Figure 5.6). The aquifer is a thin layer, so pressure, $p(x, y)$, varies only in the (x, y) plane. The pressure is measured in N wells, each located at (x_i, y_i). These measurements constitute the data, $\mathbf{d}$. The problem is to fill in the data gaps, that is, to estimate the pressure on an evenly spaced grid, A_{ij}, of (x_i, y_j) points. These gridded pressure values constitute an $I \times J$ matrix, $\mathbf{A}$, which can be unwrapped into a column vector of model parameters, $\mathbf{m}$, as described above. The prior information is the belief that the pressure satisfied a known differential equation, in this case, the diffusion equation $\partial^2 p/\partial x^2 + \partial^2 p/\partial y^2 = 0$. This equation is appropriate when the fluid flow obeys Darcy's law and the hydraulic properties of the aquifer are spatially uniform.

The problem is structured in a manner similar to the previous one-dimensional gapfilling problem. Once again, the equation, $\mathbf{Gm} = \mathbf{d}$, reduces to $d_i = m_j$, where index, j, matches up the location, (x_i, y_i), of the i-th data to the location, (x_j, y_j), of the j-th model parameter. The differential equation contains only second derivatives, which have been discussed earlier (Equation 5.26). The only nuance is that one derivative is taken in the x-direction and the other in the y-direction so that the value of pressure at five neighboring grid points is needed. (Figure 5.7):

$$\left.\frac{\partial^2 m}{\partial x^2}\right|_{x_i, y_j} = \frac{[A_{i+1,j} - 2A_{i,j} + A_{i-1,j}]}{(\Delta x)^2} \text{ and } \left.\frac{\partial^2 m}{\partial y^2}\right|_{x_i, y_j} \approx \frac{[A_{i,j+1} - 2A_{i,j} + A_{i,j-1}]}{(\Delta y)^2}$$

$$(5.30)$$

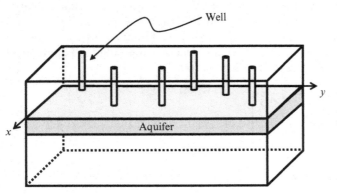

Figure 5.6 Scenario for aquifer example. Ground water is flowing through the permeable aquifer (shaded layer), driven by variations in pore pressure. The pore pressure is measured in N wells (cylinders).

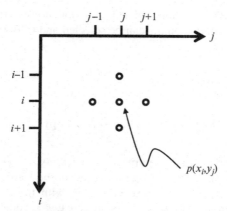

Figure 5.7 The expression, $\partial^2 p/\partial x^2 + \partial^2 p/\partial y^2$, is calculated by summing finite difference approximations for $\partial^2 p/\partial x^2$ and $\partial^2 p/\partial y^2$. The approximation for $\partial^2 p/\partial x^2$ involves the column of three grid points (circles) parallel to the i-axis and the approximation for $\partial^2 p/\partial y^2$ involves the row of three grid points parallel to the j-axis.

Note that the central point, A_{ij}, is common to the two derivatives, so five, and not six, grid points are involved. While these five points are neighboring elements of **A**, they do not correspond to neighboring elements of **m**, once **A** is unwrapped.

Once again, a decision needs to be made about what to do on the edges of the grid. One possibility is to assume that the pressure derivative in the direction perpendicular to the edge of the grid is zero (which is the two-dimensional analog to the previously discussed one-dimensional case). This corresponds to the choice, $\partial p/\partial y = 0$, on the left and right edges of the grid, and $\partial p/\partial x = 0$ on the top and bottom edges. Physically, these equations imply that the pore fluid is not flowing across the edges of the grid (an assumption that may or may not be sensible, depending on the circumstances). The four corners of the grid require special handling, as two edges are coincident at these points. One possibility is to compute the first derivative along the grid's diagonals at these four points.

In the exemplary *MatLab* script, `eda05_08`, the equation, **Fm = f**, is built up row-wise, in a series of steps: (1) the N "data" rows; (2) the $(I-2)(J-2)$ Laplace's equation rows; (3) the $(J-2)$ rows of first derivatives top-of-the-grid rows; (4) the $(J-2)$ rows of first derivatives bottom-of-the-grid rows; (5) the $(I-2)$ rows of first derivatives left-of-the-grid rows; (6) the $(I-2)$ rows of first derivatives right-of-the-grid rows; and (7) the *four* rows of first derivatives at grid-corners. When debugging a script such as this, a few exemplary rows of, **F** and **f**, from each section should be displayed and carefully examined, to ensure correctness.

The script, `eda05_08`, is set up to work on a file of test data, `pressure.txt`, that is created with a separate script, `eda05_09`. The data are simulated or *synthetic* data, meaning that they are calculated from a formula and that no actual measurements are involved. The test script evaluates a known solution of Laplace's equation

$$p(x,y) = P_0 \sin(\kappa x) \exp(-\kappa y) \quad \text{where } \kappa \text{ and } P_0 \text{ are constants} \tag{5.31}$$

on N randomly selected grid points and adds a random noise to them to simulate measurement error. Random grid points can be selected and utilized as follows:

```
rowindex=unidrnd(I,N,1);
xobs = x(rowindex);
colindex=unidrnd(J,N,1);
yobs = y(colindex);
kappa = 0.1;
dtrue = 10*sin(kappa*xobs).*exp(-kappa*yobs);
```
(MatLab eda05_09)

Here, the function, unidrnd(), returns an $N \times 1$ array, rowindex, of random integers in the range $(1, I)$. A column vector, xobs, of the x-coordinates of the data is then created from the grid coordinate vector, x, with the expression, xobs = x (rowindex). A similar pair of expressions creates a column vector, yobs, of the y-coordinates of the data. Finally, the (x, y) coordinates of the data are used to evaluate Equation (5.31) to construct the "true" synthetic data, dtrue.

Normally distributed random noise can be added to the true data to simulate measurement error:

```
sigmad = 0.1;
dobs = dtrue + random(normal,0.0,sigmad,N,1);
```
(MatLab eda05_09)

Here, the random() function returns an $N \times 1$ column vector of random numbers with zero mean and variance, σ_d^2.

Results of the eda05_08 script are shown in Figure 5.8. Note that the predicted pressure is a good match to the synthetic data, as it should be, for the data are, except for noise, exactly a solution of Laplace's equation. This step of testing a script on synthetic data should *never* be omitted. A series of tests with synthetic data are more

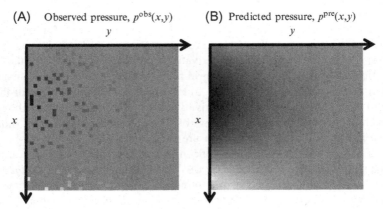

Figure 5.8 Filling in gaps in pressure data, $p(x, y)$, using the prior information that the pressure satisfies Laplace's equation, $\partial^2 p/\partial x^2 + \partial^2 p/\partial y^2 = 0$. (A) Observed pressure, $p^{obs}(x, y)$. (B) Predicted pressure, $p^{pre}(x, y)$. *MatLab* script eda05_08.

likely to reveal problems with a script than a single application to real data. Such a series of tests should vary a variety of parameters, including the grid spacing, the parameter, κ, and the noise level.

Problems

5.1 The first paragraph of Section 5.2 mentions one type of prior information that cannot be implemented by a linear equation of the form, $\mathbf{Hm} = \bar{\mathbf{h}}$. What is it?

5.2 What happens in the eda05_05 script if it is applied to an inconsistent dataset (meaning a dataset containing multiple points with the same xs but different ds)? Modify the script to implement a test of this scenario. Comment on the results.

5.3 Modify the eda05_05 script to fill in the gaps of the *cleaned* version of the Black Rock Forest temperature dataset. Make plots of selected data gaps and comment on how well the method filled them in. Suggestions: First create a short version of the dataset for test purposes. It should contain a few thousand data points that bracket one of the data gaps. Do not run your script on the complete dataset until it works on the short version. Only the top part of the script needs to be changed. First, the data must be read using the load() function. Second, you must check whether all the times are equally spaced. Missing times must be inserted and the corresponding data set to zero. Third, a vector, rowindex, that gives the row index of the good data but excludes the zero data, hot spikes, and cold spikes must be computed with the find() function.

5.4 Run the eda05_07 script in a series of tests in which you vary the parameter, κ, and the noise level, σ_d. You will need to edit the eda05_08 script to produce the appropriate file, pressure.txt, of synthetic data. Comment on your results.

5.5 Suppose that the water in a shallow lake flows only horizontally (i.e., in the (x, y) plane) and that the two components of fluid velocity, v_x and v_y are measured at a set of N observation points, (x_i, y_i). Water is approximately incompressible, so a reasonable type of prior information is that the divergence of the fluid velocity is zero; that is $\partial v_x/\partial x + \partial v_y/\partial y = 0$. Furthermore, if the grid covers the whole lake, then the condition that no water flows across the edges is a reasonable one, implying that the perpendicular component of velocity is zero at the edges. (A) Sketch out how scripts eda05_07 and eda05_08 might be modified to fill in the gaps of fluid velocity data.

5.6 In the example shown in Fig. 5.8, a two-dimensional pressure field $p(x, y)$ is reconstructed using sparse data and the prior information that the field satisfies Laplace's equation. Consider an alternative scenario in which the pressure is believed to vary smoothly but the equation that it satisfies is unknown. In that case, we might opt to use a combination of flatness:

$$\frac{dp}{dx} \approx 0 \quad \text{and} \quad \frac{dp}{dy} \approx 0$$

(say with variance σ_s^2) and smallness $p \approx 0$ (say with variance σ_m^2) to create a smooth solution. Modify eda05_06.m to solve this problem and adjust σ_s^2 and σ_m^2 by trial and error to produce a solution that is a reasonable compromise between smoothness and goodness of fit. How much worse is the error, compared to the solution that employs Laplace's equation?

References

Menke, W., 1989. Geophysical Data Analysis: Discrete Inverse Theory, Revised Edition. Academic Press, Inc., New York.

Tarantola, A., Valette, B., 1982. Inverse problems = quest for information. J. Geophys. 50, 159–170.

6 Detecting periodicities

6.1 Describing sinusoidal oscillations

Sinusoidal oscillations—those involving sines and cosines—are very common in the environmental sciences. We encountered them in the Neuse River Hydrograph and the Black Rock Forest datasets, where they were associated with seasonal variations in river discharge and air temperature, respectively. This chapter examines oscillatory behavior in more detail, developing systematic methods for detecting and quantifying periodicities.

Periodicities can be both temporal and spatial in character, with a somewhat different nomenclature used for each. In both cases, the height of the oscillation is called its *amplitude*. Temporal periodicities have a *period*, T, the time between successive cycles. Spatial periodicities have a *wavelength*, λ, the distance between successive cycles. The rate at which temporal cycles occur is called the *frequency* and the rate at which spatial cycles occur is called *wavenumber*. Frequency can be measured in units of cycles per unit time, in which case it is given the symbol, f, or it can be measured in units of radians per unit time, in which case it is given the symbol, ω. The units of cycles per second are called *Hertz*, abbreviated Hz. Similarly, wavenumber can be measured in either cycles per unit distance or radians per unit distance. Unfortunately, both units of wavenumber tend to be given the same symbol, k, in the literature. In this book, we use k exclusively to mean radians per unit distance. These quantities are related as follows:

$$f = \frac{1}{T} = \frac{\omega}{2\pi} \quad \text{and} \quad \frac{1}{\lambda} = \frac{k}{2\pi} \tag{6.1}$$

Generic temporal, $d(t)$, and spatial, $d(x)$, cosine oscillations of amplitude, C, can be written as

$$d(t) = C \cos\left\{\frac{2\pi t}{T}\right\} = C \cos\{2\pi ft\} = C \cos\{\omega t\} \quad \text{and}$$

$$d(x) = C \cos\left\{\frac{2\pi x}{\lambda}\right\} = C \cos\{kx\} \tag{6.2}$$

In nature, oscillations rarely "start" at time (or distance) zero. A cosine wave with amplitude, C, that starts (peaks) at time, t_0, is given by (Figure 6.1)

$$d(t) = C \cos\left\{\frac{2\pi(t - t_0)}{T}\right\} = C \cos\{\omega(t - t_0)\} = C \cos\{\omega t - \phi\} \tag{6.3}$$

The quantity, $\phi = \omega t_0$, is called the *phase*. The rule,

$$\cos(a-b) = \cos(a)\cos(b) + \sin(a)\sin(b) \tag{6.4}$$

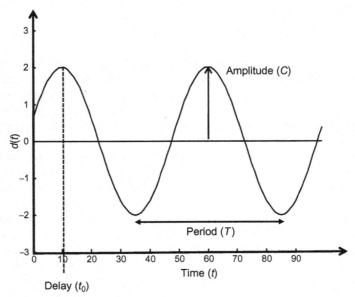

Figure 6.1 The sinusoidal function, $d(t) = C\cos\{2\pi(t - t_0)/T\}$, has amplitude, $C = 2$, period, $T = 50$, and delay, $t_0 = 10$. *MatLab* script eda06_01.

when applied to Equation (6.3), yields

$$d(t) = C \cos\{\omega(t - t_0)\}$$
$$= C \cos(\omega t_0) \cos(\omega t) + C \sin(\omega t_0) \sin(\omega t) = A \cos(\omega t) + B \sin(\omega t)$$

with

$$A = C \cos(\omega t_0) \quad \text{and} \quad B = C \cos(\omega t_0)$$

and

$$C = \sqrt{A^2 + B^2} \quad \text{and} \quad t_0 = \omega^{-1} \tan^{-1}(B/A) \tag{6.5}$$

See Note 6.1 for a discussion of how the arc-tangent is to be correctly computed. A time-shifted cosine can be represented as the sum of a sine and a cosine. An explicit time-shift variable such as t_0 is unnecessary in formulas describing periodicities, as long as sines and cosines are paired up with one another.

6.2 Models composed only of sinusoidal functions

Suppose that we have a dataset in which a variable, such as temperature, is sampled at evenly spaced intervals of time, t_i (a *time series*), say with sampling interval, Δt. One extreme is a dataset composed of only sinusoidal oscillations:

the data $=$ sum of sines and cosines

$$d(t_i) = A_1 \cos(\omega_1 t_i) + B_1 \sin(\omega_1 t_i) + A_2 \cos(\omega_2 t) + B_2 \sin(\omega_2 t) + \cdots$$
$$\tag{6.6}$$

This formula is sometimes referred to as a *Fourier series* or an *inverse discrete Fourier transform* and the column vector containing the As and Bs is called the *discrete Fourier transform (DFT)* of **d**. The As and Bs are the model parameters and the corresponding frequencies are taken to be auxiliary variables. Note that sines and cosines of a given frequency, ω_i, are paired, as was discussed in the previous section. Two key questions involve the number of frequencies that ought to be included in this representation and what their values should be. The answers to these questions involve a surprising fact about time series:

$$\text{frequencies higher than} f_{ny} = \frac{1}{2\Delta t} \text{ cannot be detected} \tag{6.7}$$

This is called *Nyquist's Sampling Theorem*. It says that the periods shorter than two time increments cannot be detected in time series with evenly spaced samples. Furthermore, as we will see below, any frequencies in the data that are higher than the *Nyquist frequency*, f_{ny}, are erroneously mapped into the $(0, f_{ny})$ frequency range. Choosing the frequency range $(0, f_{ny})$ in Equation (6.5) is, therefore, natural. For simplicity, we assume that the number N of data is an even integer. It can always be made so by shortening an odd-length time series by one point. Then the number of frequencies is $N/2 + 1$ and the number M of unknown coefficients of cosines and sines exactly equals the number N of data; that is $M = N$. These choices imply

$$f_{ny} = \frac{1}{2\Delta t} \quad \text{and} \quad \omega_{ny} = \frac{\pi}{\Delta t} \quad \text{and} \quad \Delta\omega = \frac{2\pi}{N\Delta t} \quad \text{and} \quad \Delta f = \frac{1}{N\Delta t} \qquad (6.8)$$

The reason that this relationship calls for $\frac{1}{2}N + 1$, as contrasted to $\frac{1}{2}N$ of them, is that the first and last sine term is identically zero; that is,

$$B_1 \sin(0\, t_i) = 0 \quad \text{and} \quad B_{(N/2)+1} \sin\left(\frac{2\pi}{2\Delta t} t_i\right) = B_{(N/2)+1} \sin(n\pi) = 0 \qquad (6.9)$$

where, $n = t_i/\Delta t$, is an integer. Hence, these two terms are omitted from the sum. Thus, the Fourier series contains the frequencies $[0, \Delta f, 2\Delta f, \ldots, \frac{1}{2}N\Delta f]^{\mathrm{T}}$. The lowest frequency is zero. It corresponds to a cosine of zero period, that is, to a constant. The next highest frequency is Δf. It corresponds to a sine and a cosine that have one complete oscillation over the length of the data. The next highest frequency is $2\Delta f$. It corresponds to a sine and a cosine that have two complete oscillations over the length of the data. The highest frequency is $f_{ny} = \frac{1}{2}N\Delta f = 1/(2\Delta t)$. It corresponds to a highly oscillatory cosine that reverses sign at every sample; that is $[1, -1, 1, -1, \ldots]^{\mathrm{T}}$ (Figure 6.2).

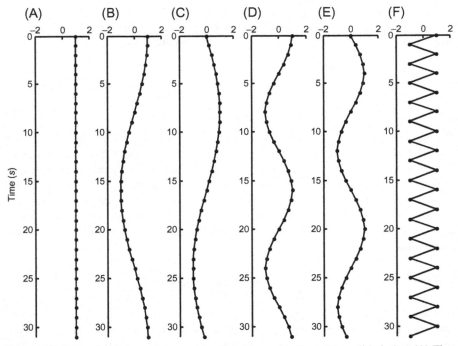

Figure 6.2 Plots of columns of the matrix, **G**. with rows indicated by solid circles. (A) First column, the constant 1. (B, C) Next two columns are, $\cos(\Delta\omega t)$ and $\sin(\Delta\omega t)$, respectively. They have one period of oscillation over the time interval of the data. (D, E) Next two columns are $\cos(2\Delta\omega t)$ and $\sin(2\Delta\omega t)$, respectively. They have two periods of oscillation over the time interval of the data. F) Last column switches between 1 and -1 every row. *MatLab* script eda06_02.

In *MatLab*, frequency-related quantities are defined as follows:

```
Nf=N/2+1;
fmax = 1/(2*Dt);
Df = fmax/(N/2);
f = Df*[0:Nf-1]';
Nw=Nf;
wmax = 2*pi*fmax;
Dw = wmax/(N/2);
w = Dw*[0:Nw-1]'
```
(*MatLab* script eda06_02)

Here, Dt is the sampling interval and N is the length of the data. N is assumed to be an even integer.

The problem that arises with frequencies higher than the Nyquist frequency can be seen by examining a pair of cosines and sines from Equation (6.6), written out for a particular time, t_k, and frequency, ω_n.

$$\cos(\omega_n t_k) = \cos((n-1)(k-1)\,\Delta\omega\Delta t) = \cos\left(\frac{2\pi\,(n-1)(k-1)}{N}\right)$$

$$\sin(\omega_n t_k) = \sin((n-1)(k-1)\,\Delta\omega\Delta t) = \sin\left(\frac{2\pi\,(n-1)(k-1)}{N}\right) \quad (6.10)$$

Here, we have used the rule, $\Delta\omega\Delta t = 2\pi/N$. Note that we index time and frequency so that t_1 and ω_1 correspond to zero time and frequency, respectively; that is, $t_k = (k-1)\Delta t$ and $\omega_n = (n-1)\Delta\omega$. Now let us examine a frequency, ω_m, that is higher than the Nyquist frequency, say $m = n + N$, where N is the number of data:

$$\cos(\omega_m t_k) = \cos((n-1+N)(k-1)\Delta\omega\Delta t) = \cos\left(\frac{2\pi(n-1+N)(k-1)}{N}\right)$$

$$= \cos\left(\frac{2\pi(n-1)(k-1)}{N}\right)\cos(2\pi(k-1)) - \sin\left(\frac{2\pi(n-1)(k-1)}{N}\right)\sin(2\pi(k-1))$$

$$= \cos\left(\frac{2\pi(n-1)(k-1)}{N}\right) + 0$$

$$\sin(\omega_m t_k) = \sin((n-1+N)(k-1)\Delta\omega\Delta t) = \sin\left(\frac{2\pi(n-1+N)(k-1)}{N}\right)$$

$$= \sin\left(\frac{2\pi(n-1)(k-1)}{N}\right)\cos(\pi(k-1)) + \cos\left(\frac{2\pi(n-1)(k-1)}{N}\right)\sin(\pi(k-1))$$

$$= \sin\left(\frac{2\pi(n-1)(k-1)}{N}\right) + 0 \quad (6.11)$$

Here, we have used the rules, $\cos(a+b) = \cos(a)\cos(b) - \sin(a)\sin(b)$ and $\sin(a+b) = \sin(a)\cos(b) + \cos(b)\sin(a)$, together with the rule that $\cos(2\pi(k+1)) = 1$ and $\sin(2\pi(k+1)) = 0$, for any integer, k. Comparing Equations (6.10) and (6.11), we conclude

$$\cos(\omega_n t_k) = \cos(\omega_m t_k) \quad \text{and} \quad \sin(\omega_n t_k) = \sin(\omega_m t_k) \tag{6.12}$$

The frequencies, ω_{n+N} and ω_n, are *equivalent*, in the sense that they yield exactly the same sinusoids (Figure 6.3). A similar calculation, omitted here, shows that ω_{N-n} and $-\omega_n$ are equivalent in this same sense. But sines and cosines with negative frequencies have the same shape, up to a sign, as those with corresponding positive frequencies; that is, $\cos(-\omega t) = \cos(\omega t)$ and $\sin(-\omega t) = -\sin(\omega t)$. Thus, only sinusoids in the frequencies in the range ω_1 (zero frequency) to $\omega_{N/2+1}$ (the Nyquist frequency) have unique shapes (Figure 6.3C). In a digital world, no frequencies are higher than the Nyquist.

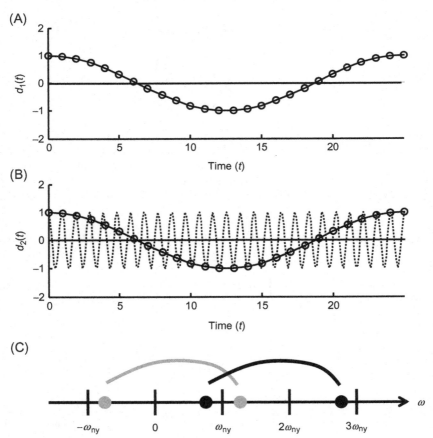

Figure 6.3 Example of aliasing. (A) Low-frequency oscillation, $d_1(t) = \cos(\omega_1 t)$, with $\omega_1 = 2\Delta\omega$, evaluated every Δt (circles). Bottom) High-frequency oscillation, $d_2(t) = \cos\{\omega_2 t\}$, with $\omega_2 = (2 + N)\Delta\omega$, evaluated every Δt (circles). Note that both the true curve (dashed) and low-frequency curve (solid) pass through the data points. (C) Schematic representation of aliasing, showing pairs of points on the frequency-axis that are equivalent. *MatLab* script eda06_03.

This limitation implies that special care must be taken when observing a time series to avoid recording any frequencies that are higher than the Nyquist frequency. This goal is usually achieved by creating a data recording system that attenuates—or eliminates—high frequencies *before* the data are converted into digital samples. If any high frequencies persist, then the dataset is said to be *aliased*, and spurious low frequencies will appear.

The linear equation, $\mathbf{d} = \mathbf{Gm}$, form of the Fourier series is as follows:

the data = sum of sines and cosines

$$
\begin{bmatrix} d_1 \\ d_2 \\ d_3 \\ \vdots \\ d_N \end{bmatrix} = \begin{bmatrix} 1 & \cos(\omega_2 t_1) & \sin(\omega_2 t_1) & \ldots & \cos(\omega_{N/2} t_1) & \sin(\omega_{N/2} t_1) & 1 \\ 1 & \cos(\omega_2 t_2) & \sin(\omega_2 t_2) & \ldots & \cos(\omega_{N/2} t_2) & \sin(\omega_{N/2} t_2) & -1 \\ 1 & \cos(\omega_2 t_3) & \sin(\omega_2 t_3) & \ldots & \cos(\omega_{N/2} t_3) & \sin(\omega_{N/2} t_3) & 1 \\ 1 & \vdots & \vdots & \vdots & \vdots & \vdots & \vdots \\ 1 & \cos(\omega_2 t_N) & \sin(\omega_2 t_N) & \ldots & \cos(\omega_{N/2} t_N) & \sin(\omega_{N/2} t_N) & -1 \end{bmatrix} \begin{bmatrix} A_1 \\ A_2 \\ B_2 \\ \vdots \\ A_{N/2} \\ B_{N/2} \\ A_{(N/2)+1} \end{bmatrix}
$$

$$(6.13)$$

In *MatLab*, the data kernel, **G**, is created as follows:

```
% set up G
G=zeros(N,M);

% zero frequency column
G(:,1)=1;

% interior M/2-1 columns
for i = [1:M/2-1]
    j = 2*i;
    k = j+1;
    G(:,j)=cos(w(i+1).*t);
    G(:,k)=sin(w(i+1).*t);
end

% nyquist column
G(:,M)=cos(w(Nw).*t);                    (MatLab eda06_02)
```

Here, the number of model parameters, M, equals the number of data, N, and the frequency values are in a column vector, w.

Remarkably, the matrices, $[\mathbf{G}^T\mathbf{G}]$ and $[\mathbf{G}^T\mathbf{G}]^{-1}$, which play such an important role in the least-squares solution, can be shown to be diagonal:

$$[\mathbf{G}^T\mathbf{G}] = N \, \text{diag}(1, \tfrac{1}{2}, \tfrac{1}{2}, \ldots, \tfrac{1}{2}, \tfrac{1}{2}, 1)$$

and

$$[\mathbf{G}^T\mathbf{G}]^{-1} = \frac{1}{N} \text{diag}(1, 2, 2, \ldots, 2, 2, 1) \tag{6.14}$$

Note that we have used the rule that the inverse of a diagonal matrix, $\mathbf{M}$, is the matrix of reciprocals of the elements of the main diagonal; that is $[\mathbf{M}^{-1}]_{ii} = 1/M_{ii}$. Equation (6.14) can be understood by noting that the elements of $\mathbf{G}^T\mathbf{G}$ are just the dot products of the columns, $\mathbf{c}$, of the matrix, $\mathbf{G}$. Many of these dot products are clearly zero (meaning that $[\mathbf{G}^T\mathbf{G}]_{ij} = \mathbf{c}^{(i)T}\mathbf{c}^{(j)} = 0$ when $i \neq j$). For instance, the second column, $\cos(\Delta\omega t)$, is symmetric about its midpoint, while the third column, $\sin(\Delta\omega t)$, is antisymmetric about it, so their dot product is necessarily zero (see Figure 6.2B and C). What is not so clear—but nevertheless true—is that every column is orthogonal to every other. We will not derive this result here, for it requires rather nitty-gritty trigonometric manipulations (but see Note 6.2). Its consequence is that the least-squares solution, $\mathbf{m} = [\mathbf{G}^T\mathbf{G}]^{-1}\mathbf{G}^T\mathbf{d}$ requires only matrix multiplication, and not matrix inversion. Furthermore, when the data are uncorrelated, then the model parameters, $\mathbf{m}$, which represent the amplitudes of the sines and cosines, are also uncorrelated, as $\mathbf{C}_m = \sigma^2_d[\mathbf{G}^T\mathbf{G}]^{-1}$ is diagonal. Furthermore, all but the first and last model parameters have equal variance.

In *MatLab*, the least-squares solution is computed as follows:

```
gtgi = 2* ones(M,1)/N;
gtgi(1)=1/N;
gtgi(M)=1/N;
mest=gtgi .* (G'*d);                                    (MatLab eda06_04)
```

The column vectors,

$$[A_1^2, A_2^2 + B_2^2, \ldots, A_{N/2}^2 + B_{N/2}^2, A_{N/2+1}^2]^T$$

and   (6.15)

$$\left[\sqrt{A_1^2}, \sqrt{A_2^2 + B_2^2}, \ldots, \sqrt{A_{N/2}^2 + B_{N/2}^2}, \sqrt{A_{N/2+1}^2}\right]^T$$

are called the *power spectral density* and *amplitude spectral density* of the time series, respectively. Either can be used to quantify the overall amount of oscillations at any given frequency, irrespective of the phase. In *MatLab*, the power spectral density, $\mathbf{s}$, is computed from mest as follows:

```
% zero frequency
s=zeros(Nw,1);
s(:,1)=mest(1)^2;

% interior points
for i = [1:M/2-1]
    j = 2*i;
    k = j+1;
    s(i+1) = mest(j)^2 + mest(k)^2;
end

% Nyquist frequency
s(Nw) = mest(M)^2;                                      (MatLab eda06_04)
```

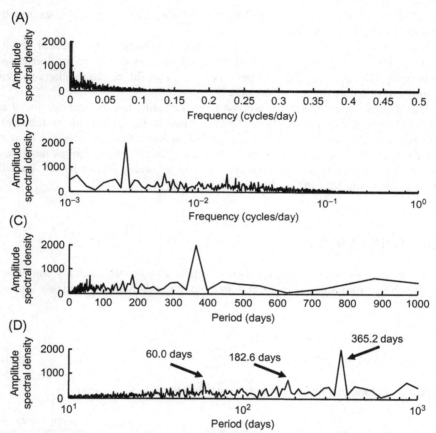

Figure 6.4 Amplitude spectral density of the Neuse River discharge dataset. (A) Plotted as a function on frequency (linear frequency axis). (B) Plotted as a function on frequency (logarithmic frequency axis). (C) Plotted as a function of period (linear horizontal axis). (D) Plotted as a function of period (logarithmic horizontal axis) *MatLab* script eda06_04.

When applied to the Neuse River discharge data, this method produces an amplitude spectral density that is largest at low frequencies (Figure 6.4A). In such cases, a graph with a logarithmic frequency axis (Figure 6.4B) allows one to see details that are squeezed near the origin of a linear plot (Figure 6.4A). In some applications, plotting amplitude spectral density as a function of period, as contrasted to frequency, may be preferable (as in this case, where the annual period of 365.2 days is more recognizable than the equivalent frequency of 0.002738 cycles/day). The amplitude spectral density consists of several peaks, superimposed on a "noisy background" that gradually declines with period. The three largest peaks have periods of 365.2, 182.6, and 60.2, (one, one-half, and one-sixth years) and are associated with oscillations in stream flow caused by seasonal fluctuations in rainfall.

A common practice is to omit the zero-frequency element of the spectral density from the plots (as was done here). It only reflects the mean value of the time series and is not really relevant to the issue of periodicities. Large values can require a vertical scaling that obscures the rest of the plot.

The *MatLab* script, eda06_04, worked fine calculating the spectral density of an $N \approx 4000$ length time series. However, the $N \times N$ matrix, $\mathbf{G}$, will become prohibitively large for larger datasets. Fortunately, a very efficient algorithm, called the *fast Fourier transform* (*FFT*), has been discovered for solving for $\mathbf{m}^{est}$ that requires much less storage and much less computation than "brute force" multiplication by $\mathbf{G}^{\mathrm{T}}$. We will return to this issue later in the chapter.

6.3 Going complex

Many of the formulas of the previous section can be substantially simplified by switching from sines and cosines to complex exponentials, utilizing *Euler's formulas:*

$$\exp(i\omega t) = \cos(\omega t) + i\sin(\omega t)$$
$$\exp(-i\omega t) = \cos(\omega t) - i\sin(\omega t) \tag{6.16}$$

Here, i is an imaginary unit. These formulas imply

$$\cos(\omega t) = \frac{\exp(i\omega t) + \exp(-i\omega t)}{2} \quad \text{and} \quad \sin(\omega t) = \frac{\exp(i\omega t) - \exp(-i\omega t)}{2i} \tag{6.17}$$

The main complication (besides the need to use complex numbers) is that both positive and negative frequencies are needed in the Fourier series. Previously, we paired sines and cosines; now we will pair complex exponentials with positive and negative frequencies:

$$d(t) = \cdots A\cos(\omega t) + B\sin(\omega t)\cdots = \cdots C^{-}\exp(-i\omega t) + C^{+}\exp(i\omega t)\cdots \tag{6.18}$$

Here, C^{-} and C^{+} are the coefficients of the negative-frequency and positive-frequency terms, respectively. The requirement that these two different representations be equal implies a relationship between the As and Bs and the Cs. We write out the Cs in terms of their real and imaginary parts, $C^{-} = C_{R}^{-} + iC_{I}^{-}$ and $C^{+} = C_{R}^{+} + iC_{I}^{+}$ and perform the multiplication explicitly:

$$C^- \exp(-i\omega t) + C^+ \exp(i\omega t) = (C_R^- + iC_I^-)\{\cos(\omega t) - i\sin(\omega t)\}$$
$$+ (C_R^+ + iC_I^+)\{\cos(\omega t) + i\sin(\omega t)\}$$
$$= (C_R^- + C_R^+)\cos(\omega t) + (C_I^- - C_I^+)\sin(\omega t)$$
$$+ i(C_I^- + C_I^+)\cos(\omega t) + i(C_R^+ - C_R^-)\sin(\omega t)$$

$$(6.19)$$

As the time series, $d(t)$, is real, the two imaginary terms must be zero. This happens when C^- and C^+ are complex conjugates of each other: $C^- = C_R - iC_I$ and $C^+ = C_R + iC_I$. Comparing Equations (6.17) and (6.18), we find

$$A = (C_R^- + C_R^+) = 2C_R^+ \quad \text{and} \quad B = (C_I^- - C_I^+) = -2C_I^+ \tag{6.20}$$

The power spectral density is now computed from $C_R^2 + C_I^2$, which is equivalent to C times its complex conjugate: $C_R^2 + C_I^2 = |C|^2 = C^*C$. Here, the asterisk means complex conjugation.

We can represent a real function as a Fourier series that involves sines and cosines of real amplitudes, A and B, respectively, or alternatively, as a Fourier series that involves complex exponentials with complex coefficients, C:

$$d_k = \sum_{n=1}^{\frac{N}{2}+1} \{A_n \cos(\omega_n t_k) + B_n \sin(\omega_n t_k)\} \quad \text{with} \quad \omega_n = (0, \Delta\omega, 2\Delta\omega, \dots, \tfrac{1}{2}N\Delta\omega)$$

$$(6.21a)$$

$$d_k = \frac{1}{N}\sum_{n=1}^{N} C_n \exp(i\omega_n t_k) \quad \text{with} \quad \omega_n = (0, \Delta\omega, 2\Delta\omega, \dots, \tfrac{1}{2}N\Delta\omega,$$
$$- (\tfrac{1}{2}N - 1)\Delta\omega, \dots, -2\Delta\omega, -\Delta\omega)$$

$$(6.21b)$$

Note the nonintuitive ordering of the frequencies in the summation of the complex exponentials, which we will discuss in detail below. The factor of N^{-1} has been added to the complex summation in order to match *MatLab*'s convention so that now $(2/N)C_i = A_i - iB_i$.

The solution of Equation (6.21b) for the coefficients, C_n, requires the complex version of least squares (see Note 4.1). We will not discuss it in any detail here, except to note that the matrix, $[G^H G]^{-1} = N^{-1}I$, is diagonal (see Note 6.2) so that, like the sine and cosine version of the DFT, matrix inversion is not required. The complex coefficients, C_j, are calculated from the time series, $d(t_n)$, by

$$C_j = \sum_{n=1}^{N} d(t_n) \exp(-i\omega_j t_n) \tag{6.22}$$

In the sine and cosine case, the sum has $N/2 + 1$ pairs of sines and cosines, but the coefficients of the first and last pairs are zero, so the total number of unknowns is N. In the complex exponential case, the sum has N complex coefficients, but except for the first and last, they occur in complex conjugate pairs, so only the $N/2 + 1$ co-efficients of the nonnegative frequencies count for the unknowns. Each of these has a real and imaginary part, except for the first and the last, which are purely real, so the number of unknowns is once again $2(N/2 + 1) - 2 = N$. Note that if the data were complex, the coefficients would all be different; that is, N complex coefficients are needed to represent N complex data.

We return now to the issue of the ordering of the frequencies, which is to say, the order of the model parameters, **m**. The ordering presented in Equation (6.21b) has the nonnegative frequencies first and the negative frequencies last. This ordering, while nonintuitive, is actually more useful than a strictly ascending ordering, because the nonnegative frequencies are really the only ones needed when dealing with real time series. The negative frequencies, being complex conjugates of the positive ones, are redundant. However, the ordering has a more subtle rationale related to aliasing. The negative frequencies correspond exactly to the positive frequencies that one would get, if the positive ordering were extended past the Nyquist frequency. For example, the last frequency, $-\Delta\omega$, is exactly equivalent to $+(N-1)\Delta\omega$, in the sense that both correspond to the same complex exponential.

In *MatLab*, the arrays of frequencies are created as follows:

```
M=N;
tmax=Dt*(N-1);
t=Dt*[0:N-1]';
fmax=1/(2.0*Dt);
df=fmax/(N/2);
f=df*[0:N/2,-N/2+1:-1]';
Nf=N/2+1;
dw=2*pi*df;
w=dw*[0:N/2,-N/2+1:-1]';
Nw=Nf;                                          (MatLab eda06_05)
```

Here, N is the length of the data, again presumed to be an even integer. The quantities, Nf and Nw, are the numbers of nonnegative frequencies. *MatLab*'s fft() (for *fast Fourier transform*) function solves for the complex Fourier coefficients very efficiently, and should always be used in preference to the least-squares procedure. The amplitude spectral density is computed as follows:

```
% compute Fourier coefficients
mest = fft(d);
% compute amplitude spectral density
s=abs(mest(1:Nw));                              (MatLab eda06_05)
```

Note that the amplitude spectral density is computed from the complex absolute value of the coefficients, C. The results are, of course, identical to least-squares, but they are computed with orders of magnitude less time and storage requirements (as can be seen by comparing the run times of scripts eda06_04 and eda06_05). The time series can be rebuilt from its Fourier coefficients by using the ifft() function:

```
dnew = ifft(mest);                              (MatLab eda06_05)
```

Crib Sheet 6.1 Experimental design for spectra

Questions to ask yourself

What is the highest frequency f^{high} that you need to detect?
What is the minimum frequency separation δf of spectral peaks
that you need to resolve?
How much data N can you record and analyze?

Important quantities

duration of recording, T
sampling interval, Δt
number of samples (amount of data), $N = T/\Delta t$
Nyquist (maximum) frequency, $f^{max} = 1/(2\Delta t)$
frequency spacing, $\Delta f = f^{max}/(N/2)$

Design principles

The Nyquist frequency must be greater than the highest frequency you need to
detect, and preferably at least twice as high

$$f^{max} \approx 2f^{high}$$

The frequency sampling must be smaller than the minimum frequency separation
you need to resolve, and preferably ten times smaller

$$\Delta f \approx \delta f/10$$

However, you need to verify that you have the ability to record and analyze the
corresponding amount of data

Design formulas

$$\Delta t = 1/(2f^{max}) \approx 1/f^{high}$$

$$N = 2f^{max}/\Delta f \approx 40f^{high}/\delta f$$

$$T = N\Delta t \approx 40/\delta f$$

6.4   Lessons learned from the integral transform

The Fourier series is very closely related to the *Fourier transform*, as can be seen by a
side-by-side comparison of the two:

$$\tilde{d}(\omega) = \int_{-\infty}^{+\infty} d(t)\exp(-i\omega t)\mathrm{d}t \quad \text{and} \quad C(\omega_j) = \sum_{k=1}^{N} d(t_k)\exp(-i\omega_j t_k) \tag{6.23}$$

The integral converts the function, $d(t)$, into its *Fourier transform*, the function, $\tilde{d}(\omega)$.
Similarly, the summation converts a time series, $d(t_j)$, into its *DFT*, the column vector,
$C(\omega_j)$. If we assume that the data are nonzero only between 0 and t_{max}, then the integral
can be approximated by its Riemann sum:

$$\int_{0}^{t_{max}} d(t)\exp(-i\omega_j t)\mathrm{d}t \approx \Delta t \sum_{k=1}^{N} d(t_k)\exp(-i\omega_j t_k) \tag{6.24}$$

Comparing Equations (6.22) and (6.23), we deduce that $\tilde{d}(\omega_j) \approx \Delta t\, C(\omega_j)$, that is, the Fourier transform and DFT coefficients differ only by the constant, Δt. A similar relationship holds between the inverse transform and the Fourier summation as well:

$$d(t_j) = \frac{1}{2\pi} \int_{-\infty}^{+\infty} \tilde{d}(\omega)\, \exp(i\omega t_j)\, d\omega \approx \frac{\Delta\omega}{2\pi} \sum_{k=1}^{N} \tilde{d}(\omega_k)\, \exp(i\omega_k t_j) \qquad (6.25)$$

This again gives $\tilde{d}(\omega_j) \approx \Delta t\, C(\omega_j)$ (as $\Delta\omega \Delta t = 2\pi/N$).

In the context of this book, Fourier transforms are of interest because they can be more readily manipulated than Fourier series. Many of the relationships that are true for Fourier transforms will also be true—or approximately true—for Fourier series. We summarize a few of the most useful relationships below.

6.5 Normal curve

The transform of the Normal function, $d(t) = \exp(-a^2 t^2)$, is

$$\tilde{d}(\omega) = \int_{-\infty}^{+\infty} \exp(-a^2 t^2)\exp(-i\omega t) dt = 2 \int_{0}^{+\infty} \exp(-a^2 t^2)\cos(\omega t) dt$$

$$= \frac{\sqrt{\pi}}{a} \exp\left(-\frac{\omega^2}{4a^2}\right) \qquad (6.26)$$

Here, we have expanded $\exp(\omega t)$ into $\cos(\omega t) + i\sin(\omega t)$. The integrand, $\sin(\omega t)\exp(-a^2 t^2)$, consists of an odd function (the sine) of time multiplied by an even function (the exponential) of time. It is, therefore, odd and so its integral over $(-\infty, +\infty)$ is zero. The product, $\cos(\omega t)\exp(-a^2 t^2)$, is an even function, so its integral on $(-\infty, +\infty)$ is twice its integral on $(0, \infty)$. A standard table of integrals (e.g., integral 679 of the CRC Standard Mathematical Tables) is used to evaluate the final integral.

If we write $a^2 = 1/2\sigma^2$, then the exponential has the form of a Normal curve centered at time zero:

$$d(t) = \frac{1}{\sqrt{2\pi}\sigma} \exp\left(-\frac{t^2}{2\sigma^2}\right) \quad \text{and} \quad \tilde{d}(\omega) = \exp\left(-\frac{\omega^2}{2\sigma^{-2}}\right) \qquad (6.27)$$

Thus, up to an overall normalization, the transform of a Normal curve with variance, σ^2, is a Normal curve with variance, σ^{-2}.

This result is extremely important because it quantifies how the widths of functions and the widths of their Fourier transforms are related. When $d(t)$ is a wide function, $\tilde{d}(\omega)$ is narrow, and vice-versa. A spiky function, such as a narrow Normal curve, has a transform that is rich in high frequencies. A very smooth function, such as a wide Normal curve, has a transform that lacks high frequencies (Figure 6.5).

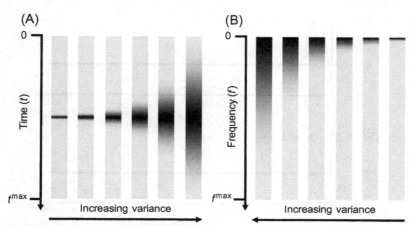

Figure 6.5 (A) Shaded column-vectors of a series of Normal functions with increasing variance. (B) Corresponding amplitude spectral density *MatLab* script eda06_06.

6.6 Spikes

The relationship between the width of a function and its Fourier transform can be pursued further by defining the *Dirac delta function*—a Normal curve in the limit of vanishing small variance:

$$\delta(t - t_0) = \lim_{\sigma \to 0} \frac{1}{\sqrt{2\pi}\sigma} \exp\left(-\frac{(t - t_0)^2}{2\sigma^2}\right) \tag{6.28}$$

This *generalized function* is zero everywhere, except at the point, t_0, where it is singular. Nevertheless, it has unit area. When the product, $\delta(t - t_0)f(t)$, is integrated, the result is just $f(t)$ evaluated at the point, $t = t_0$:

$$\int_{-\infty}^{+\infty} \delta(t - t_0)f(t)dt = f(t_0) \tag{6.29}$$

This result can be understood by noting that the Dirac function is nonzero only in a vanishingly small interval of time, t_0. Within this interval, the function, $f(t)$, is just the constant, $f(t_0)$. No error is introduced by replacing $f(t)$ with $f(t_0)$ everywhere and taking it outside the integral, which then integrates to unity.

The Fourier transform of a spike at $t_0 = 0$ is unity (Figure 6.6):

$$\tilde{d}(\omega) = \int_{-\infty}^{+\infty} \delta(t)\exp(-i\omega t)\, dt = \exp(-i\omega t)|_{t=0} = 1 \tag{6.30}$$

This is the limiting case of an indefinitely narrow Normal function (see Equation 6.27).

The Dirac function, being "infinitely spiky," has a transform that contains all frequencies in equal proportions. The transform with a spike at $t = t_0$, is

$$\tilde{d}(\omega) = \int_{-\infty}^{+\infty} \delta(t - t_0)\exp(-i\omega t)dt = \exp(-i\omega t)|_{t=t_0} = \exp(-i\omega t_0) \tag{6.31}$$

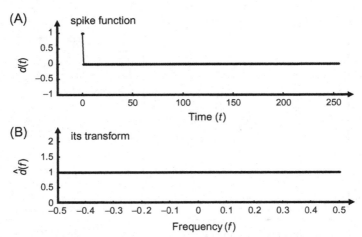

Figure 6.6 (A) Spike function is zero except at time, $t = 0$. (B) Corresponding transform is unity. *MatLab* script eda06_07.

Although it is an oscillatory function of time, its power spectral density is constant: $s(\omega) = \tilde{d}^*(\omega)\tilde{d}(\omega) = \exp(+i\omega t_0)\exp(-i\omega t_0) = 1$.

The Dirac function can appear in a function's Fourier transform as well. The transform of $\cos(\omega_0 t)$ is

$$\tilde{d}(\omega) = \pi\left(\delta(\omega - \omega_0) + \delta(\omega + \omega_0)\right) \tag{6.32}$$

This formula can be verified by inserting it into the inverse transform:

$$d(t) = \frac{1}{2\pi}\int_{-\infty}^{+\infty} \pi(\delta(\omega - \omega_0) + \delta(\omega + \omega_0))\exp(i\omega t)\mathrm{d}\omega$$

$$= \frac{\exp(i\omega_0 t) + \exp(-i\omega_0 t)}{2} = \cos(\omega_0 t) \tag{6.33}$$

As one might expect, the transform of the pure sinusoid, $\cos(\omega_0 t)$, contains only two frequencies, $\pm\omega_0$.

6.7 Area under a function

The area, A, under a function, $d(t)$, is its Fourier transform evaluated at zero frequency; that is, $A = \tilde{d}(\omega = 0)$:

$$\tilde{d}(\omega = 0) = \int_{-\infty}^{+\infty} d(t)\exp(0)\mathrm{d}t = \int_{-\infty}^{+\infty} d(t)\mathrm{d}t = A \tag{6.34}$$

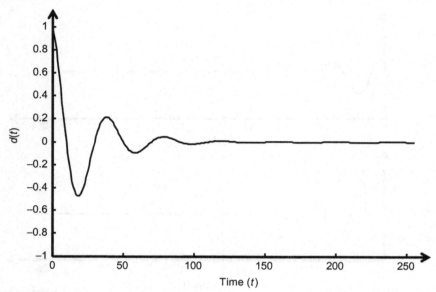

Figure 6.7 The area under the exemplary function, $d(t)$, is computed by the `sum()` and `fft()` functions. Both give the same value, 2.0014. *MatLab* script eda06_08.

as $\exp(0) = 1$. In *MatLab*, the area is computed as (Figure 6.7)

```
dt=fft(d);
area = real(dt(1));                                    (MatLab eda06_08)
```

6.8 Time-delayed function

Multiplying the transform, $\tilde{d}(\omega)$, by the factor, $\exp(-i\omega t_0)$, time-delays the function by a time interval, t_0:

$$\tilde{d}_{\text{delayed}}(\omega) = \int_{-\infty}^{+\infty} d(t - t_0)\exp(-i\omega t)\mathrm{d}t = \int_{-\infty}^{+\infty} d(t')\exp(-i\omega(t' + t_0))\mathrm{d}t'$$

$$= \exp(-i\omega t_0)\int_{-\infty}^{+\infty} d(t')\exp(-i\omega t')\mathrm{d}t' = \exp(-i\omega t_0)\tilde{d}(\omega) \qquad (6.35)$$

Here, we use the transformation of variables, $t' = t - t_0$, noting that $dt' = dt$ and that $t' \to \pm\infty$ as $t' \to \pm\infty$. In the literature, the process of modifying a Fourier transform by multiplication with a factor, $\exp(-i\omega t_0)$, is sometimes referred to as introducing a *phase ramp*, as it changes the phase by an amount proportional to frequency (i.e., by a ramp-shaped function):

$$\varphi(\omega) = \omega t_0 \qquad (6.36)$$

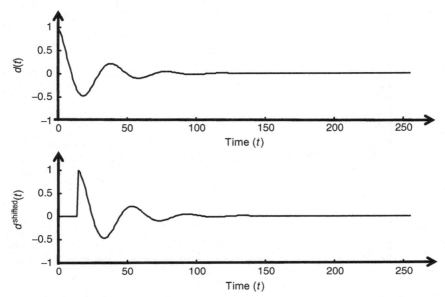

Figure 6.8 The exemplary function, $d(t)$, is time shifted by an interval, t_0. *MatLab* script eda06_09.

The time-shift result appeared previously, when we were calculating the transform of a time-delayed spike (Equation 6.31). The transform of the time-shifted spike differed from the transform of a spike at time zero by a factor of $\exp(-i\omega t_0)$. In *MatLab*, the time delay is accomplished as follows (Figure 6.8).:

```
t0 = t(16);
ds=ifft(exp(-i*w*t0).*fft(d));            (MatLab eda06_09)
```

where t0 is the delay. Note that the symbol i is being used as the imaginary unit. This is the *MatLab* default, but one must be careful not to reset its value to something else, for example, by using it as a counter in a for loop (see Note 1.1).

6.9 Derivative of a function

Multiplying the transform, $\tilde{d}(\omega)$, by the factor, $i\omega$, computes the transform of the derivative, dd/dt:

$$\int_{-\infty}^{+\infty} \frac{dd}{dt}\exp(-i\omega t)dt = d(t)\exp(-i\omega t)\Big|_{-\infty}^{+\infty} - (-i\omega)\int_{-\infty}^{+\infty} d(t)\exp(-i\omega t)dt$$

$$= 0 + (i\omega)\tilde{d}(\omega) = i\omega\tilde{d}(\omega) \qquad (6.37)$$

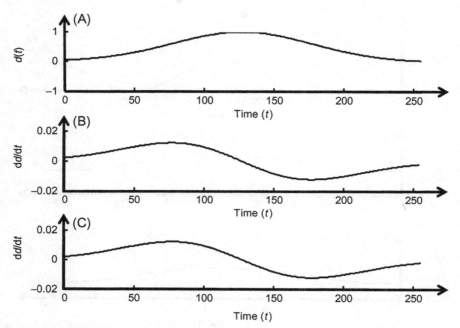

Figure 6.9 (A) exemplary function, $d(t)$. (B) Derivative of $d(t)$ as computed by finite differences. (C) Derivative as computed by fft(). *MatLab* script eda06_10.

Here, we have used integrations by parts, $\int u dv = uv - \int v du$, together with the limit, $\exp(-i\omega t) \rightarrow 0$ as $t \rightarrow \pm\infty$. In *MatLab*, the derivative can be computed as follows (Figure 6.9):

```
dddt = ifft(i*w.*fft(d));                    (MatLab eda06_10)
```

6.10 Integral of a function

Dividing the transform, $\tilde{d}(\omega)$, by the factor, $i\omega$, computes the transform of the integral, $\int d(t) \, dt$:

$$\int_{-\infty}^{+\infty} \int_0^t d(t')dt' \exp(-i\omega t)dt = \int_0^t d(t')dt' \frac{\exp(-i\omega t)}{-i\omega}\Big|_{-\infty}^{+\infty} - \frac{1}{-i\omega}\int_{-\infty}^{+\infty} d(t)\exp(-i\omega t)dt$$

$$= 0 + \frac{1}{i\omega}\,\tilde{d}(\omega) = \frac{1}{i\omega}\,\tilde{d}(\omega) \tag{6.38}$$

Here, we have used integrations by parts, $\int u dv = uv - \int v du$ together with the limit, $\exp(-i\omega t) \rightarrow 0$ as $t \rightarrow \pm\infty$. Note, however, that the zero-frequency value is undefined (as 1/0 is undefined). As shown in Equation (6.34), this value is the total area under the curve, so it functions as integration constant and must be set to an appropriate value. In *MatLab*, with the zero-frequency value set to zero, the integral is calculated as follows (Figure 6.10):

```
integral=ifft(-i*fft(d).*[0,1./w(2:N)']');   (MatLab eda06_11)
```

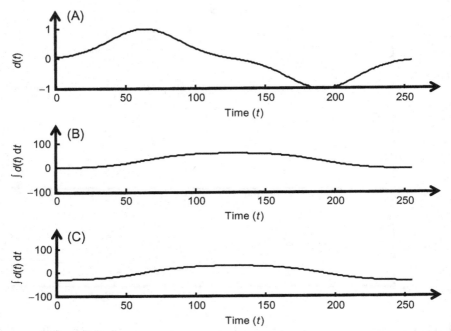

Figure 6.10 (A) exemplary function, $d(t)$. (B) Integral of $d(t)$ as computed by Riemann sums. (C) Integral as computed by $\mathtt{fft()}$. Notice that the two integrals differ by a constant offset. *MatLab* script eda06_11.

6.11 Convolution

Finally, we derive a result that pertains to the *convolution* operation, which will be developed further and utilized heavily in subsequent chapters. Given two functions, $f(t)$ and $g(t)$, their convolution (written $f*g$) is defined as

$$f(t) * g(t) = \int_{-\infty}^{+\infty} f(\tau)g(t - \tau)d\tau \tag{6.39}$$

The transform of the convolution is

$$\int_{-\infty}^{-\infty} \int_{-\infty}^{+\infty} f(\tau)g(t - \tau)d\tau \exp(-i\omega t)dt =$$

$$= \int_{-\infty}^{-\infty} f(\tau) \int_{-\infty}^{+\infty} g(t - \tau) \exp(-i\omega t)dt d\tau$$

$$= \int_{-\infty}^{-\infty} f(\tau) \int_{-\infty}^{+\infty} g(t') \exp(-i\omega(t' + \tau))dt' d\tau$$

$$= \int_{-\infty}^{-\infty} f(\tau) \exp(-i\omega\tau)d\tau \int_{-\infty}^{+\infty} g(t') \exp(-i\omega t')dt'$$

$$= \tilde{f}(\omega)\tilde{g}(\omega) \tag{6.40}$$

Here, we have used the transformation of variables, $t' = t - \tau$. Thus, the transform of the convolution of two functions is the product of their transforms.

Crib Sheet 6.2 Exemplary pairs of time series $d(t)$ and its amplitude spectral density $s(f)$ (*MatLab* script eda06_12)

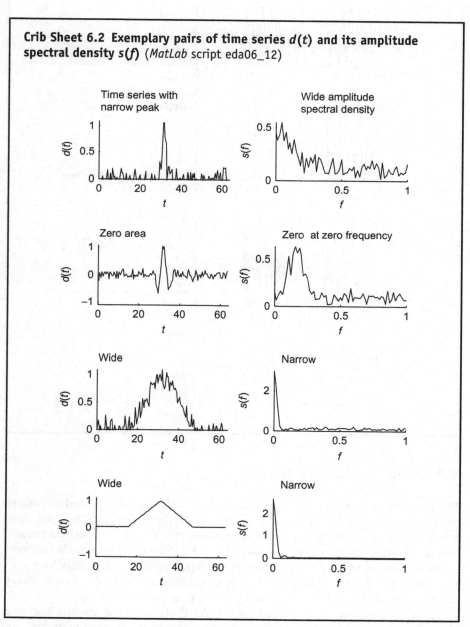

Continued

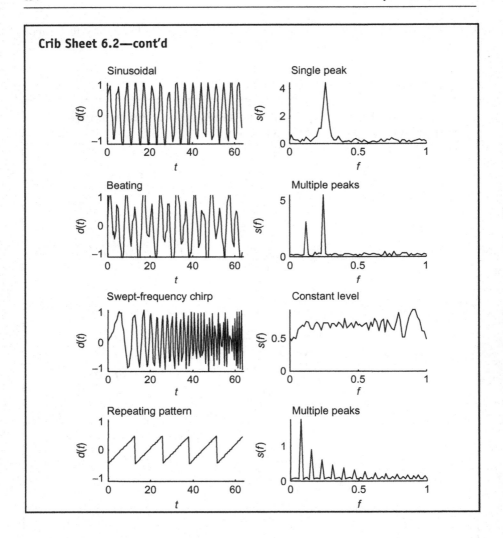

6.12   Nontransient signals

Previously, in developing the relationship between the Fourier integral and its discrete analog, we assumed that the function of interest, $d(t)$, was zero outside of the time window of observation. Only *transient* signals have this property; we can theoretically record the whole phenomenon, as it lasts only for a finite time. An equally common scenario is one in which the data represent just a portion of an indefinitely long phenomenon that has no well-defined beginning or end. Both the Neuse River Hydrograph and Black Rock Forest temperature datasets are of this type.

Many nontransient signals do not vary dramatically in overall pattern from one time window of observation to another (meaning that their statistical properties are *stationary*; that is, constant with time). One parameter that is independent of the window length is the *power*, P:

$$P = \frac{1}{T}\int [d(t)]^2 dt \approx \frac{\Delta t}{N\Delta t}\sum_{i=1}^{N}[d_i(t)]^2 = \frac{1}{N}\mathbf{d}^T\mathbf{d} \tag{6.41}$$

In some cases, such as when $d(t)$ represents velocity, P literally is *power*, that is energy per unit time. Usually, however, the word is understood in the more abstract sense to mean the overall size of a signal that is fluctuating about zero. Note that when the data have zero mean, $P = N^{-1}\mathbf{d}^T\mathbf{d}$ is the formula for the variance of $\mathbf{d}$. In this special case, the power is equivalent to the variance of the signal, $d(t)$.

The power in a time series has a close relationship to its power spectral density. A time series is related to its Fourier transform by the linear rule, $\mathbf{d} = \mathbf{Gm}$, where $\mathbf{m}$ is a column vector of Fourier amplitudes. If we were to use the sines and cosines representation of the Fourier transform, then the matrix, $\mathbf{G}$, is given in Equation (6.13). Substituting $\mathbf{d} = \mathbf{Gm}$ into Equation (6.41) yields

$$P = \frac{1}{N}\mathbf{d}^T\mathbf{d} = \frac{1}{N}(\mathbf{Gm})^T(\mathbf{Gm}) = \frac{1}{N}\mathbf{m}^T[\mathbf{G}^T\mathbf{G}]\mathbf{m} \tag{6.42}$$

However, according to Equation (6.14), $\mathbf{G}^T\mathbf{G} = (N/2)\mathbf{I}$ (except for the first and last coefficient, which we shall ignore). Hence,

$$P = \frac{1}{N}\frac{N}{2}\mathbf{m}^T\mathbf{m} = \frac{1}{2}\sum_{i=1}^{\frac{N}{2}+1}(A_i^2 + B_i^2) = \frac{1}{2}\sum_{i=1}^{\frac{N}{2}+1}\frac{4}{N^2}|C_i|^2 = \frac{2}{(\Delta t)^2 N^2 \Delta f}\int_{0}^{f_{ny}}|\tilde{d}(f)|^2 df \tag{6.43}$$

This result is called *Parseval's Theorem*. Here, the As and Bs are the coefficients in the cosines and sines representation of the Fourier series (Equation 6.13) and the Cs are the coefficients of the *MatLab*'s version of the DFT (Equation 6.21b). The two representations are related by $(2/N)C_i = A_i - iB_i$. The Fourier transform is approximately, $\tilde{d}(f) = \Delta t C_i$. If we *define* the power spectral density, $s^2(f)$, of a nontransient signal as

$$s^2(f) = \frac{2}{T}|\tilde{d}(f)|^2 \quad \text{then} \quad P = \int_0^{f_{ny}} s^2(f)df \tag{6.44}$$

Here, we have used the relations $T = N\Delta t$ and $\Delta t\Delta f = N^{-1}$ to reduce $2/[(\Delta t)^2 N^2 \Delta f]$ to $2/T$. The power is the integral of the power spectral density from zero frequency to the Nyquist frequency. In *MatLab*, the power spectral density is computed as follows:

```
tmax=N*Dt;
C=fft(d);
dbar=Dt*C;
thepsd = (2/tmax) * abs(dbar(1:Nf)).^2;      (MatLab eda06_13)
```

and the total power (variance) is:

    P = Df * sum(thepsd);   (*MatLab* eda06_13)

Here, the data, d, are presumed to have sampling interval, Dt, and length, N.

This simple generalization of the idea of spectral density is closely connected to the limitation that we cannot take the Fourier transform of the whole phenomenon, for it is indefinitely long, but only a portion of it. Nevertheless, we would like for our results to be relatively insensitive to the length of the time window. The factor of $1/T$ in Equation (6.44) normalizes for window length.

Suppose the units of $d(t)$ are u. Then, the Fourier transform, $\tilde{d}(\omega)$, has units of u-s and the power spectral density has units of $u^2s^2/s = u^2s = u^2/Hz$. For example, for discharge data with units of cubic feet per second, $u = ft^3/s$, and power spectral density has units of ft^3, or $ft^3/s \; Hz^{-1}$ (in the Neuse river hydrograph shown in Figure 6.11, we use the mixed units of ft^3/s per cycle/day).

The amplitude spectral density has units of the square root of the power spectral density; that is, $u \; Hz^{-1/2}$.

We have yet to discuss two important elements of working with spectra: First, we have made no mention of confidence limits, yet these are important in determining whether an observed periodicity (i.e., an observed spectral peak) is statistically significant. The reason we have omitted it so far is that the power spectral density is not a linear function of the model parameters, but instead depends on the sum of squares of the model parameters. We lack the tools to propagate error between the data and results in this case. Second, we have made no mention of the *tapering* process that is

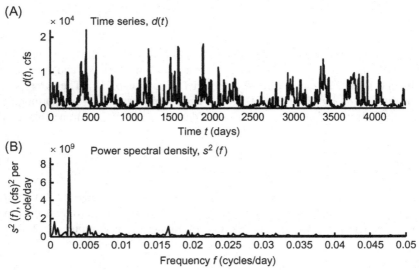

Figure 6.11 (A) Neuse River hydrograph, $d(t)$. (B) Its power spectral density, $s^2(f)$. *MatLab* script eda06_14.

often used to prepare data before computing its Fourier transform. We will address these important issues in Chapters 12 and 9, respectively.

Problems

6.1. Write a *MatLab* script that uses the `fft()` function to differentiate the Neuse River Hydrograph dataset. Plot the results.

6.2. What is the Fourier transform of $\sin(\omega_0 t)$? Compare it to the transform of $\cos(\omega_0 t)$.

6.3. The *MatLab* script, `eda06_15`, creates a file, `noise.txt`, containing normally distributed random time series, $d(t)$, with zero mean and unit variance. (A) Compute and plot the power spectral density of this time series. (B) Create a second time series, $a(t)$, that is a moving window average of $d(t)$; that is, each point in $a(t)$ is the average of, say L, neighboring points in $d(t)$. (C) Compute and plot the power spectral density of $a(t)$ for a suite of values of L. Comment on the results.

6.4. Suppose that you needed to compute the DFT of the function

$$d(t) = \exp\left(-\frac{t^2}{2\sigma^2}\right)$$

using *MatLab's* `fft()` function. This function is centered on $t = 0$, and therefore has nonnegligible values for points to the left of the origin. Unfortunately, we have defined the time and data column-vectors, **t** and **d**, to start at time zero, so there seems to be no place to put these data values. One solution to this problem is to shift the function to the center of the time window, say by an amount, t_0, compute its Fourier transform, and then multiply the transform by a phase factor, $\exp(i\omega t_0)$, that shifts it back. Another solution relies on the fact that, in discrete transforms, *both* time and frequency suffer from aliasing. Just as the last frequencies in the transform were large positive frequencies and small negative frequencies, the last points in the time series are simultaneously

$$\mathbf{t} = [\ldots \quad (N-3)\Delta t \quad (N-2)\Delta t \quad (N-1)\Delta t]^{\mathrm{T}}$$

and

$$\mathbf{t} = [\cdots -3\Delta t \quad -2\Delta t \quad -\Delta t]^{\mathrm{T}}$$

Therefore, one simply puts the negative part of $d(t)$ at the right-hand end of **d**. Write a *MatLab* script to try both methods and check that they agree.

6.5. Compute and plot the amplitude spectral density of a cleaned version of the Black Rock Forest temperature dataset. (A) What are its units? (B) Interpret the periods of any spectral peaks that you find.

References

Zwillinger, D., 1996. CRC Standard Mathematical Tables and Formulae, 30th ed. CRC Press.

7 The past influences the present

7.1   Behavior sensitive to past conditions

Consider a scenario in which a thermometer is placed on the flame emitted by a burner (Figure 7.1A). The gas supplied to the burner varies with time, causing the heat, $h(t)$, of the flame, measured in Watts (W), to fluctuate. An ideal thermometer would respond instantaneously to changes in heat, so the temperature, $\theta(t)$, measured in Kelvin (K), would track heat exactly, that is, $\theta(t) \propto h(t)$. In this case, the past history of the heat has no effect on the temperature. The temperature *now* depends only on the heat at *this* moment. If, however, a metal plate is inserted into the flame between the burner and the thermometer (Figure 7.1B), the situation changes radically. The plate takes time to warm up or cool down. The thermometer does not immediately detect a change in heat; it now takes time to *respond* to changes. Even if the flame were turned off, the thermometer would not immediately detect the event. For a while, it would continue to register high temperatures that reflected the heat supplied by the flame in the past. In this case, $\theta(t)$ would not be proportional to $h(t)$, although the two would still be related in some fashion.

The relationship between $\theta(t)$ and $h(t)$ might be *linear*. Suppose that the time history of heat, $h(t)$, caused temperature, $\theta(t)$. If the heat were doubled to $2h(t)$, we might find that the temperature also doubled, to $2\theta(t)$. Moreover, if two flames were operated individually, with $h_1(t)$ causing $\theta_1(t)$ and $h_2(t)$ causing $\theta_2(t)$, we might find that the combined heat of the two flames, $h_1(t) + h_2(t)$, caused temperature, $\theta_1(t) + \theta_2(t)$.

Another aspect of this scenario is that only *relative time* matters. If we turn on the burner at 9 AM *today*, varying its heat, $h(t)$, and recording temperature, $\theta(t)$, we might expect that the same experiment, if started at 9 AM tomorrow with the same $h(t)$,

Environmental Data Analysis with MATLAB®. http://dx.doi.org/10.1016/B978-0-12-804488-9.00007-0

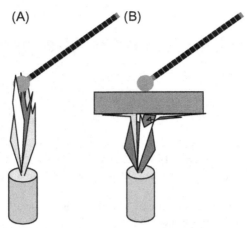

Figure 7.1 (A) Instantaneous response of a thermometer to heat emitted by burner. (B) Delayed response, due to plate that takes time to warm and cool.

would lead to exactly the same $\theta(t)$ (as long as we measured time from the start of each experiment). The underlying notion is that the physics of heat transport through the plate does not depend on day of the week.

This linear behavior can be quantified by assuming that both the heat, $h(t)$, and temperature, $\theta(t)$, are time series; then writing the temperature, θ_i, at time, t_i, as a linear function of past and present heat, h_j, with $j \leq I$,

the data = a linear function of past values of another time series
$$\theta_i = g_1 h_i + g_2 h_{i-1} + g_3 h_{i-2} + g_4 h_{i-3} + \cdots$$
$$\theta_{i+1} = g_1 h_{i+1} + g_2 h_i + g_3 h_{i-1} + g_4 h_{i-2} + \cdots$$
$$\theta_{i+2} = g_1 h_{i+2} + g_2 h_{i+1} + g_3 h_i + g_4 h_{i-1} + \cdots$$
$$\text{or} \quad \theta_i = \sum_{j=1}^{\infty} g_j h_{i-j+1} \quad \text{or} \quad \boldsymbol{\theta} = \mathbf{g} * \mathbf{h} \tag{7.1}$$

Note that each formula involves only current and past values of the heat. This is an expression of *causality*—the future cannot affect the present. The relationship embodied in Equation (7.1) is called a *convolution*, and is denoted by the asterisk, * (which does *not* mean multiplication when used in this context).

The gs in Equation (7.1) are coefficients that express the linear proportionality between heat and temperature. Note that exactly the same coefficients, g_1, g_2, g_3, g_4 ... are used in all the formulas in Equation (7.1). This pattern implements the *time-shift invariance* that we discussed earlier—only the relative time between the application of the heat and the measurement of the temperature matters. Of course, this is only an idealization. If, for instance, the plate were oxidizing during the experiment, then its thermal properties would change with time and the set of coefficients that linked $\theta(t)$ to $h(t)$ at one time would be different from that at another.

For Equation (7.1) to be of any practical use, we need to assume that only the *recent* past affects the present, that is, the coefficients, g_1, g_2, g_3, g_4, ..., eventually diminish in magnitude, sufficiently for the sequence to be approximated as having a finite length. This notion implies that the underlying physics has a characteristic time scale over which equilibration occurs. Heat that was supplied prior to this characteristic *response time* has negligible effect on the current temperature. In the case of the flame in the laboratory, we would not expect that yesterday's experiment would affect today's experiment. The temperature of the plate has adequate time to equilibrate overnight.

As Equation (7.1) is linear, it can be arranged into a matrix equation of the form $\mathbf{d} = \mathbf{Gm}$:

temperature = linear, causal, time-shift invariant function of heating

or

$$
\begin{bmatrix} \theta_1 \\ \theta_2 \\ \theta_3 \\ \cdots \\ \theta_N \end{bmatrix} = \begin{bmatrix} g_1 & 0 & 0 & \cdots & 0 \\ g_2 & g_1 & 0 & \cdots & 0 \\ g_3 & g_2 & g_1 & \cdots & 0 \\ \cdots & \cdots & \cdots & \cdots & 0 \\ g_N & g_{N-1} & g_{N-2} & \cdots & g_1 \end{bmatrix} \begin{bmatrix} h_1 \\ h_2 \\ h_3 \\ \cdots \\ h_N \end{bmatrix}
$$

or

$$
\theta = \mathbf{Gh} \tag{7.2}
$$

Note, however, that a problem arises because of the lack of observation before time, t_1. Here, we have simply ignored those values (or equivalently, assumed that they are zero). At time t_1, the best that we can write is $\theta_1 = g_1 h_1$, as we have no information about heating at earlier times.

When grouped together in a column vector, $\mathbf{g}$, the coefficients are called a *filter*. A filter has a simple and important interpretation as the temperature time series that one would observe if a *unit impulse* (spike) of heat were applied: inserting $\mathbf{h} = [1, 0, 0, 0, \ldots, 0]^T$ into Equation (7.2) yields $\theta = [g_1, g_2, g_3, \ldots, g_N]^T$. Thus, the time series, $\mathbf{g}$, is often called the *impulse response* of the *system* (Figure 7.2). The equation, $\theta = \mathbf{g} * \mathbf{h}$, can then be understood to mean that the output—the data, θ—equals the input—the heat, $\mathbf{h}$—convolved with the impulse response of the system.

Each column, $\mathbf{c}^{(i)}$, of the matrix, $\mathbf{G}$, in Equation (7.2) is the time series, $\mathbf{g}$, which is shifted so that g_1 aligns with row i; that is, it is the response of an impulse at time t_i. If we read the matrix equation as $\theta = h_1 \mathbf{c}^{(1)} + h_2 \mathbf{c}^{(2)} + h_3 \mathbf{c}^{(3)} +, \ldots$, then we can see that the temperature, θ, is built up by adding together many time-shifted versions of the impulse response, each with an amplitude governed by the amount of heat. In this view, the time series of heat is composed of a sequence of spikes and the time series for temperature is constructed by *mixing* together scaled versions of the impulse responses of each of those spikes (Figure 7.3).

So far, we have been examining a scenario in which the data vary with time. The same sort of behavior occurs in spatial contexts as well. Consider a scenario in which a spatially variable mass, $h(x)$, of a pollutant is accidentally spilled on a

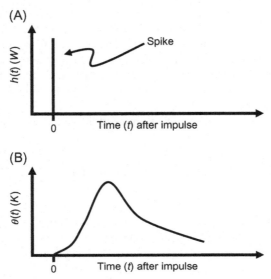

Figure 7.2 Hypothetical impulse response of the hot plate scenario. (A) An impulse (spike) of heat, h, is applied to the bottom of the plate at time, $t = 0$. (B) The temperature, θ, of the top surface of the plate first increases, as heat begins to diffuse through plate. It then decreases, as the plate cools.

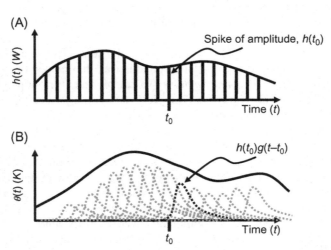

Figure 7.3 Interpretation of the response to heating. (A) The heat, $h(t)$, is viewed as consisting of a sequence of impulses (spikes). (B) The temperature, $\theta(t)$, is viewed as consisting of a sum of scaled and delayed impulse responses (dashed curves). A spike in heat of amplitude, $h(t_0)$ at time, $t = t_0$, makes a contribution, $h(t_0)g(t-t_0)$, to the overall temperature.

highway. If, a year later, its concentration, $\theta(x)$, on the road surface is measured, it will not obey $\theta(x) \propto h(x)$, because weather and traffic will have spread out the deposit. This type of scenario can also be represented with filters, but in this case they are noncausal, that is, the concentration at position x_0 is affected by deposition at both $x \leq x_0$ and $x > x_0$:

concentration $=$ linear, spatial-shift invariant function of mass at all positions

or

$$\theta_i = \cdots + g_{-2}h_{i+3} + g_{-1}h_{i+2} + g_0h_{i+1} + g_1h_i + g_2h_{i-1} + g_3h_{i-2} + g_4h_{i-3} + \cdots$$

We need to assume that the filter coefficients (the gs) die away at both the left and the right so that the filter can be approximated by one of finite length.

Crib Sheet 7.1 Filter basics

Definition of a filter

A filter is a time series $\mathbf{f}$ (of length L) that expresses a linear relationship between an output time series $\mathbf{d}$ (of length N) and an input time series $\mathbf{m}$ (of length $M = N$) through an operation called a *convolution*.

Convolution notation

$$\mathbf{d} = \mathbf{f} * \mathbf{m} \quad \text{which is equivalent to} \quad \mathbf{d} = \mathbf{m} * \mathbf{f}$$

(Here $*$ denotes the convolution operation, *not* multiplication).

When to use a filter

Use a filter when $\mathbf{d}$ depends on the past history of $\mathbf{m}$.

Length of a filter

L expresses the duration of the past that influences the present. Filters can be of any length, but short filters ($L \ll N$) usually behave better than long filters.

Matrix forms of the convolution

$$
\begin{bmatrix} d_1 \\ d_2 \\ d_3 \\ \vdots \\ d_N \end{bmatrix}
=
\begin{bmatrix}
f_1 & 0 & 0 & \cdots & 0 \\
\vdots & f_1 & 0 & \cdots & \vdots \\
f_L & \vdots & f_1 & \cdots & \vdots \\
0 & f_L & \vdots & \cdots & \vdots \\
0 & 0 & f_L & & \vdots \\
0 & 0 & 0 & & 0 \\
\vdots & \vdots & \vdots & \vdots & \vdots \\
0 & 0 & 0 & 0 & f_1
\end{bmatrix}
\begin{bmatrix} m_1 \\ m_2 \\ m_3 \\ \vdots \\ m_N \end{bmatrix}
$$

```
d = toeplitz([f',zeros(1,N-L)]',[f(1),zeros(1,N-1)]) * m;
```

Continued

Crib Sheet 7.1—cont'd

$$
\begin{bmatrix} d_1 \\ d_2 \\ d_3 \\ \vdots \\ d_N \end{bmatrix} = \begin{bmatrix} m_1 & 0 & 0 & \cdots & 0 \\ m_2 & m_1 & 0 & \cdots & 0 \\ m_3 & m_2 & m_1 & \cdots & 0 \\ \vdots & \vdots & \vdots & \vdots & \vdots \\ m_N & m_{N-1} & m_{N-2} & \cdots & m_{N-L+1} \end{bmatrix} \begin{bmatrix} f_1 \\ f_2 \\ f_3 \\ \vdots \\ f_L \end{bmatrix}
$$

```
d = toeplitz(m,[m(1),zeros(1,L-1)]) * f;
```

Computing a convolution

```
d = conv(f,m); % perform convolution
d = d[1:N]; % truncate result to length N (optional)
```

7.2 Filtering as convolution

Equations (7.1) and (7.2) involve the discrete time series $\boldsymbol{\theta}$, $\mathbf{h}$, and $\mathbf{g}$. They can readily be generalized to the corresponding continuous case, for functions $\theta(t)$, $h(t)$, and $g(t)$, by viewing Equation (7.1) as the Riemann sum approximation of an integral. Then the heat, $h(t)$, is related to the temperature, $\theta(t)$, by the *convolution integral* introduced in Section 6.11:

$$
\theta(t) = \int_{-\infty}^{+\infty} h(t-\tau)\,g(\tau)\,d\tau \rightarrow \theta(t_i) \approx \Delta t \sum_{j=1}^{\infty} h(t_i - \tau_j)\,g(\tau_j) \approx \Delta t \sum_{j=1}^{\infty} h_{i-j+1}g_j
$$

$$(7.3)$$

Note that the function, $g(t)$ (as in Equation 7.3), and the time series, $\mathbf{g}$ (as in Equation 7.1) differ by a scaling factor: $g_i = \Delta t g(t_i)$, which arises from the Δt in the Riemann summation.

The integral formulation is useful because it will allow us to deduce properties of filters that might not be so obvious from the matrix equation (Equation 7.2) that describes the discrete case. As an example, we note that an alternative form of the convolution can be derived from Equation (7.3) by the transformation of variables, $t' = (t-\tau)$:

$$
\theta(t) = \int_{-\infty}^{+\infty} h(t-\tau)\,g(\tau)\,d\tau \rightarrow \theta(t) = \int_{-\infty}^{+\infty} g(t-\tau)\,h(\tau)\,d\tau
$$

or

$$
\theta(t) = h(t) * g(t) = g(t) * h(t)
$$

$$(7.4)$$

For causal filters, the function, $g(t)$, is zero for times $t < 0$. In the first integral, $g(t)$ is *forward in time* and $h(t)$ is *backward in time*, and in the second integral, it is vice-versa. Just as with the discrete version of the convolution, the integral version is denoted by the asterisk: $\theta(t) = h(t) * g(t)$. The integral convolution is symmetric, in the sense that $h(t) * g(t) = g(t) * h(t)$.

In Section 7.1, we interpreted the discrete convolution, $\boldsymbol{\theta} = \mathbf{g} * \mathbf{h}$, to mean that the output, $\boldsymbol{\theta}$, is the input, $\mathbf{h}$, convolved with the impulse response of the system, $\mathbf{g}$. We interpreted the column vector, $\mathbf{g}$, as the impulse response because $\boldsymbol{\theta} = \mathbf{g}$ when the input is a spike, $\mathbf{h} = [1, 0, 0, 0, \ldots, 0]^{\mathrm{T}}$. The same rationale carries over to the integral convolution. The equation $\theta(t) = g(t) * h(t)$ means that the output, $\theta(t)$, is the input, $h(t)$, convolved with the impulse response of the system, $g(t)$. We represent the spike with the Dirac function, $\delta(t)$ (see Section 6.6). Then $\theta(t) = g(t)$ when $h(t) = \delta(t)$ (as can be verified by inserting $h(t) = \delta(t)$ into Equation 7.4).

The alternative form of the integral convolution (Equation 7.4), when converted to a Riemann summation, yields the following matrix equation:

output = linear function of impulse response

or

$$
\begin{bmatrix}
\theta_1 \\
\theta_2 \\
\theta_3 \\
\cdots \\
\theta_N
\end{bmatrix}
=
\begin{bmatrix}
h_1 & 0 & 0 & \cdots & 0 \\
h_2 & h_1 & 0 & \cdots & 0 \\
h_3 & h_2 & h_1 & \cdots & 0 \\
\cdots & \cdots & \cdots & \cdots & 0 \\
h_N & h_{N-1} & h_{N-2} & \cdots & h_1
\end{bmatrix}
\begin{bmatrix}
g_1 \\
g_2 \\
g_3 \\
\cdots \\
g_N
\end{bmatrix}
\tag{7.5}
$$

In Equation (7.2), the heat plays the role of the model parameters. If solved by least squares, this equation allows the reconstruction of the heat, using observations of temperature (with the impulse response assumed known, perhaps by deriving it from the fundamental physics). In Equation (7.4), the impulse response plays the role of the model parameters. If solved by least squares, this equation allows the reconstruction of the impulse response, using observations of temperature. In this case, the heat is assumed to be known, perhaps by measuring it, too. (Note, however, that the data kernel would then contain potentially noisy observations, which violates one of the underlying assumptions of least-squares methodology.)

As another example of the utility of the integral formulation of the convolution, note that we had previously derived a relation pertaining to the Fourier transform of a convolution (see Section 6.11). In this context, the Fourier transform of temperature, $\tilde{\theta}(\omega)$, is equal to the product of the transforms of the heat and the impulse response, $\tilde{\theta}(\omega) = \tilde{h}(\omega)\tilde{g}(\omega)$. As we will see later in the chapter, this relationship will prove useful in some computational scenarios.

7.3 Solving problems with filters

Here, we consider a more environmentally relevant temperature scenario, which, however, shares a structural similarity with the burner example. Suppose a thin underground layer generates heat (for example, because of a chemical reaction) and the temperature of the surrounding soil or sediment is measured. As with the burner example, the heat production, $h(t)$, of the layer is unknown and the temperature, $\theta(t)$, of the soil at a distance, $z = 1$ meter, away from the layer is the observation.

(Perhaps direct observation of the layer is contraindicated owing to concerns about release of hazardous chemicals). The impulse response, $g(t)$, of such a layer is predicted on the basis of the physics of heat flow (Menke and Abbott, 1990, their Equation 6.3.4):

$$g(t) = \frac{1}{\rho c_p} \frac{1}{\sqrt{2\pi}\sqrt{2\kappa t}} \exp\left(-\frac{1}{2}\frac{z^2}{2\kappa t}\right) \tag{7.6}$$

Note that the impulse response is proportional to a Normal curve with variance, $2\kappa t$, centered about $z = 0$. The equation involves three material constants, the density, ρ, of the soil, its heat capacity, c_p, and its thermal diffusivity, κ. Typical values for soil are $\rho \approx 1500 \text{ kg/m}^3$, $c_p \approx 800 \text{ J/kg K}$, and $\kappa \approx 1.25 \times 10^{-6} \text{ m}^2/\text{s}$. Note that time is divided by the quantity $(2\kappa)^{-1} \approx 4 \times 10^5$ s, which defines a time scale of about 4.6 days for the heat transport process. Thus, observations made every few hours are sufficient to adequately measure variations of temperature. The quantity, $\rho c_p \approx 1.2 \times 10^6 \text{ J/kg K}$, is the amount of heat, 1.2 million Joules in this case, that needs to be supplied to raise the temperature of a cubic meter of rock by 1 K. Thus, for example, a chemical reaction supplying heat at a rate of 1 W/m^2 will warm up a cubic meter of soil by 0.08 K in 1 day.

The impulse response (Figure 7.4A) is problematical in two respects: (1) Although it rapidly rises to its maximum value, it then decays only slowly, so a large number of

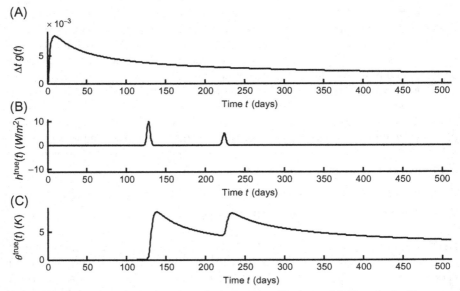

Figure 7.4 (A) Impulse response, $g(t)$, of a heat-generating layer. (B) Hypothetical heat production, $h^{\text{true}}(t)$. (C) Corresponding temperature, $\theta^{\text{true}}(t)$, at 1 m distance from the layer. *MatLab* script eda07_01.

samples are needed to describe it accurately. (2) It contains a factor of $t^{-1/2}$ that is singular at time, $t = 0$ although the function itself is zero, as the exponential factor tends to zero faster than the $t^{-1/2}$ factor blows up. Thus, the $t = 0$ value must be coded separately in *MatLab*:

```
M=1024;
N=M;
Dtdays = 0.5;
tdays = Dtdays*[0:N-1]';
Dtseconds = Dtdays*3600*24;
tseconds = Dtseconds*tdays;
------
g = zeros(M,1);
g(1)=0.0; % manually set first value
g(2:M) = (1/(rho*cp)) * (Dtseconds/sqrt(2*pi)) .* ...
    (1./sqrt(2*kappa*tseconds(2:M))) .* ...
    exp(-0.5*(z^2)./(2*kappa*tseconds(2:M)));
```

(MatLab eda07_01)

Note that two column vectors of time are being maintained: tdays, with units of days and sampling of ½ day, which is used for plotting, and tseconds, with units of seconds, which is used in formulas that require SI units. Note also that the formula for the impulse response contains a factor of Δt not present in Equation (7.6), the scaling factor between the continuous and discrete forms of the convolution.

Given a heat production function, $h(t)$ (Figure 7.4B), the temperature can be predicted by performing the convolution $\theta(t) = g(t) * h(t)$. In *MatLab*, one can build the matrix, **G** (Equation 7.2) and then perform the matrix multiplication:

```
G=toeplitz([g',zeros(N-M,1)']',[g(1),zeros(1,M-1)]);
qtrue2 = G*htrue;                           (MatLab eda07_01)
```

Here we use the character, q, instead of θ, as *MatLab* variables need to have Latin names. An alternative (and preferable) way to perform the convolution is to use *MatLab*'s conv() (for *convolve*) function, which performs the convolution summation (Equation 7.1) directly:

```
tmp = conv(g, htrue);
qtrue=tmp(1:N);
```

The conv() function returns a column vector that is $N + M - 1$ samples long, which we truncate here to N samples.

When we attempt to solve this problem with least squares, *MatLab* reports that the solution,

```
hest1=(G'*G)\(G'*qobs);
```

is ill-behaved. Apparently, more than one heat production, $\mathbf{h}^{est}$, predicts the same—or nearly the same—observations. A quick fixup is to use the damped least squares solution, which adds prior information that $\mathbf{h}^{est}$ is small. The damped least squares solution is easy to implement. We merely add a factor of $\epsilon^2 \mathbf{I}$ to $\mathbf{G}^T\mathbf{G}$:

```
GTG=G'*G;
GTGmax = max(max(abs(GTG)));
e2=1e-4*GTGmax;
hest1=(GTG+e2*eye(M,M))\(G'*qobs);        (MatLab eda07_01)
```

The function $eye()$ returns the identity matrix, $\mathbf{I}$. In principle, the size of ϵ^2 should be governed by the ratio of the variance of the data to the variance of the prior information. However, we make no pretense here of knowing what this ratio might be. Instead, we take a more pragmatic approach, choosing a value for ϵ^2 ($e2$ in the script) by trial-and-error. In order to expedite the trial-and-error tuning, we write ϵ^2 as proportional to the largest element in $\mathbf{G}^T\mathbf{G}$, so that a small damping factor (10^{-4}, in this case) corresponds to the case where $\epsilon^2\mathbf{I}$ is smaller than $\mathbf{G}^T\mathbf{G}$.

The damped least squares solution returns a heat production, $\mathbf{h}^{est}$, that closely resembles the true heat production, $\mathbf{h}^{true}$, when run on synthetic data with a low (0.1%) noise level (Figure 7.5). The situation is less favorable as the noise level is increased to 1% (Figure 7.6). Although the two peaks in heat production, $\mathbf{h}^{true}$, can still be discerned in $\mathbf{h}^{est}$, they are superimposed on a background of high amplitude, high-frequency fluctuations. The problem is that the impulse response, $g(t)$, is a slowly varying function that tends to smooth out high-frequency variations in heat production. Thus, while the low-frequency components of the heat production are well constrained by the data, $\boldsymbol{\theta}^{obs}$, the high-frequency components are not. The damped least squares implements the prior information of *smallness*, but smallness does not discriminate between frequencies and, so, does little to correct this problem. As an alternative, we try prior information of *smoothness*, implemented with generalized least squares. As a smooth solution is one

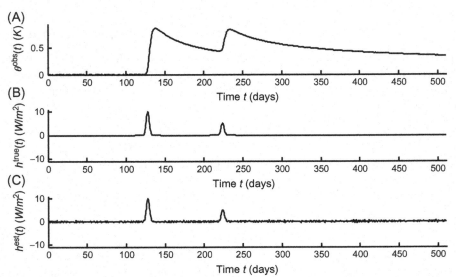

Figure 7.5 (A) Synthetic temperature data, $\theta^{obs}(t)$, constructed from the true temperature plus random noise. (B) True heat production, $h^{true}(t)$. (C) Estimated heat production, $h^{est}(t)$, calculated with damped least squares. *MatLab* script eda07_01.

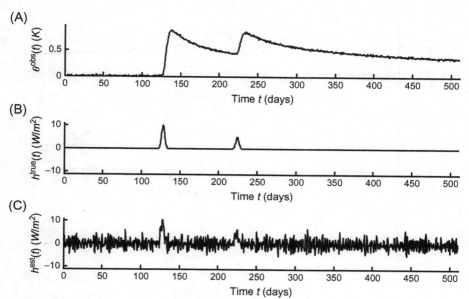

Figure 7.6 (A) Synthetic temperature data, $\theta^{obs}(t)$, constructed from the true temperature plus a higher level of random noise than in Figure 7.5. (B) True heat production, $h^{true}(t)$. (C) Estimated heat production, $h^{est}(t)$, calculated with damped least squares. *MatLab* script eda07_02.

lacking high-frequency fluctuations, this method does much better (Figure 7.7). The *MatLab* code that created and solved the generalized least squares equation, $\mathbf{Fm} = \mathbf{f}$, is as follows:

```
e=10*max(max(abs(G)));
F=zeros(N+M,M);
f=zeros(N+M,1);
F(1:N,1:M)=G;
f(1:N)=qobs2;
H=toeplitz(e*[-2, 1, zeros(1,M-2)]',...
    e*[-2, 1, zeros(1,M-2)]);
F(N+1:N+M,:)=H;
f(N+1:N+M)=0;
hest3 = (F'*F)\(F'*f);                          (MatLab eda07_02)
```

Here the matrix equation, $\mathbf{f} = \mathbf{Fm}$, is built up from its two component parts, the data equation, $\mathbf{d} = \mathbf{Gm}$ and the prior information equation, $\mathbf{h} = \mathbf{Hm}$. As before, the quantity ϵ is in principle controlled by the square-root of the ratio of the variances of the data and the prior information. However, we again take a pragmatic approach and tune it by trial-and-error.

The above code makes rather inefficient use of computer memory. The matrix, $\mathbf{G}$, is constructed from exactly one column-vector, $\mathbf{g}$, so that almost all of the elements of the matrix are redundant. Similarly, the matrix, $\mathbf{H}$, has only three nonzero diagonals, so most of its elements are zero. This wastefulness is not a problem for short filters,

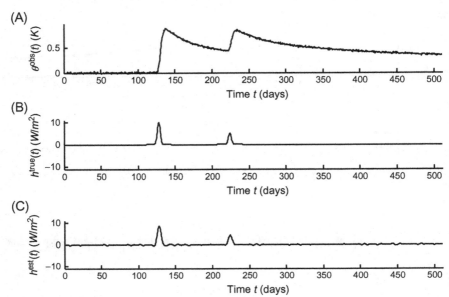

Figure 7.7 (A) Synthetic temperature data, $\theta^{obs}(t)$, constructed from the true temperature plus the same level of random noise as in Figure 7.6. (B) True heat production, $h^{true}(t)$. (C) Estimated heat production, $h^{est}(t)$, calculated with generalized least squares using prior information of smoothness. *MatLab* script eda07_02.

but the storage requirements and computation time quickly becomes unmanageable for long ones. The solution can be computed much more efficiently using *MatLab*'s bicg() function. As was described in Section 5.8, this function requires a user-supplied function that efficiently computes $\mathbf{F^T(Fv)} = \mathbf{G^T(Gv)} + \mathbf{H^T(Hv)}$, where $\mathbf{v}$ is an arbitrary vector. This function, which we call filterfun(), is analogous to (but more complicated than) the afun() function discussed in Section 5.8:

```
function y = filterfun(v,transp_flag)
global g H;
% get dimensions
N = length(g);
M = length(v);
[K, M2] = size(H);
temp1 = conv(g,v); % G v is of length N
a=temp1(1:N);
b=H*v; % H v is of length K
temp2=conv(flipud(g),a); % GT (G v) is of length M
a2 = temp2(N:N+M-1);
b2 = H'*b; % HT (H v) is of length M
% FT F v = GT G v+HT H v
y = a2 + b2;
return
```
                                                                    (*MatLab* filterfun)

The quantities, $\mathbf{g}$ and $\mathbf{H}$, are defined as global variables, in both this function and the main script, so *MatLab* recognizes these variables as referring to the same quantities in both scripts. The required vector, $\mathbf{G}^T(\mathbf{Gv}) + \mathbf{H}^T(\mathbf{Hv})$, is built up in five steps: (1) The quantity $\mathbf{a} = \mathbf{Gv}$ is just the convolution of $\mathbf{g}$ and $\mathbf{v}$, and is computed using the `conv()` function. (2) The quantity $\mathbf{b} = \mathbf{Hv}$ is performed with a simple matrix multiplication and presumes that $\mathbf{H}$ is a sparse matrix. (3) The quantity $\mathbf{a}_2 = \mathbf{G}^T\mathbf{a}$ is a convolution, but the matrix $\mathbf{G}^T$ differs from the normal convolution matrix $\mathbf{G}$ in three respects. First, the ordering of $\mathbf{g}$ in the columns of $\mathbf{G}^T$ is reversed ('flipped') with respect to its ordering in $\mathbf{G}$. Second, $\mathbf{G}^T$ is upper-triangular, while $\mathbf{G}$ is lower-triangular. Third, $\mathbf{G}^T$ is $M \times N$ whereas $\mathbf{G}$ is $N \times M$. Thus, modifications in procedure are required to compute $\mathbf{a}_2 = \mathbf{G}^T\mathbf{a}$ using the `conv()` function. The order of the elements of $\mathbf{g}$ must be flipped before computing its convolution with $\mathbf{a}$. Moreover, the results must be extracted from the last M elements of the column vector returned by `conv()` (not, as with $\mathbf{Gv}$, from the first N elements). (4) The quantity $\mathbf{b}_2 = \mathbf{H}^T\mathbf{b}$ is computed via normal matrix multiplication, again presuming that $\mathbf{H}$ is sparse. (5) The quantity $\mathbf{a}_2 + \mathbf{b}_2$ is computed by normal vector addition. In the main script, the quantities $\mathbf{H}$, $\mathbf{h}$, and $\mathbf{f}$ are created as follows:

```
clear g H;
global g H;
------
e=10*max(abs(g));
K=M;
L=N+K;
H=spalloc(K,M,3*K);
for j = [2:K]
    H(j,j-1)=e;
end
for j = [1:K]
    H(j,j)=-2*e;
end
for j = [1:K-1]
    H(j,j+1)=e;
end
h=zeros(K,1);
f(1:N)=qobs2;
f(N+1:N+K)=h;
```
                                                                              (*MatLab* eda07_03)

The `spalloc()` function creates a $K \times M$ sparse matrix $\mathbf{H}$, capable of holding $3K$ nonzero elements. Its three nonzero diagonals are set by the three for loops. The column vector $\mathbf{h}$ is zero. The column vector $\mathbf{f}$ is built up from θ^{est} and $\mathbf{h}$. The quantity $\mathbf{F}^T\mathbf{f}$ is computed using the same methodology as in `filterfun()` and then both `filterfun()` and $\mathbf{F}^T\mathbf{f}$ are passed to the biconjugate gradient function, `bicg()`:

```
temp=conv(flipud(g),qobs2);
FTfa = temp(N:N+M-1);
FTfb = H'*h;
FTf=FTfa+FTfb;
hest3=bicg(@filterfun,FTf,1e-10,3*L);
```
                                                        (*MatLab* eda07_03)

The solution provided by eda07_03 is the same as that provided by eda07_02, but the memory requirements are much less and the execution speed is much faster.

7.4 An example of an empirically-derived filter

At least during the summer when precipitation occurs as rain, we might expect that the discharge of a river is proportional to the amount of rain that falls on its watershed. However, rain does not instantaneously cause discharge but rather is delayed behind it because the rainwater needs time to work its way through soils and tributaries before it can register on a hydrographic station. This scenario is well suited to being described by a filter. River discharge $d(t)$ is related to precipitation $g(t)$ through the convolution relationship $d(t) = g(t) * r(t)$, where the filter $r(t)$ respresents the response of the river's drainage network to a pulse of rain.

We apply this idea to the Hudson River, using a dataset with daily measurements of river discharge and precipitation, both measured in the Albany, New York area:

> *The file precip_discharge_data.txt contains three columns of data, time in days starting January 1, 2002, Hudson River discharge in cubic feet per second measured at the Green Island hydrographic station near Albany, New York and precipitation in inches per day measured at the Albany Airport. I download the discharge data from the USGS's Surface-Water Daily Data for the Nation database and the precipitation data from NOAA's lobal Surface Summary of Day database and merged the two together. A few of the precipitation values were listed as 99.99; I reset them to zero.*

We focus on a 101 day long time period in Summer 2004, which contains four signifcant storms and about a dozen days with more moderate precipitation (Figure 7.8 A). As expected, the observed river discharge (Figure 7.8 B, sold curve) rises after each storm and varies more smoothly than does precipitation.

We use the same Generalized Least Squares procedure described in Section 7.3 to estimate the filter $r(t)$. Note, however, that in Section 7.3 we assumed that the output and filter were known and solved for the input, whereas here we are assuming that the input and output are known and are solving for the filter. The underlying equations are the same but the interpretation is different. In the current case, the data kernel $\mathbf{G}$ in the standard equation $\mathbf{Gm} = \mathbf{d}$ is built from the precipitation $g(t)$, the model parameters $\mathbf{m}$ represent the unknown response $r(t)$ and the data $\mathbf{d}$ respresent the discharge $d(t)$. We expect that the response of the river to precipitation is relatvely short-lived, since a pulse of rain should drain away over the course of a few weeks, so this is a case where $M < N$. We also implement prior information of the form $\mathbf{Hm} = \mathbf{h}$. Motivated by the notion that $r(t)$ decays smoothly with time, we use the prior information that $dr/dt \approx 0$, so $\mathbf{H}$ is a first difference matrix and $\mathbf{h} = 0$. The results depend on the relative error in the data and prior inforation, $\epsilon = \sigma_d/\sigma_h$. We tune both M and ϵ by trial and error until we achieve a filter that is a good compromise between being smooth and fitting the data well.

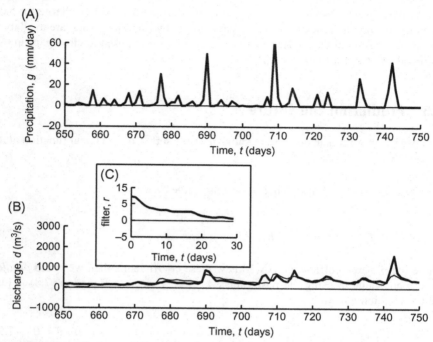

Figure 7.8 River discharge $d(t)$ predicted from precipitation $g(t)$ using an empirically-derived filter $r(t)$. (A) Precipitation $g(t)$ observed in Albany, New York for a 101 day period in summer 2004. (B) Observed Hudson River discharge $d^{obs}(t)$ for the same time period (bold) and predicted discharge $d^{pre}(t) = r(t)*g(t)$ (solid) for a filter $r(t)$ estimated using generalized least squares. (C) The estimated filter $r(t)$, which is 30 days long. *MatLab* script eda07_04.

As expected, the estimated filter $r(t)$ (Figure 7.8 C) declines with time, reaching half of its starting value after about 10 days. The predicted discharge matches the observations, with two exceptions: the prediction misses the peak at day 707; and the prediction overestimates the peak at day 742. These errors may be due to our use of data from a single meteorological station to represent precipitation in the whole Hudson River watershed. A storm may miss the station or dump a disproportionate amount of rain on it. A more complicated model that allows for several stations might have worked better:

$$d(t) = g^{(1)}(t) * r^{(1)}(t) + g^{(2)}(t) * r^{(2)}(t) + \dots \qquad (7.7)$$

Here $g^{(i)}(t)$ is the precipitation observed at the i-th meterological station and $r^{(i)}(t)$ is the resonse of the river discharge to a pulse of precipitation that falls near that station. This problem can be solved using Generalized Least Squares, but the bookkeeping needed to construct the equations $\mathbf{Gm} = \mathbf{d}$ and $\mathbf{Hm} = \mathbf{h}$ is challenging.

The river's estimated response $r(t)$ can be used to predict river discharge for past time periods in which meterterological data, but no hydrographic data, are available. This process, sometimes referred to as *retrodiction*, is often used to close gaps in the historical record of a river's dischage.

7.5   Predicting the future

Suppose we approximate the present value of a time series, **d**, as a linear function of its past values:

present value = linear function of past values

or

$$d_i = p_2 d_{i-1} + p_3 d_{i-2} + p_4 d_{i-3} + \cdots + p_M d_{i-M-1} \qquad (7.8)$$

where the p's are coefficients. If we knew these coefficients, then we could *predict* the current value d_i using the past values $d_{i-1}, d_{i-2}, d_{i-3}, \ldots$ Equation (7.8) is just the convolution equation:

$$p_1 d_i + p_2 d_{i-1} + p_3 d_{i-2} + p_4 d_{i-3} + \cdots = 0 \quad \text{with } p_1 = -1 \quad \text{so} \quad \mathbf{p} * \mathbf{d} = 0 \quad (7.9)$$

Here, the *prediction error filter* **p** is the unknown. The convolution equation, **p** * **d** = 0, has the same form as the heat production equation, **g** * **h** = **θ**, that we encountered in the previous section, and the same generalized least squared methodology can be used to solve it. The prior information, $p_1 = (-1)$, is assumed to be extremely certain and given much smaller variance than **p** * **d** = 0. In most practical cases, the future, insofar as it can be predicted, is a function primarily of the *recent* past. Thus, the prediction error filter is short and *MatLab*'s standard matrix algebra can be used in the computation of **p** (although the bicongugate gradient method remains a good alternative).

As an example, we compute the prediction error filter of the Neuse River Hydrograph (Figure 7.9) using both methods, obtaining, as expected, identical results in the two cases. We choose $M = 100$ as the length of **p**. The filter has most of its high amplitude in the first 3 of 4 days, suggesting that a shorter filter might produce a similar prediction. The small feature at a time of 42 days is surprising, because it indicates some dependence of the present on times more than a month in the past. However, degree to which this feature actually *improves* the prediction needs to be investigated before its significance can be understood.

The prediction error, **e** = **p** * **d**, is an extremely interesting quantity (Figure 7.10), because it contains the part of the present that *cannot* be predicted from the past; that is, the new information. In this case, we expect that the unpredictable part of the Neuse hydrograph is the pattern of storms, while the predictable part is response of the river to those storms. The narrow spikes in the error, **e**, which correlate to peaks in the discharge, seem to bear this out.

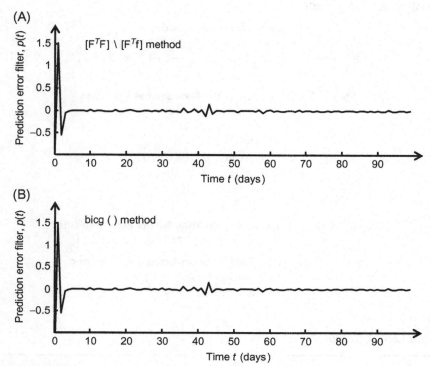

Figure 7.9 Prediction error filter for the Neuse River Hydrograph data, with length of $M = 100$. (A) Filter computed by standard matrix algebra $[F^T F] \backslash [F^T f]$. (B) Filter computed with the biconugate gradient function, `bicg()`. *MatLab* script eda07_05.

Crib Sheet 7.2 Least squares estimation of a filter

Step 1: Indentify the time series and their lengths
The filter equation $\mathbf{d} = \mathbf{g} * \mathbf{m}$ relates input $\mathbf{g}$ to output $\mathbf{d}$ (both of length N) via convolution with a filter $\mathbf{m}$ (of length M).

Step 2: Choose the length *M* of the filter
In general, $M \leq N$. Filters are most useful (and best behaved) when $M \ll N$.

Step 3: Choose the form of the prior information (if any)
Prior information $\epsilon\mathbf{h} = \epsilon\mathbf{H}\mathbf{m}$ (where $\mathbf{h}$ has K rows and $\epsilon = \sigma_d/\sigma_d$) can improve the performance of the estimation process. For instance, smoothness (with $K = M - 2$, $\mathbf{h} = \mathbf{0}$ and $\mathbf{H}$ the second derivative matrix) can reduce short-period fluctuations caused by noise in the data. In practice, the value of ϵ is set by trial and error.

Step 4: Build the prior information equation
Declare globals at top of the script:
```
clear all;
global g H;
```

Continued

Make **H** a sparse matrix:

```
e = epsilon*ones(M,1)/(Dt^2);
H = spdiags([e,-2*e,e],[-1:1],M-2,M);
h = zeros(M-2,1);
```

Step 5: Build the r.h.s. of the least squares equations

```
L=N+K;
f(1:N)=d;
f(N+1:N+K)=h;
temp=conv(flipud(g),d);
FTfa = temp(N:N+M-1);
FTfb = H'*h;
FTf=FTfa+FTfb;
```

Step 6: Solve the least squares equation for the estimated filter

```
mest=bicg(@filterfun,FTf,1e-10,3*L);
```

Step 7: Calculate (and plot) the predicted data and error

```
dpre = conv(g,mest);
dpre = dpre(1:N);
e = d-dpre;
E=e'*e;
```

MatLab eda07_06

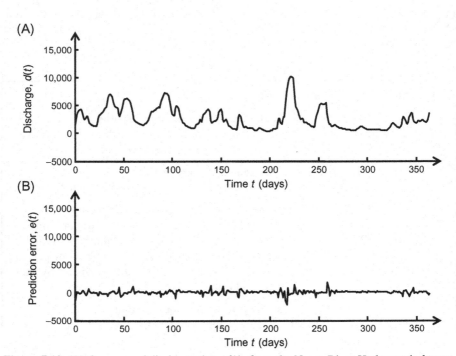

Figure 7.10 (A) One year of discharge data, $d(t)$, from the Neuse River Hydrograph dataset. The prediction error, $e(t)$, based on the $M = 100$ length prediction error filter shown in Figure 7.9. The filter is computed for the whole dataset. A shorter time segment is shown here for visual clarity. *MatLab* script eda07_05.

7.6   A parallel between filters and polynomials

Being able to perform convolutions of short time series by hand is very useful, so we describe here a simple method of organizing the calculation in the convolution formula (Equation 7.1). Suppose we want to calculate $\mathbf{c} = \mathbf{a} * \mathbf{b}$, where both $\mathbf{a}$ and $\mathbf{b}$ are of length 3. We start by writing down $\mathbf{a}$ and $\mathbf{b}$ as row vectors, with $\mathbf{a}$ written backward and time and $\mathbf{b}$ written forward in time, with one sample of overlap. We obtain c_1 by multiplying column-wise, treating blanks as zeros:

$$
\begin{array}{cccc}
a_3 & a_2 & a_1 & \\
& b_1 & b_2 & b_3 \xrightarrow{\text{yields}} c_1 = a_1 b_1 \\
\times & & & \\
& a_1 b_1 & &
\end{array}
\tag{7.10}
$$

To obtain c_2, we shift the top time series one sample to the right, multiply column-wise, and add the results:

$$
\begin{array}{cccc}
a_3 & a_2 & a_1 & \\
& b_1 & b_2 & b_3 \xrightarrow{\text{yields}} c_2 = a_2 b_1 + a_1 b_2 \\
\times & & & \\
& a_2 b_1 & a_1 b_2 &
\end{array}
\tag{7.11}
$$

To obtain c_3, we shift, multiply, and add again.

$$
\begin{array}{ccc}
a_3 & a_2 & a_1 \\
b_1 & b_2 & b_3 \xrightarrow{\text{yields}} c_3 = a_3 b_1 + a_2 b_2 + a_1 b_3 \\
\times & & \\
a_3 b_1 & a_2 b_2 & a_1 b_3
\end{array}
\tag{7.12}
$$

And so forth. The last nonzero element is c_5:

$$
\begin{array}{cccc}
& & a_3 & a_2 & a_1 \\
b_1 & b_2 & b_3 & & \xrightarrow{\text{yields}} c_5 = a_3 b_3 \\
\times & & & & \\
& & a_3 b_3 & &
\end{array}
\tag{7.13}
$$

An astute reader might have noticed that this is exactly the same pattern of coefficients that one would obtain if one multiplied the polynomials:

$$
a(z) = a_1 + a_2 z + a_3 z^2 \quad \text{and} \quad b(z) = b_1 + b_2 z + b_3 z^2
$$

so

$$
c(z) = a(z)b(z) = a_1 b_1 + (a_2 b_1 + a_1 b_2)z + \cdots + a_3 b_3 z^4
\tag{7.14}
$$

We can perform a convolution by converting the time series to polynomials, as above, multiplying the polynomials, and forming a time series from the coefficients of the product. The process of forming the polynomial from a time series is trivial: multiply the first element by z^0, the second by z^1, the third by z^2, and so forth, and add. The process of forming a time series from a polynomial is equally trivial: the first element is the coefficient of the z^0 term, the second of the z^1 term, the third of the z^2 term, and so forth. Yet, while it is trivial to perform, this process turns out to be extremely important, because it allows us to apply a very large body of knowledge about polynomials to issues associated with filters and convolutions. Because of its importance, the process of forming a polynomial from a time series is given a special name, the *z-transform*.

7.7   Filter cascades and inverse filters

Consider a polynomial $g(z)$ of order $N - 1$ that represents a filter, **g**, of length, N. According to the *Fundamental Theorem of Algebra*, the polynomial can be written as the product of its $N - 1$ factors:

$$g(z) = g_1 + g_2 z + \cdots g_N z^{N-1} = g_N(z - r_1)(z - r_2) \cdots (z - r_{N-1}) \tag{7.15}$$

where the rs are the roots of the polynomial (that is, the zs for which the polynomial is zero) and the factor of g_N acts as an overall normalization. Thus the filter, **g**, can be written as a *cascade* of convolutions of $N - 1$ length-two filters:

$$\mathbf{g} = g_N \mathbf{g}_1 * \mathbf{g}_2 * \cdots \mathbf{g}_{N-1} = g_N \begin{bmatrix} -r_1 \\ 1 \end{bmatrix} * \begin{bmatrix} -r_2 \\ 1 \end{bmatrix} * \cdots * \begin{bmatrix} -r_{N-1} \\ 1 \end{bmatrix} \tag{7.16}$$

Thus, any long filtering operation can be broken down into a sequence of many small ones. (Note, however, that some of the length-two time series may be complex, as the roots of a polynomial are, in general, complex).

The goal of the temperature scenario in Section 7.3 was to solve an equation of the form $\mathbf{g} * \mathbf{m} = \mathbf{d}^{obs}$ for **m**. In that section, we used generalized least squares to solve the equivalent matrix equation $\mathbf{Gm} = \mathbf{d}$. Another way to solve the problem is to construct an *inverse filter* $\mathbf{g}^{inv}$ such that $\mathbf{g}^{inv} * \mathbf{g} = [1, 0, 0, \dots, 0]^T$. Then $\mathbf{g} * \mathbf{m} = \mathbf{d}^{obs}$ can be solved by convolving each side of the equation by the inverse filter: $\mathbf{m}^{est} = \mathbf{g}^{inv} * \mathbf{d}^{obs}$.

The z-transform of the inverse filter is evidently $g^{inv}(z) = 1/g(z)$, because then $g(z)g^{inv}(z) = 1$. Note, however, that the function $1/g(z)$ is not a polynomial, so the method of inverting the z-transform and thereby converting the function, $1/g(z)$, back to the time series, $\mathbf{g}^{inv}$, is unclear. Suppose, however, that $g(z)$ was written as a product of its factors $(z - r_i)$, as in Equation (7.15). Then, the inverse filter is the product of the reciprocal of each of these binomials:

$$g^{inv}(z) = \frac{1}{g_N}(z - r_1)^{-1}(z - r_2)^{-1} \cdots (z - r_{N-1})^{-1} \tag{7.17}$$

Each of these reciprocals can now be expanded using the binomial theorem:

$$(z - r_i)^{-1} = (-r_i^{-1})\left(1 - \frac{z}{r_i}\right)^{-1} = (-r_i^{-1})(1 + r_i^{-1}z + r_i^{-2}z^2 + r_i^{-3}z^3 \cdots)$$

(7.18)

Writing the same result in terms of time series

$$\begin{bmatrix} -r_i \\ 1 \end{bmatrix}^{\text{inv}} = (-r_i^{-1}) \begin{bmatrix} 1 \\ r_i^{-1} \\ r_i^{-2} \\ \cdots \end{bmatrix}$$

(7.19)

Thus, the inverse of a length-two filter is infinite in length. This is not a problem, as long as the elements of the inverse filter die away quickly, which happens when $|r_i^{-1}| < 1$ or, equivalently, $|r_i| > 1$.

Each length-two filter in the cascade for **g** turns into an infinite length filter in the cascade for the inverse filter, $\mathbf{g}^{\text{inv}}$. Therefore, while **g** is a length-N filter, $\mathbf{g}^{\text{inv}}$ is an infinite length filter. Any attempt (as in Section 7.3) to find a finite-length version of $\mathbf{g}^{\text{inv}}$ is at best approximate, and can really succeed only when all the roots of $g(z)$ satisfy $|r_i| > 1$. Nevertheless, the approximation is quite good in some cases. In the lingo of filter theory, the roots, r_i, must all lie *outside the unit circle*, $r_i^2 = 1$, for the inverse filter to exist. In this case, the filter, **g**, is said to be *minimum phase*.

This method can be used to construct inverse filters of short filters (Figure 7.11). The first step is to find the roots of the polynomial associated with the filter, **g**:

```
% find roots of g
r = roots(flipud(g));                          (MatLab eda07_07)
```

Here, the filter, g, is of length N. Note that the order of elements in g are flipped, because *MatLab*'s root() function expects the highest order coefficient first. Then a length, Ni, approximation of the inverse filter is computed:

```
% construct inverse filter, gi, of length Ni
Ni = 50;
gi = zeros(Ni,1);
gi(1)=1/gN;
% filter cascade, one filter per loop
for j = [1:N-1]
    % construct inverse filter of a length-2 filter
    tmp = zeros(Ni,1);
    tmp(1) = 1;
```

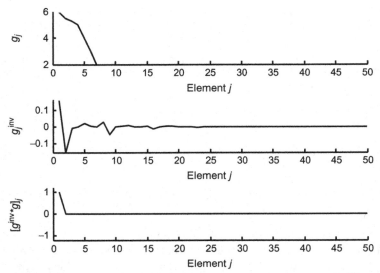

Figure 7.11 A filter, **g**, its inverse filter, **g**^{inv}, and the convolution of the two, **g**^{inv}***g**. *MatLab* script eda07_07.

```
for k = [2:Ni]
    tmp(k) = tmp(k-1)/r(j);
end
tmp = -tmp/r(j);
gi = conv(gi,tmp);
gi=gi(1:Ni);
end
% delete imaginary part (which should be zero)
gi = real(gi);                              (MatLab eda07_07)
```

First, the inverse filter is initialized to a spike of amplitude $1/g$N. Then, each component filter (Equation 7.18) of the cascade (Equation 7.17) is constructed and convolved into gi. After each convolution, the results are truncated to length Ni. Finally, the imaginary part of the inverse filter, which is zero up to round-off error, is deleted.

As an aside, we mention that Fourier transforms can also be used to solve the equation **g** * **m** = **d** and to understand the inverse filter. As the Fourier transform of a convolution is the product of the Fourier transforms, we have

$$g(t) * m(t) = d^{\text{obs}}(t) \rightarrow \tilde{g}(\omega)\,\tilde{m}(\omega) = \tilde{d}^{\text{obs}}(\omega) \quad \text{so} \quad \tilde{m}^{\text{est}}(\omega) = \frac{1}{\tilde{g}(\omega)}\tilde{d}^{\text{obs}}(\omega)$$

$$(7.20)$$

This equation elucidates the problem of the nonuniqueness of the solution, $m^{\text{est}}(t)$. If the Fourier transform of the impulse response, $\tilde{g}(\omega_0)$, is zero for any value of frequency, ω_0, then this *spectral hole* hides the corresponding value of $\tilde{m}(\omega_0)$, in the sense that the data, $\tilde{d}(\omega_0)$, does not depend on it. Thus, $\tilde{m}^{\text{est}}(\omega)$ is nonunique; its value at frequency ω_0 is arbitrary. In practice, even nonzero but small values of $\tilde{g}(\omega)$ are problematical, as corresponding large values of $\tilde{g}^{\text{inv}}(\omega)$ amplify noise in the data, $\tilde{d}^{\text{obs}}(\omega)$, and lead to a noisy solution, $m^{\text{est}}(t)$. We encountered this problem in the heat production scenario of Section 7.3. As the impulse response (Figure 7.4A) is a very smooth function, its Fourier transform has low amplitude at high frequencies. This leads to high-frequency noise present in the data being amplified during the solution process.

Equation (7.20) also implies that the Fourier transform of the inverse filter is $\tilde{g}^{\text{inv}}(\omega) = 1/\tilde{g}(\omega)$; that is, the Fourier transform of the inverse filter is the reciprocal of the Fourier transform of the filter. A problem arises, however, with spectral holes, as $\tilde{g}^{\text{inv}}(\omega)$ is singular at those frequencies. Because of problems associated with spectral holes, the *spectral division* method defined by Equation (7.20) is not usually a satisfactory method for computing the solution to $\mathbf{g} * \mathbf{m} = \mathbf{d}$ or for constructing the inverse filter. The generalized least squares method, based on solving the matrix form of the equation $\mathbf{g} * \mathbf{g}^{\text{inv}} = [1, 0, 0, \ldots, 0]^{\text{T}}$, usually performs much better, as prior information can be used to select a well-behaved solution.

7.8   Making use of what you know

In the standard way of evaluating a convolution equation (e.g., $\boldsymbol{\theta} = \mathbf{g} * \mathbf{h}$, as in Equation 7.1), we compute the elements of $\boldsymbol{\theta}$ in sequence, $\theta_1, \theta_2, \theta_3, \ldots$ but *independently* of one another, even though we know the value of θ_1 before we start to calculate θ_2, know the values of θ_1 and θ_2 before we start to calculate θ_3, and so forth. The known but unused values of $\boldsymbol{\theta}$ are a source of information that can be put to work.

Suppose that the convolution equation (Equation 7.1) is modified by adding a second summation:

$$\theta_i = \sum_{j=1}^{\infty} g_j \, h_{i-j+1} = \sum_{j=1}^{N} u_j \, h_{i-j+1} - \sum_{j=2}^{M} v_j \, \theta_{i-j+1} \tag{7.21}$$

Here, $\mathbf{u}$ and $\mathbf{v}$ are filters of length N and M, respectively, whose relationships to $\mathbf{g}$ are yet to be determined. Note that the last summation starts at $j = 2$, so that only previously calculated elements of $\boldsymbol{\theta}$ are employed. The introduction of past values of $\boldsymbol{\theta}$ into the convolution equation is called *recursion* and filters that include recursion (i.e., include the last term in Equation 7.21) are called *Infinite Impulse Response* (*IIR*) filters. Filters that omit recursion (i.e., omit the last term in Equation 7.21) are called *Finite Impulse Response* (*FIR*) filters. If we define $v_1 = 1$, then we can rewrite Equation (7.21) as:

$$\sum_{j=1}^{N} u_j \ h_{i-j+1} = \sum_{j=1}^{M} v_j \ \theta_{i-j+1} \quad \text{or} \quad \mathbf{u} * \mathbf{h} = \mathbf{v} * \boldsymbol{\theta} \tag{7.22}$$

Recall that we began this discussion by seeking an efficient way to evaluate $\boldsymbol{\theta} = \mathbf{g} * \mathbf{h}$. Equation (7.22) implies $\boldsymbol{\theta} = \mathbf{v}^{\mathrm{inv}} * \mathbf{u} * \mathbf{h}$ and so, if we could find filters $\mathbf{u}$ and $\mathbf{v}$ so that $\mathbf{g} = \mathbf{v}^{\mathrm{inv}} * \mathbf{u}$, then Equation (7.21) would be equivalent to $\boldsymbol{\theta} = \mathbf{g} * \mathbf{h}$. However, it will only improve efficiency if the two filters, $\mathbf{u}$ and $\mathbf{v}$, are shorter than $\mathbf{g}$. What makes this possible is the fact that even a very short filter, $\mathbf{v}$, has an infinitely long inverse filter, $\mathbf{v}^{\mathrm{inv}}$.

In order to illustrate how an IIR filter can be designed, we examine the following simple case:

$$\mathbf{g} = \frac{1}{2} \left[1, \frac{1}{2}, \frac{1}{4}, \frac{1}{8}, \ldots \right]^{T} \tag{7.23}$$

Here, $\mathbf{g}$ is a causal *smoothing* filter. It has only positive coefficients that rapidly decrease with time and the sum of its elements is unity. Each element of $\boldsymbol{\theta}$ is dependent on just the current value and recent past of $\mathbf{h}$. The choices

$$\mathbf{u} = [\frac{1}{2}, 0]^{T} \quad \text{and} \quad \mathbf{v} = [1, -v_2]^{T} \quad \text{with } v_2 = \frac{1}{2} \tag{7.24}$$

work in this case, as $\mathbf{v}^{\mathrm{inv}} = [1, v_2, v_2^2, \ldots]^{T} = [1, \frac{1}{2}, \frac{1}{4}, \ldots]^{T}$ (see Equation 7.19). The generalized convolution equation (Equation 7.21) then reduces to

$$\theta_i = \frac{1}{2} \ h_i + \frac{1}{2} \ \theta_{i-1} \tag{7.25}$$

which involves very little computation, indeed! This filter is implemented in *MatLab* as follows Figure 7.11):

```
q1=zeros(N,1);
q1(1)=0.5*h1(1);
for j=[2:N]
    q1(j)=0.5*(h1(j)+q1(j-1));
end                                          (MatLab eda07_08)
```

Here, $h1$ is the original time series and $q1$ is the smoothed version. Both are of length N. *MatLab* provides a function, $filter()$, that implements Equation (7.20) and can be used as an alternative to the for loop (Figure 7.12):

```
u=[0.5,0]';
v=[1.0,-0.5];
q1 = filter(u, v, h1);                       (MatLab eda07_09)
```

We will return to the issue of IIT filter design in Chapter 9.

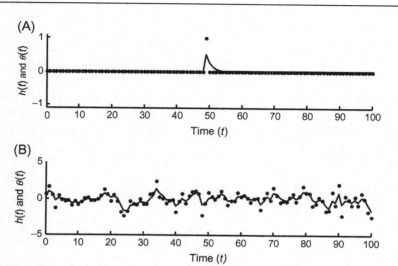

Figure 7.12 Recursive smoothing filter applied to two time series. (A) Time series, $h(t)$ (dots), is a unit spike. The smoothed version, $\theta(t)$ (*solid line*), a decaying exponential, is the impulse response of the smoothing filter. (B) Time series, $h(t)$ (dots), consists of random noise with zero mean and unit variance. Smoothed version, $\theta(t)$ (solid line), averages out extremes of variation. *MatLab* scripts eda07_08 and eda07_09.

Problems

7.1 Calculate by hand the convolution of $\mathbf{a} = [1, 1, 1, 1]^{\mathrm{T}}$ and $\mathbf{b} = [1, 1, 1, 1]^{\mathrm{T}}$. Comment on the shape of the function $\mathbf{c} = \mathbf{a} * \mathbf{b}$.

7.2 Plot the prediction error, E, of the prediction of the Neuse River hydrograph as a function of the length, N, of the prediction error filter. Is a point of diminishing returns reached?

7.3 What is the z-transform of a filter that delays a time series by one sample?

7.4 Note that any filter, $\mathbf{g}$, with $g_1 = 0$ is a nonstationary phase, as its z-transform is exactly divisible by z and so has a root at $z = 0$ that is not outside the unit circle. A simple way to change a stationary phase filter into one that is nonstationary phase filter is to decrease the size of its first element towards zero. Modify script eda07_07 to examine what happens to the inverse filter when you decrease the size of g_1 towards zero in a series of steps. Increasing Ni might make the behavior clearer.

7.5 Generalize the recursive filter developed at the end of Section 7.8 for the case $g(t) \propto \exp(-t/\tau)$, that is, a smoothing filter of unit area and arbitrary width, τ. Start by writing $g_j \propto [1, c, c^2, \ldots]^{\mathrm{T}}$ with $c = \exp(-\Delta t/\tau)$.

7.6 Use your result from the previous problem to smooth the Neuse River hydrograph by with a sequence of filters, $g(t) \propto \exp(-t/\tau)$, with $\tau = 5$, 15 and 40 days. Plot your results and comment on the effect of filtering on the appearance of the hydrograph.

References

Menke, W., Abbott, D., 1990. Geophysical Theory. Columbia University Press, New York. 458pp.

8 Patterns suggested by data

8.1 Samples as mixtures

We previously examined an Atlantic Rock data set that consisted of chemical analyses of rock samples. The data are organized in an $N \times M$ matrix, $\mathbf{S}$, of the form given below:

$$\mathbf{S} = \begin{bmatrix} \text{element 1 in sample 1} & \text{element 2 in sample 1} & \cdots & \text{element } M \text{ in sample 1} \\ \cdots & \cdots & \cdots & \cdots \\ \text{element 1 in sample } N & \text{element 2 in sample } N & \cdots & \text{element } M \text{ in sample } N \end{bmatrix}$$

$$(8.1)$$

In the Atlantic Rock dataset case, $N > M$; that is, the number of rock samples is larger than the number of chemical elements that were measured in each. In other cases, the situation might be reversed.

Rocks are composed of minerals, crystalline substances with distinct chemical compositions. Some advantage might be gained in viewing a rock as a mixture of minerals and then associating a chemical composition with each of the minerals, especially if the number, say P, of minerals is less than the number, M, of chemical elements in the analysis.

In the special case of $M = 3$ chemical elements, we can plot the compositions on a ternary diagram (Figure 8.1). For rocks containing just $P = 2$, the samples lie on a line connecting the two minerals. In this case, viewing the samples as a mixture of minerals provides a significant simplification, as two minerals are simpler than three elements.

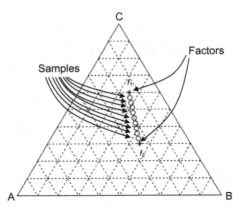

Figure 8.1 Ternary diagram for elements A, B, C. The three vertices correspond to materials composed of pure A, B, and C, respectively. A suite of samples (circles) are composed of a mixture of two factors, $\mathbf{f}_1$ and $\mathbf{f}_2$ (stars), and therefore lie on a line connecting the elemental composition two factors. *MatLab* script eda08_01.

The condition, $P < M$, can be recognized by graphing the data in this low-dimensional case. In higher dimensional cases, more analysis is required.

In the generic case, we will continue to use the term *elements* to refer to parameters that are measured for each sample, with the understanding that this word is used abstractly and may not refer to chemical elements. However, we will use the term *factors* in place of the term minerals. The notion that samples contain factors and factors contain elements is equivalent to the following equation:

samples = a linear mixture of factors

or

$$\mathbf{S} = \mathbf{CF}$$

$$(8.2)$$

The $N \times P$ matrix, $\mathbf{C}$, called the *factor loadings*, quantifies the amount of factors in each sample:

$$\mathbf{C} = \begin{bmatrix} \text{factor 1 in sample 1} & \text{factor 2 in sample 1} & \cdots & \text{factor } P \text{ in sample 1} \\ \cdots & \cdots & \cdots & \cdots \\ \text{factor 1 in sample } N & \text{factor 2 in sample } N & \cdots & \text{factor } P \text{ in sample } N \end{bmatrix}$$

$$(8.3)$$

The $P \times M$ matrix, $\mathbf{F}$, quantifies the amount of elements in each factor:

$$\mathbf{F} = \begin{bmatrix} \text{element 1 in factor 1} & \text{element 2 in factor 1} & \cdots & \text{element } M \text{ in factor 1} \\ \cdots & \cdots & \cdots & \cdots \\ \text{element 1 in factor } P & \text{element 2 in factor } P & \cdots & \text{element } M \text{ in factor } P \end{bmatrix}$$

$$(8.4)$$

Note that individual samples are *rows* of the sample matrix, **S**, and individual factors are *rows* of the factor matrix, **F**. This arrangement, while commonplace in the literature, departs from the convention of this book of exclusively using column vectors (compare with Equation 4.13). We will handle this notational inconsistency by continuing to use column vector notation for individual samples and factors, $\mathbf{s}^{(i)}$ and $\mathbf{f}^{(i)}$, respectively, and then viewing **S** and **F** as being composed of rows of $\mathbf{s}^{(i)\mathrm{T}}$ and $\mathbf{f}^{(i)\mathrm{T}}$, respectively.

Note that we have turned a problem with $N \times M$ quantities into a problem with $N \times P + P \times M$ quantities. Whether or not this change constitutes a simplification will depend on P (that is, whether $N \times P + P \times M$ is larger or smaller than $N \times M$) and on the physical interpretation of the factors. In cases where they have an especially meaningful interpretation, as in the case of minerals, we might be willing to tolerate an increase in the number of parameters.

When the matrix of factor, **F**, is known, least squares can be used to determine the coefficients, **C**. Equation (8.2) can be put into standard form, $\mathbf{Gm} = \mathbf{d}$, by first transposing it, $\mathbf{F}^{\mathrm{T}}\mathbf{C}^{\mathrm{T}} = \mathbf{S}^{\mathrm{T}}$, and then recognizing that each column of $\mathbf{C}^{\mathrm{T}}$ can be computed independently of the others; that is, with **d** a given column of $\mathbf{S}^{\mathrm{T}}$, and with **m** the corresponding column of $\mathbf{C}^{\mathrm{T}}$, and $\mathbf{G} = \mathbf{F}^{\mathrm{T}}$. However, in many instances both the number, P, of factors and the factors, **F**, themselves, are unknown.

The number of factors, P, has no upper bound. However, in general, at most $P = M$ factors are *needed* to exactly represent any set of samples (that is, one factor per element). Furthermore, as we shall describe below, methods are available for detecting the case where the data can be represented with *fewer* than M factors. However, in practice, the determination of this minimum value of P is always somewhat fuzzy because of measurement noise. Furthermore, we might choose to use a value of P *less* than the minimum value required to represent the data exactly, if the approximation, $\mathbf{S} \approx \mathbf{CF}$, is an adequate one.

Even after specifying P, the process of determining **C** and **F** is still nonunique. Given one solution, $\mathbf{S} = \mathbf{C}_1\mathbf{F}_1$, another solution, $\mathbf{S} = \mathbf{C}_2\mathbf{F}_2$, can always be constructed with $\mathbf{C}_2 = \mathbf{C}_1\mathbf{M}$ and $\mathbf{F}_2 = \mathbf{M}^{-1}\mathbf{F}_1$, where **M** is any $P \times P$ matrix that possesses an inverse. Prior information must be introduced to select the *most desirable* solution.

Two possible choices of factors in the two-factor example are illustrated in Figure 8.2. Factors $\mathbf{f}_1$ and $\mathbf{f}_2$ bound the group of samples, so that all samples can be represented by mixtures of positive amounts of each factor (Figure 8.2A). This choice is appropriate when the factors represent actual minerals, because minerals occur only in positive amounts. More abstract choices are also possible (Figure 8.2B), such as factor $\mathbf{f}_1$ representing the composition of the *typical* sample and factor $\mathbf{f}_2$ representing deviation of samples from the typical value. In this case, some samples will contain a negative amount of factor, $\mathbf{f}_2$.

8.2 Determining the minimum number of factors

A surprising amount of information on the structure of a matrix can be gained by studying how it affects a column vector that it multiplies. Suppose that **M** is an $N \times N$ square matrix and that it multiplies an *input* column vector, **v**, producing an

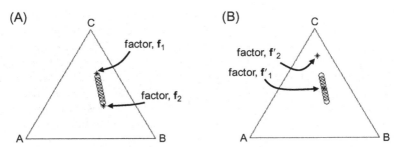

Figure 8.2 Two choice of factors (stars). (A) Factors, $\mathbf{f}_1$ and $\mathbf{f}_2$, bind the samples (circles) so that the samples are a mixture of a positive amount of each factor. (B) Factor, $\mathbf{f}'_1$, is the typical sample and factor, $\mathbf{f}'_2$, represents deviations of samples from the typical value. *MatLab* scripts eda08_02 and eda08_03.

output column vector, $\mathbf{w} = \mathbf{M}\mathbf{v}$. We can examine how the output, $\mathbf{w}$, compares to the input, $\mathbf{v}$, as $\mathbf{v}$ is varied. The following is one question of particular importance:

When is the output parallel to the input?   (8.5)

If $\mathbf{w}$ is parallel to $\mathbf{v}$, then $\mathbf{w} = \lambda \mathbf{v}$, where λ is a scalar proportionality factor. The parallel vectors satisfy the equation:

$$\mathbf{M}\mathbf{v} = \lambda \mathbf{v} \quad \text{or} \quad (\mathbf{M} - \lambda \mathbf{I})\mathbf{v} = 0 \qquad (8.6)$$

Notice that only the direction of $\mathbf{v}$, and not its length, is meaningful, because if $\mathbf{v}$ solves the equation, so does $c\mathbf{v}$, where c is an arbitrary scalar constant. We will find it convenient to use $\mathbf{v}$s that are unit vectors satisfying $\mathbf{v}^T\mathbf{v} = 1$ (or if $\mathbf{v}$ is complex, then $\mathbf{v}^{T*}\mathbf{v} = 1$, where * means complex conjugation).

    The obvious solution to Equation (8.6), $\mathbf{v} = (\mathbf{M} - \lambda \mathbf{I})^{-1}0 = 0$, is not very interesting. A nontrivial solution is possible only when the matrix inverse, $(\mathbf{M} - \lambda \mathbf{I})^{-1}$, does not exist. This is the case where the parameter λ is specifically chosen to make the determinant, $\det(\mathbf{M} - \lambda \mathbf{I})$, vanish (as a matrix with zero determinant has no inverse). Every determinant is calculated by adding together terms, each of which contains the product of N elements of the matrix. As each element of the matrix contains, at most, one instance of λ, the product will contain powers of λ up to λ^N. Thus, the equation, $\det(\mathbf{M} - \lambda \mathbf{I}) = 0$, is an N-th order polynomial equation for λ. An N-th order polynomial equation has N roots, so we conclude that there must be N different proportionality factors, say λ_i, and N corresponding column vectors, say $\mathbf{v}^{(i)}$, that solve $\mathbf{M}\mathbf{v}^{(i)} = \lambda_i \mathbf{v}^{(i)}$. The column vectors, $\mathbf{v}^{(i)}$, are called the *characteristic vectors* (or *eigenvectors*) of the matrix, $\mathbf{M}$, and the proportionality factors, λ_i, are called the *characteristic values* (or *eigenvalues*). Equation (8.6) is called the *algebraic eigenvalue problem*. As we will show below, a matrix is completely specified by its eigenvectors and eigenvalues.

In the special case where $\mathbf{M}$ is symmetric, the eigenvalues, λ_i, are real, as can be seen by calculating the imaginary part of λ_i and showing that it is zero. The imaginary part is found using the rule that, given a complex number, z, its imaginary part satisfies $2iz^{\text{imag}} = z - z^*$, where z^* is the complex conjugate of z. We first premultiply Equation (8.6) by $\mathbf{v}^{(i)\text{T}*}$:

$$\mathbf{v}^{(i)\text{T}*}\mathbf{M}\mathbf{v}^{(i)} = \lambda_i\mathbf{v}^{(i)\text{T}*}\mathbf{v}^{(i)} = \lambda_i \tag{8.7}$$

We then take its complex conjugate

$$\mathbf{v}^{(i)\text{T}}\mathbf{M}\mathbf{v}^{(i)*} = \lambda_i^* \tag{8.8}$$

using the rule, $(ab)^* = a^*b^*$, and subtract

$$2i\lambda_i^{\text{imag}} = \lambda_i - \lambda_i^* = \mathbf{v}^{(i)\text{T}*}\mathbf{M}\mathbf{v}^{(i)} - \mathbf{v}^{(i)\text{T}}\mathbf{M}\mathbf{v}^{(i)*} = 0 \tag{8.9}$$

Here, we rely on the rule that for any two vectors, $\mathbf{a}$ and $\mathbf{b}$, the quantities, $\mathbf{a}^{\text{T}}\mathbf{M}\mathbf{b}$ and $\mathbf{b}^{\text{T}}\mathbf{M}\mathbf{a}$ are equal when $\mathbf{M}$ is symmetric. Equation (8.6) will yield real eigenvectors when the eigenvalues are real.

In the special case where $\mathbf{M}$ is symmetric, the eigenvectors are mutually perpendicular, $\mathbf{v}^{(i)\text{T}}\mathbf{v}^{(j)} = 0$ for $i \neq j$ (this rule is subject to a *caveat*, discussed below). This orthogonality can be seen by premultiplying the equation, $\mathbf{M}\mathbf{v}^{(i)} = \lambda_i\mathbf{v}^{(i)}$, by $\mathbf{v}^{(j)\text{T}}$, and the equation, $\mathbf{M}^{(j)} = \lambda_j\mathbf{v}^{(j)}$, by $\mathbf{v}^{(i)\text{T}}$ and subtracting:

$$\mathbf{v}^{(j)\text{T}}\mathbf{M}\mathbf{v}^{(i)} - \mathbf{v}^{(i)\text{T}}\mathbf{M}\mathbf{v}^{(j)} = 0 = \left(\lambda_i - \lambda_j\right)\mathbf{v}^{(i)\text{T}}\mathbf{v}^{(j)} \tag{8.10}$$

Thus, the eigenvectors are orthogonal, $\mathbf{v}^{(i)\text{T}}\mathbf{v}^{(j)} = 0$, as long as the eigenvalues are *distinct* (numerically different, $\lambda_i \neq \lambda_j$). This exception is the *caveat* alluded to above. We do not discuss it further here, except to mention that while such pairs of eigenvectors are not *required* to be mutually perpendicular, they can be *chosen* to be so. Thus, the rule $\mathbf{v}^{(i)\text{T}}\mathbf{v}^{(j)} = 0$ for $i \neq j$ can be extended to all the eigenvectors of $\mathbf{M}$. We can also choose them to be of unit length so that $\mathbf{v}^{(i)\text{T}}\mathbf{v}^{(j)} = 1$ for $i = j$. Thus, $\mathbf{v}^{(i)\text{T}}\mathbf{v}^{(j)} = \delta_{ij}$, where δ_{ij} is the Kronecker delta symbol (see Section 4.7).

Customarily, the N eigenvalues are sorted into descending order. They can be arranged into a diagonal matrix, $\mathbf{\Lambda}$, whose elements are $[\mathbf{\Lambda}]_{ij} = \lambda_i\delta_{ij}$, where δ_{ij} is the Kronecker Delta. The corresponding N eigenvectors, $\mathbf{v}^{(i)}$, can be arranged as the columns of an $N \times N$ matrix, $\mathbf{V}$, which satisfies $\mathbf{V}^{\text{T}}\mathbf{V} = \mathbf{I}$. Equation (8.6) can then be succinctly written:

$$\mathbf{M}\mathbf{V} = \mathbf{V}\mathbf{\Lambda} \quad \text{or} \quad \mathbf{M} = \mathbf{V}\mathbf{\Lambda}\mathbf{V}^{\text{T}} \tag{8.11}$$

Thus, the matrix, $\mathbf{M}$, can be reconstructed from its eigenvalues and eigenvectors. Furthermore, if any of the eigenvalues are zero, the corresponding vs can be thrown out of the representation of $\mathbf{M}$:

$$
\mathbf{M} = \mathbf{V}\boldsymbol{\Lambda}\mathbf{V}^{\mathrm{T}} = \begin{bmatrix} \mathbf{v}_1 & \mathbf{v}_1 & \cdots & \mathbf{v}_P & \mathbf{v}_{P+1} & \cdots & \mathbf{v}_N \end{bmatrix} \begin{bmatrix} \lambda_1 & & & & & & \\ & \lambda_2 & & & & & \\ & & \cdots & & & & \\ & & & \lambda_P & & & \\ & & & & 0 & & \\ & & & & & \cdots & \\ & & & & & & 0 \end{bmatrix} \begin{bmatrix} \mathbf{v}_1^{\mathrm{T}} \\ \mathbf{v}_2^{\mathrm{T}} \\ \cdots \\ \mathbf{v}_P^{\mathrm{T}} \\ \mathbf{v}_{P+1}^{\mathrm{T}} \\ \cdots \\ \mathbf{v}_N^{\mathrm{T}} \end{bmatrix}
$$

$$
= \begin{bmatrix} \mathbf{v}_1 & \mathbf{v}_2 & \cdots & \mathbf{v}_P \end{bmatrix} \begin{bmatrix} \lambda_1 & & & \\ & \lambda_2 & & \\ & & \cdots & \\ & & & \lambda_P \end{bmatrix} \begin{bmatrix} \mathbf{v}_1^{\mathrm{T}} \\ \mathbf{v}_2^{\mathrm{T}} \\ \cdots \\ \mathbf{v}_P^{\mathrm{T}} \end{bmatrix} = \mathbf{V}_P \boldsymbol{\Lambda}_P \mathbf{V}_P^{\mathrm{T}}
$$

$$(8.12)$$

Returning now to the sample factorization problem, $\mathbf{S} = \mathbf{CF}$, we find that it could be solved if $\mathbf{S}$ were a square, symmetric matrix. In this special case, after computing its eigenvalues, $\boldsymbol{\Lambda}$, and eigenvectors, $\mathbf{V}$, we could write

$$
\mathbf{S} = (\mathbf{V}_P\boldsymbol{\Lambda}_P)\left(\mathbf{V}_P^{\mathrm{T}}\right) = \mathbf{CF} \quad \text{with} \quad \mathbf{C} = \mathbf{V}_P\boldsymbol{\Lambda}_P \quad \text{and} \quad \mathbf{F} = \mathbf{V}_P^{\mathrm{T}} \tag{8.13}
$$

Thus, we would have both determined the minimum number, P, of factors and divided $\mathbf{S}$ into two parts, $\mathbf{C}$ and $\mathbf{F}$. The factors, $\mathbf{f}^{(i)} = \mathbf{v}^{(i)}$ (the rows of $\mathbf{F}^{\mathrm{T}}$), are all mutually perpendicular. Unfortunately, $\mathbf{S}$ is usually neither a square nor symmetric matrix.

The solution is to first consider the matrix $\mathbf{S}^{\mathrm{T}}\mathbf{S}$, which *is* a square, $M \times M$ symmetric matrix. Calling its eigenvalue and eigenvector matrices, $\boldsymbol{\Lambda}$ and $\mathbf{V}$, respectively, we can write

$$
\mathbf{S}^{\mathrm{T}}\mathbf{S} = \mathbf{V}_P\boldsymbol{\Lambda}_P\mathbf{V}_P^{\mathrm{T}} = \mathbf{V}_P\boldsymbol{\Lambda}_P^{1/2}\boldsymbol{\Lambda}_P^{1/2}\mathbf{V}_P^{\mathrm{T}} = \mathbf{V}_P\boldsymbol{\Lambda}_P^{1/2}\mathbf{I}\boldsymbol{\Lambda}_P^{1/2}\mathbf{V}_P^{\mathrm{T}} = \mathbf{V}_P\boldsymbol{\Lambda}_P^{1/2}\mathbf{U}_P^{\mathrm{T}}\mathbf{U}_P\boldsymbol{\Lambda}_P^{1/2}\mathbf{V}_P^{\mathrm{T}}
$$
$$
= \left(\mathbf{U}_P\boldsymbol{\Lambda}_P^{1/2}\mathbf{V}_P^{\mathrm{T}}\right)^{\mathrm{T}}\left(\mathbf{U}_P\boldsymbol{\Lambda}_P^{1/2}\mathbf{V}_P^{\mathrm{T}}\right) \tag{8.14}
$$

Here, $\boldsymbol{\Lambda}_P^{1/2}$ is a diagonal matrix whose elements are the square root of the elements of the eigenvalue matrix, $\boldsymbol{\Lambda}_P$ (see Note 8.1). Note that we have replaced the identity matrix, $\mathbf{I}$, with $\mathbf{U}_P^{\mathrm{T}}\mathbf{U}_P$, where $\mathbf{U}_P$ is an as yet to be determined $N \times P$ matrix which must satisfy $\mathbf{U}_P^{\mathrm{T}}\mathbf{U}_P = \mathbf{I}$. Comparing the first and last terms, we find that

$$
\mathbf{S} = \mathbf{U}_P\boldsymbol{\Sigma}_P\mathbf{V}_P^{\mathrm{T}} \quad \text{with} \quad \boldsymbol{\Sigma}_P = \boldsymbol{\Lambda}_P^{1/2} \quad \text{and} \quad \mathbf{U}_P = \mathbf{S}\mathbf{V}_P\boldsymbol{\Lambda}_P^{-1/2} \tag{8.15}
$$

Note that the $N \times P$ matrix, $\mathbf{U}_P$, satisfies $\mathbf{U}_P^{\mathrm{T}}\mathbf{U}_P = \mathbf{I}$ and the $P \times M$ matrix, $\mathbf{V}_P'$, satisfies $\mathbf{V}_P^{\mathrm{T}}\mathbf{V}_P = \mathbf{I}$. The $P \times P$ diagonal matrix, $\boldsymbol{\Sigma}_P$, is called the matrix of *singular values* and Equation (8.15) is called the *singular value decomposition* of the matrix, $\mathbf{S}$. The sample factorization is then

$$\mathbf{S} = \mathbf{CF} = (\mathbf{U}_P \mathbf{\Sigma}_P)(\mathbf{V}_P^T) \quad \text{with} \quad \mathbf{C} = \mathbf{U}_P \mathbf{\Sigma}_P \quad \text{and} \quad \mathbf{F} = \mathbf{V}_P^T \tag{8.16}$$

Note that the factors are mutually perpendicular unit vectors. The singular values (and corresponding columns of $\mathbf{U}_P$ and $\mathbf{V}_P$) are usually sorted according to size, with the largest first. As the singular values appear in the expression for the factor loading matrix, $\mathbf{C}$, the factors are sorted into the order of contribution to the samples, with those making the largest contribution first. The first factor, $\mathbf{f}_1$, makes the largest contribution of all and usually similar in shape to the average sample.

Because of observational noise, the eigenvalues of $\mathbf{S}^T\mathbf{S}$ can rarely be divided into two clear-cut groups of P nonzero eigenvalues (the square roots of which are the singular values of $\mathbf{S}$) and $M - P$ exactly zero eigenvalues (which are dropped from the representation of $\mathbf{S}$). Much more common is the case where no eigenvalue is exactly zero, but where many are exceedingly small. In this case, the singular value decomposition has $P = M$. It is still possible to throw out eigenvectors, $\mathbf{v}^{(i)}$, corresponding to small eigenvalues, λ_i, but then the representation is only approximate; that is, $\mathbf{S} \approx \mathbf{CF}$. However, because $\mathbf{S}$ is noisy, the distinction between $\mathbf{S} = \mathbf{CF}$ and $\mathbf{S} \approx \mathbf{CF}$ may not be important. Judgment is required in choosing P, for too small a value will lead to an unnecessarily poor representation of the samples, and too large will result in retaining factors whose only purpose is to *fit the noise*. In the case of the Atlantic Rock dataset, these noise factors correspond to fictitious minerals not actually present in the rocks.

8.3 Application to the Atlantic Rocks dataset

The *MatLab* code for computing the singular value decomposition is

```
[U, SIGMA, V] = svd(S,0);
sigma = diag(SIGMA);
Ns = length(sigma);
F = V';
C = U*SIGMA;                              (MatLab eda08_04)
```

The svd() function does not throw out any of the zero (or near-zero) eigenvalues; this is left to the user. Here, U is an $N \times M$ matrix, SIGMA is an $M \times M$ diagonal matrix of singular values, and V is an $M \times M$ matrix. The diagonal of SIGMA has been copied into the column-vector, sigma, for convenience. A plot of the singular values of the Atlantic Rock data set reveals that the first value is by far the largest, values 2 through 5 are intermediate in size and values 6 through 8 are near-zero. The fact that the first singular value, Σ_{11}, is much larger than all the others reflects the composition of the rock samples having only a small range of variability. Thus, all rock samples contain a large amount of the first factor, $\mathbf{f}_1$—the typical sample. Only five factors, $\mathbf{f}_1, \mathbf{f}_2, \mathbf{f}_3, \mathbf{f}_4$, and $\mathbf{f}_5$, out of a total of eight are needed to describe the samples and their variability about the typical sample requires only four (factors 2 through 5) (Figure 8.3):

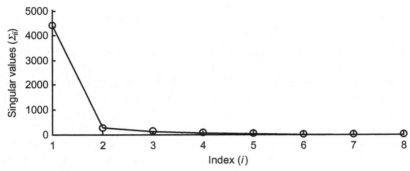

Figure 8.3 Singular values, Σ_{ii}, of the Atlantic Ocean rock dataset. *MatLab* script eda08_04.

Element	f_1	f_2	f_3	f_4	f_5
SiO_2	+0.908	+0.007	−0.161	+0.209	+0.309
TiO_2	+0.024	−0.037	−0.126	+0.151	−0.100
Al_2O_3	+0.275	−0.301	+0.567	+0.176	−0.670
FeO-total	+0.177	−0.018	−0.659	−0.427	−0.585
MgO	+0.141	+0.923	+0.255	−0.118	−0.195
CaO	+0.209	−0.226	+0.365	−0.780	+0.207
Na_2O	+0.044	−0.058	−0.0417	+0.302	−0.145
K_2O	+0.003	−0.007	−0.006	+0.073	+0.015

The role of each of the factors can be understood by examining its elements. Factor 2, for instance, increases the amount of MgO while decreasing mostly Al_2O_3 and CaO, with respect to the typical sample.

The *factor analysis* has reduced the dimensions of variability of the rock dataset from 8 elements to 4 factors, improving the effectiveness of scatter plots. *MatLab*'s three-dimensional plotting capabilities are useful in this case, as any three of the four factors can be used as axes and the resulting three-dimensional scatter plot viewed from a variety of perspectives. The following *MatLab* command plots the coefficients of factors 2 through 4 for each sample:

```
plot3(C(:,2), C(:,3), C(:,4), 'k.');        (MatLab eda08_04)
```

The plot can then be viewed from different perspectives by using the rotation controls of the figure window (Figure 8.4). Note that the samples appear to form two populations, one in which the variability is due to f_2 and another due to f_3.

8.4 Spiky factors

As mentioned earlier, the factors, **F**, of a set of samples, **S**, are nonunique. The equation, $S = CF$, can always be modified to $S = CM^{-1}MF$, where **M** is an arbitrary $P \times P$ matrix, defining a new set of factors, $F' = MF$. Singular value decomposition is useful

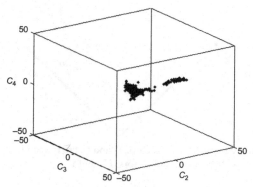

Figure 8.4 Three-dimensional perspective view of the coefficients, C_i, of factors 2, 3, and 4 in each of the rock samples (dots) of the Atlantic Ocean Rock dataset. *MatLab* script eda08_04.

because it allows the determination of a set of $P \leq M$ factors that adequately approximate **S**. However, it does not always provide the most desirable set of factors. Modifying the set of P factor by using the matrix, **M**, does not change the value of P or the quality of the fit, but can be used to produce factors with more desirable properties than those produced by singular value decomposition.

One possible guiding principle is the prior information that the factors should be *spiky*; that is, they should have just a few large elements, with the other elements being near-zero. Minerals, for example, obey this principle. While a rock can contain upward of twenty chemical elements, typically it will be composed of minerals such as fosterite (Mg_2SiO_4), anorthite ($CaAl_2Si_2O_8$), rutile (TiO_2), etc., each of which contains just a few elements. Spikiness is more or less equivalent to the idea that the elements of the factors should have *high variance*. The usual formula for the variance, $\sigma_d{}^2$, of a data set, **d**, is

$$\sigma_d^2 = \frac{1}{N}\left(\sum_{i=1}^{N}(d_i - \bar{d})^2\right) = \frac{1}{N^2}\left(N\sum_{i=1}^{N}d_i^2 - \left(\sum_{i=1}^{N}d_i\right)^2\right) \tag{8.17}$$

Its generalization to a factor, f_i, is

$$\sigma_f^2 = \frac{1}{M^2}\left(M\sum_{i=1}^{M}f_i^4 - \left(\sum_{i=1}^{M}f_i^2\right)^2\right) \tag{8.18}$$

Note that this is the variance of the *squares* of the elements of the factors. Thus, a factor, **f**, has a large variance, σ_f^2, if the absolute values of its elements have high variation. The signs of the elements are irrelevant.

The *varimax* procedure is a way of constructing a matrix, **M**, that increases the variance of the factors while preserving their orthogonality. It is an iterative procedure, with each iteration operating on only one pair of factors, with other pairs being

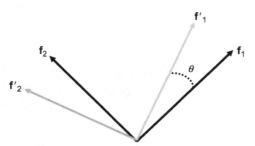

Figure 8.5 Two mutually perpendicular factors, $\mathbf{f}_1$ and $\mathbf{f}_2$, are rotated in their plane by an angle, θ, creating two new mutually orthogonal vectors, $\mathbf{f'}_1$ and $\mathbf{f'}_2$.

operated upon in subsequent iterations. The idea is to view the factors as vectors, and to rotate them in their plane (Figure 8.5) by an angle, θ, chosen to maximize the sum of their variances. The rotation changes only the two factors, leaving the other $P - 2$ factors unchanged, as in the following example:

$$\begin{bmatrix} \mathbf{f}_1^T \\ \mathbf{f}_2^T \\ \cos(\theta)\mathbf{f}_3^T + \sin(\theta)\mathbf{f}_5^T \\ \mathbf{f}_4^T \\ -\sin(\theta)\mathbf{f}_3^T + \cos(\theta)\mathbf{f}_5^T \\ \mathbf{f}_6^T \end{bmatrix} = \begin{bmatrix} 1 & 0 & 0 & 0 & 0 & 0 \\ 0 & 1 & 0 & 0 & 0 & 0 \\ 0 & 0 & \cos(\theta) & 0 & \sin(\theta) & 0 \\ 0 & 0 & 0 & 1 & 0 & 0 \\ 0 & 0 & -\sin(\theta) & 0 & \cos(\theta) & 0 \\ 0 & 0 & 0 & 0 & 0 & 1 \end{bmatrix} \begin{bmatrix} \mathbf{f}_1^T \\ \mathbf{f}_2^T \\ \mathbf{f}_3^T \\ \mathbf{f}_4^T \\ \mathbf{f}_5^T \\ \mathbf{f}_6^T \end{bmatrix} \quad \text{or } \mathbf{F'} = \mathbf{MF}$$

$$(8.19)$$

Here, only the pair, $\mathbf{f}_3$ and $\mathbf{f}_5$, are changed.

In Equation (8.19), the matrix, $\mathbf{M}$, represents a rotation of *one pair* of vectors. The rotation matrix for many such rotations is just the product of a series of pair-wise rotations. Note that the matrix, $\mathbf{M}$, obeys the rule, $\mathbf{M}^{-1} = \mathbf{M}^T$ (that is, $\mathbf{M}$ is a *unary* matrix). For a given pair of factors, $\mathbf{f}^A$ and $\mathbf{f}^B$, the rotation angle, θ, is determined by minimizing $\Phi(\theta) = M^2(\sigma_{fA}{}^2 + \sigma_{fB}{}^2)$ with respect to θ (i.e., by solving $d\Phi/d\theta = 0$).

The minimization requires a substantial amount of algebraic and trigonometric manipulation, so we omit it here. The result is (Kaiser, 1958) as follows:

$$\theta = \frac{1}{4}\tan^{-1}\frac{2M\sum_i u_i v_i - \sum_i u_i \sum_i v_i}{M\sum_i \left(u_i^2 - v_i^2\right) - \left(\left(\sum_i u_i\right)^2 - \left(\sum_i v_i\right)^2\right)}$$

with

$$u_i = \left(f_i^A\right)^2 - \left(f_i^B\right)^2 \text{ and } v_i = 2f_i^A f_i^B \quad (8.20)$$

By way of example, we note that the two vectors $\mathbf{f}^A = \frac{1}{2}[1, \ 1, \ 1, \ 1]^T$ and $\mathbf{f}^B = \frac{1}{2}[1, \ -1, \ 1, \ -1]^T$ are extreme examples of two *nonspiky* orthogonal vectors, because all their elements have the same absolute value. When applied to them, the varimax procedure returns $\mathbf{f}^{A'} = (1/\sqrt{2})[1, 0, 1, 0]^T$ and $\mathbf{f}^{B'} = (1/\sqrt{2}) [0, -1, 0, -1]^T$, which are significantly spikier than the originals (see *MatLab* script eda08_05). The *MatLab* code is as follows:

```
u = fA.^2 - fB.^2;
v = 2* fA.* fB;

A = 2*M*u'*v;
B = sum(u)*sum(v);
top = A - B;

C = M*(u'*u-v'*v);
D = (sum(u)^2) - (sum(v)^2);
bot = C - D;

q = 0.25 * atan2(top,bot);

cq = cos(q);
sq = sin(q);

fAp = cq*fA + sq*fB;
fBp = - sq*fA + cq*fB;                                    (MatLab eda08_05)
```

See Note 6.1 for a discussion of the atan2() function. Here, the original pair of factors fA and fB, and the rotated pair are fAp and fBp.

We apply this procedure to factors $\mathbf{f}_2$ through $\mathbf{f}_5$ of the Atlantic Rock dataset (that is, the factors related to deviations about the typical rock). The varimax procedure is applied to all pairs of these factors and achieves convergence after several such iterations. The *MatLab* code for the loops is as follows:

```
FP = F;

% spike these factors using the varimax procedure
k = [2, 3, 4, 5]';
Nk = length(k);

for iter = [1:3]
for ii = [1:Nk]
for jj = [ii+1:Nk]

% spike factors i and j
i=k(ii);
j=k(jj);

% copy factors from matrix to vectors
fA = FP(i,:)';
fB = FP(j,:)';

% standard varimax procedure to determine rotation angle q
------
```

(A)   (B)

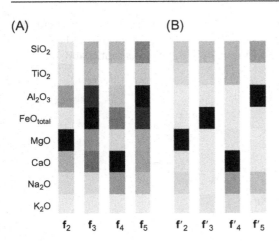

SiO₂

TiO₂

Al₂O₃

FeO_total

MgO

CaO

Na₂O

K₂O

$\mathbf{f}_2$ $\mathbf{f}_3$ $\mathbf{f}_4$ $\mathbf{f}_5$ $\mathbf{f'}_2$ $\mathbf{f'}_3$ $\mathbf{f'}_4$ $\mathbf{f'}_5$

Figure 8.6 (A) Factors, $\mathbf{f}_2$ through $\mathbf{f}_5$, of the Atlantic Rock data set, as calculated by singular value decomposition. (B) Factors, $\mathbf{f'}_2$ through $\mathbf{f'}_5$, after application of the varimax procedure. *MatLab* script eda08_06.

```
% copy rotated factors back to matrix
FP(i,:) = fAp';
FP(j,:) = fBp';

end
end
end                                      (MatLab eda08_06)
```

Here the rotated matrix of factors, FP, is initialized to the original matrix of factors, F, and then modified by the varimax procedure (omitted and replaced with a "------"), with each pass through the inner loop rotating one pair of factors. The procedure converges very rapidly, with three iterations of the outside loop being sufficient.

The resulting factors (Figure 8.6) are much spikier than the original ones. Each now involves mainly variations in one chemical element. For example, $\mathbf{f'}_2$ mostly represents variations in MgO and $\mathbf{f'}_5$ mostly represents variations in Al₂O₃.

8.5 Weighting of elements

An element can be important even though its concentration is extremely low. Gold, for instance, is economic in ore at 5 parts per million and arsenic is unhealthy in drinking water at 10 parts per billion. When a data set contains both high- and low-concentration elements, the procedure described in Section 8.3 for choosing a set of P important factors will lead to an approximation $\mathbf{S}_P \approx \mathbf{S}^{obs}$ that only poorly describes patterns in the low-concentration elements, because the overall quality of the fit is dominated by the much larger errors of the high-concentration elements.

This problem can be avoided by weighting each element by an amount, say w_i, that reflects its importance. The singular value decomposition is performed on the matrix $\mathbf{S}^{obs}\mathbf{W}$, where $\mathbf{W}$ is a diagonal matrix with elements $W_{ii} = w_i$:

$$\mathbf{S}^{obs}\mathbf{W} = \mathbf{U}\mathbf{\Sigma}\mathbf{V}^{T} \quad \text{so} \quad \mathbf{C} = \mathbf{U}_P\mathbf{\Sigma}_P \quad \text{and} \quad \mathbf{F} = \mathbf{V}_P{}^{T}\mathbf{W}^{-1} \tag{8.21}$$

The weights can be chosen in any of several ways. One possibility is to set them intuitively, to reflect the qualitative importance of each element. Another possibility is to scale them according to the certainty of the measurements; that is, to set $w_i = 1/\sigma_i$, where σ_i^2 is the prior variance of the i-th element. This choice leads to each element being fit to similar relative error (Figure 8.7).

Crib Sheet 8.1 Steps in factor analysis

Step 1: Organize the data as a sample matrix S

$$\mathbf{S}^{obs} = \begin{bmatrix} \text{element 1 in sample 1} & \cdots & \text{element } M \text{ in sample 1} \\ \cdots & \cdots & \cdots \\ \text{element 1 in sample } N & \cdots & \text{element } M \text{ in sample } N \end{bmatrix}$$

Step 2: Establish weights that reflect the importance of the elements
$$w_i \quad \text{for} \quad i = 1, \cdots, M$$

one possibility: $w_i = 1/\sigma_i$ with σ_i^2 the prior variance of element i

Step 3: Perform singular value decomposition and form the factor matrix F and loading matrix C

$$\mathbf{U}\boldsymbol{\Sigma}\mathbf{V}^T = \mathbf{S}^{obs}\mathbf{W} \quad \text{with} \quad W_{ij} = \delta_{ij}w_i$$

$$\mathbf{F} = \mathbf{V}^T\mathbf{W}^{-1} \quad \text{and} \quad \mathbf{C} = \mathbf{U}\boldsymbol{\Sigma}$$

```
[U2,SIGMA2,V2] = svd(Sobs*diag(w),0);
   Fpre2 = V2'*diag(1./w); % factors
   Cpre2 = U2*SIGMA2;  % loadings
```

Step 4: Determine the number P of important factors
Plot the diagonal of $\boldsymbol{\Sigma}$ as a function of row index i and choose P to include all rows with "large" Σ_{ii}

Step 5: Reduce the number of factors from M to P

$$\mathbf{F} = \begin{bmatrix} \text{element 1 in factor 1} & \cdots & \text{element } M \text{ in factor 1} \\ \cdots & \cdots & \cdots \\ \text{element 1 in factor } P & \cdots & \text{element } M \text{ in factor } P \end{bmatrix}$$

$$\mathbf{C} = \begin{bmatrix} \text{factor 1 in sample 1} & \cdots & \text{factor } P \text{ in sample 1} \\ \cdots & \cdots & \cdots \\ \text{factor 1 in sample } N & \cdots & \text{factor } P \text{ in sample } N \end{bmatrix}$$

Continued

Crib Sheet 8.1—cont'd

```
Fpre2P = Fpre2(1:P,:);
Cpre2P = Cpre2(:,1:P);
```

Step 6: Predict the data
$$\mathbf{S}^{pre} = \mathbf{CF}$$
```
Spre2P = Cpre2P*Fpre2P;        MatLab eda08_07
```

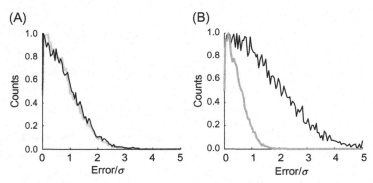

Figure 8.7 Histograms of relative error (error normalized by measurement error σ) for two elements in a 10,000 sample, 7-element synthetic dataset with exactly P = 3 factors. (A) For a high-concentration element, unweighted (black curve) and weighted (grey curve) factor analyis yield similar distributions of errors. (B) For a low-concentration element, the weighted method yields smaller errors. *MatLab* eda08_07.

8.6   Q-mode factor analysis and spatial clustering

The eigenvectors $\mathbf{U}$ and $\mathbf{V}$ play completely symmetric roles in the singular value decomposition of the sample matrix $\mathbf{S} = \mathbf{U\Sigma V}^T$. We introduced an asymmetry when we grouped them as $\mathbf{S} = (\mathbf{U\Sigma})(\mathbf{V}^T) = \mathbf{CF}$, to define the loadings $\mathbf{C}$ and factors $\mathbf{F}$.

This grouping is associated with the term *R-mode factor analysis*. The R-mode factors $\mathbf{F} = \mathbf{V}^T$ express patterns of variability among elements. Thus, for example, the table in Section 8.3 indicates that the elements in the Atlantic rock dataset contains a pattern, quantified by factor $\mathbf{f}^{(2)}$, in which Al_2O_3 and MgO are strongly and negatively correlated and another pattern, quantified by factor $\mathbf{f}^{(3)}$, in which Al_2O_3 and FeO_{total} are strongly and negatively correlated. These correlations reduce the effective number of elements; that is, they to allow us to substitute a small number of factors for a large number of elements.

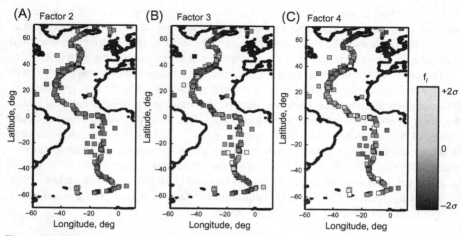

Figure 8.8 Maps of three Q-mode factors of the Atlantic rock dataset, which express patterns of geographical variability. (A) The i-th component of factor $\mathbf{f}^{(2)}$, plotted at the geographical location of sample i (squares) and shaded according to its numberical value. The range of the shading is scaled according to the standard deviation σ of the components. (B) The factor $\mathbf{f}^{(3)}$. (C) The factor $\mathbf{f}^{(4)}$. See text for further discussion. *MatLab* eda08_08.

Alternately, we could have grouped the singular value decomposition as $\mathbf{S} = (\mathbf{U})(\mathbf{\Sigma V}^T)$, an approach associated with the term *Q-mode factor analysis*. The equivalent transposed form $\mathbf{S}^T = (\mathbf{V\Sigma})(\mathbf{U}^T)$ is more frequently encountered in the literature, and is also more easily understood, since it can be interpreted as 'normal factor analysis' applied to the matrix $\mathbf{S}^T$. The transposition has reversed the sense of samples and elements, so the Q-mode factors $\mathbf{F} = \mathbf{U}^T$ quantify patterns of variability among samples, in the same way that the R-mode $\mathbf{V}^T$ quantifies patterns of variability among elements.

The sampling in the Atlantic rock dataset is geographical, so the Q-mode factors express patterns of spatial variability and can be used to detect geographical clustering. These spatial patterns can be brought out by plotting the components of each factor on a map, with the i-th component plotted at the location of sample i and with its numberial value depicted by the color (or grey shade) or the symbo. When applied to the Atlantic rock datset, this technique demonstrated that that the chemistry of the mid-Atlantic ridge (north-south bands of symbols in Figure 8.8) is strongly segmented, with ~ 10 degree long sections of failry uniform chemistry punctuated by spatially-sharp jumps.

8.7 Time-Variable functions

The samples in the Atlantic Rock dataset do not appear to have an intrinsically-meaningful order. As far as we know, their order in the file could merely reflect the order that the samples were entered into the database by the personnel who compiled the data set. However, one might imagine a similar dataset in which sample order is significant. One example would be a set of chemical analyses made in the same place as a sequence of times. The time sequence could be used to characterize the chemical evolution of the system. Sample, $\mathbf{s}^{(i)}$, quantifies the chemical composition at time, t_i.

The ordering of the elements in the Atlantic Rock dataset does not appear to have any special significance either. It does not reflect their abundance in a typical rock. It is not even alphabetical. However, one might imagine a similar dataset in which the order of chemical constituents *is* significant. One example would be a set of analyses of the concentration of alkanes (methane (CH_4), ethane (C_2H_8), propane (C_3H_8), etc.) in a hydrocarbon mixture, as the chemical properties of these carbon-chain molecules are critically dependent on the number of carbon atoms in the chain. In this case, ordering the elements by the increasing length, x, of the carbon chain would be appropriate. The standard equation for factors (Equation 8.2) could then be interpreted in terms of variation in x and t:

$$\mathbf{s}^{(i)} = \sum_{k=1}^{P} C_{ki}\mathbf{f}^{(k)} \quad \text{or} \quad s(x_j, t_i) = \sum_{k=1}^{P} C_k(t_i)f_k(x_j) \tag{8.22}$$

Note that the analysis has broken out the (x, t) dependence into dependence on, x and t, separately. The factors, $f_k(x_j)$, each describe a pattern in x and the factor loadings, $C_k(t_i)$, describe the temporal, t, variation of those patterns. When used in this context, the factors are called *empirical orthogonal functions*, or *EOFs*. In the hydrocarbon example, above, a plot of the elements of a factor, $f_k(x)$, against x would display the distribution of alkane chain length within the k-th factor. A plot of $C_k(t)$ against time, t, would display the time-dependent amplitude the k-th factor. One might imagine a chemical evolution process in which chain length of alkanes decreased systematically with time. This behavior would be manifested in a temporal evolution of the factor loadings, with the mix becoming increasingly rich in those factors that contained larger fractions of short-length alkanes.

A sample is just a collection of numbers arranged in column vector format. While the alkane data set has a one-dimensional organization that makes the use of vectors *natural*, a one-dimensional organization is not required by the factor analysis method. Samples could, for instance, have the natural organization of a two-dimensional or three-dimensional grid. The grid merely would need to be rearranged into a vector for the method to be applied (see Section 5.9).

The Climate Analysis Center (CAC) Equatorial Pacific Ocean Sea Surface Temperature data set is one such example. It is a time sequence of two-dimensional grids of the surface temperature of a patch of the equatorial Pacific Ocean, a part of the world important to the study of the El Niño/La Niña climate oscillation. Bill Menke, who retrieved the data, provides the following report:

I downloaded this data set from the web site of the International Research Institute (IRI) for Climate and Society at Lamont-Doherty Earth Observatory. They call it CAC (for Climate Analysis Center) and describe it as containing "climatological, smoothed and raw sea surface temperature of the tropical Pacific Ocean". I retrieved a text file, cac_sst.txt, *of the entire "smoothed sea-surface temperature anomaly".*

It contains deviations of sea surface temperature, in K, from the average value for a region of the Pacific Ocean (29°S to 29°N, 124°E to 70°W, 2° grid spacing) for each month from January 1970 through March 2003. I made one set of minor changes to the file using a text editor, replacing the words for months, "Jan", "Feb" ... with the numbers, 1, 2 ..., to make it easier to read in MatLab. You have to skip past a monthly header line when you read the file – I wrote a MatLab script that does this. The data center gave two references for the data set, Reynolds and Smith (1994) and Woodruff et al. (1993).

A 6-year portion of the CAC dataset is shown (Figure 8.9). Many of the monthly images show an east-west band of cold (white) temperatures that is characteristic

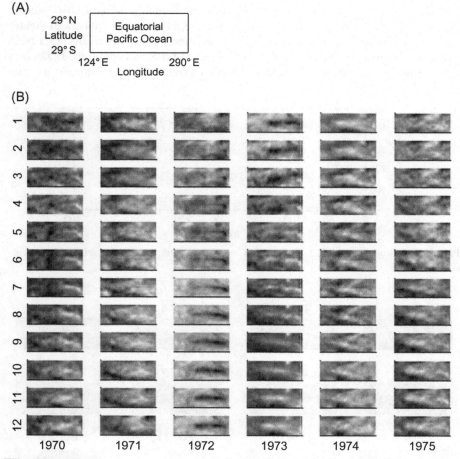

Figure 8.9 Sea surface temperature anomaly for the equatorial Pacific Ocean. (A) Index map. (B) Maps for each month of the year for the 1970–1975 time period are shown, but the dataset continues through March 2003. Darker shades correspond to warmer temperatures.
Source: CAC smoothes sea surface temperature anomaly dataset, IRI Data Library. *MatLab* script eda08_09.

of the La Niña phase of the oscillation. A few (e.g., late 1972) show a warm (black) band, most prominent in the eastern Pacific that is characteristic of the El Niño phase.

The CAC data set comprises $N = 399$ monthly images, each with $M = 2520$ grid points (30 in latitude, 84 in longitude). Each factor (or empirical orthogonal function, EOF) is a 2520-length vector that folds into a 30×84 spatial grid of temperature values. The total number of EOFs is M = 399, but as is shown in Figure 8.10, many have exceedingly small singular values and can be discarded. As the data represent a temperature *anomaly* (that is, temperature minus its mean value), the first EOF will not resemble the mean temperature, but rather will characterize the maximum amplitude spatial variation. A visual inspection (Figure 8.11) reveals that it consists of an east-west cold (white) band crossing the equatorial region and is thus a La Niña pattern.

Plots of the factor loadings, the time-dependent amplitude of the EOFs, indicate that the first five factors have time-variation with significant power at periods greater than 1 year (Figure 8.12). The coefficient of the first EOF is essentially a La Niña index, as the shape of the first EOF is diagnostic of La Niña conditions. Peaks in it indicate times when the La Niña pattern is particularly strong, and troughs indicate

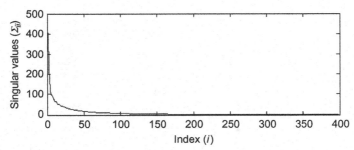

Figure 8.10 Singular values, Σ_{ii}, of the CAC sea surface temperature dataset. *MatLab* script eda08_10.

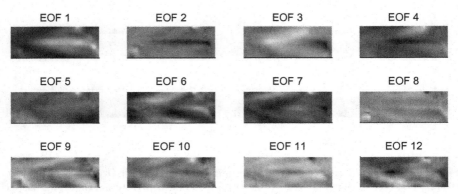

Figure 8.11 First 12 empirical orthoginal functions (EOFs) of the CAC sea surface temperature dataset. *MatLab* Script eda08_11.

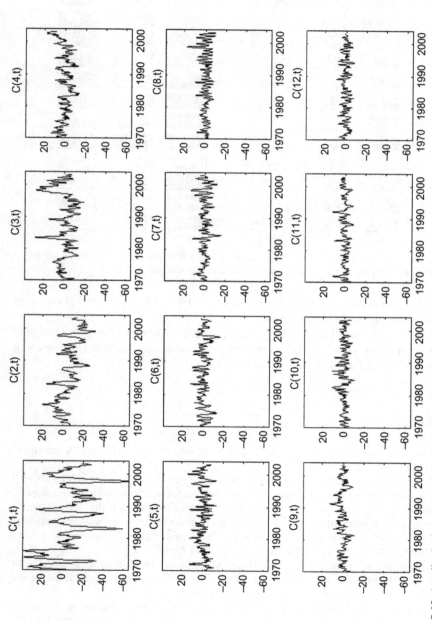

Figure 8.12 Amplitude time series, $C_i(t)$, of the First 12 EOFs of the CAC sea surface temperature dataset for the time period January 1970 to March 2003. *MatLab* script eda08_11.

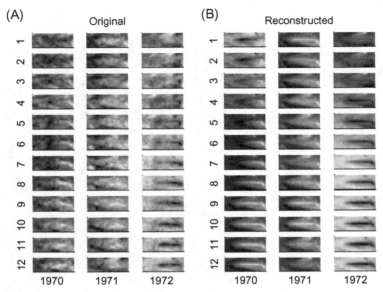

Figure 8.13 (A) First 3 years of the CAC sea surface temperature dataset. (B) Reconstruction using first five empirical orthogonal functions. *MatLab* script eda08_12.

when it is particularly weak. The El Niño years of 1972, 1983, and 1997 show up as prominent troughs in this time series.

The EOFs can be used as a method of smoothing the temperature data in a way that preserves the most important spatial and temporal variations. A reconstruction using just the first five EOFs is shown in Figure 8.13.

Problems

8.1. Write the matrix, $\mathbf{SS}^T$, in terms of $\mathbf{U}$, $\mathbf{V}$, and $\mathbf{\Sigma}$ and show that the columns of $\mathbf{U}$ are eigenvectors of $\mathbf{SS}^T$.

8.2. The varimax procedure uses one type of prior information, spikiness, to build a set of P "improved" factors, $\mathbf{f}'_i$, out of the set of P significant factors, $\mathbf{f}_i$, computed using singular value decomposition. Another, different type of prior information is that the factors, $\mathbf{f}'_i$, are close in shape to some other set of P factors, $\mathbf{f}_i^s$, that are *specified*. Find linear mixtures of the factors, $\mathbf{f}_i$, computed using singular value decomposition that comes as close as possible to $\mathbf{f}_i^s$, in the sense of minimizing, $\Sigma_i \, (\mathbf{f}'_i - \mathbf{f}_i^s) \, (\mathbf{f}'_i - \mathbf{f}_i^s)^T$.

8.3. Compute the power spectra of each of the EOF amplitude (factor loading) time series for the CAC dataset. Which one has the most power at a period of exactly 1 year? Describe and interpret the spatial pattern of the corresponding EOF.

8.4. Cut the Black Rock Forest temperature dataset up into a series of one-day segments.
Discard any segments that contain hot or cold spikes or data dropouts. Subtract out the mean of each segment so that they reflect only the temperature changes through the course of the day, and not the seasonal cycle of temperatures. Consider each segment a sample and

analyze the dataset with empirical orthogonal function analysis, making and interpreting plots that are analogous to Figure 8.9-8.13.

8.5. Invent a scenario in which the factor loadings (amplitudes of EOF's) are a function of two spatial coordinates, (x, y) instead of time, t, and in which they could be used to solve a nontrivial problem.

References

Kaiser, H.F., 1958. The varimax criterion for analytic rotation in factor analysis. Psychometrika 23, 187–200.

Reynolds, R.W., Smith, T.M., 1994. Improved global sea surface temperature analyses. J. Climate 7.

Woodruff, S.D., Lubker, S.J., Wolter, K., Worley, S.J., Elms, J.D., 1993. Comprehensive Ocean-Atmosphere Data Set (COADS) Release 1a: 1980–1992, NOAA Earth System Monitor 4, September.

9 Detecting correlations among data

9.1 Correlation is covariance

When we create a *scatter* plot of observations, we are treating the data as random variables. The underlying idea is that two data types (or elements), say d_i and d_j, are scattering about their typical values. Sometimes the scatter is due to measurement noise. Sometimes it is due to an unmodeled natural process that we can only treat probabilistically. But in either case, we are viewing the cloud of data points as being drawn from a joint probability density function, $p(d_i, d_j)$. The data are correlated if the covariance of this function is nonzero. Thus, the covariance matrix, $\mathbf{C}$, is extremely useful in quantifying the degree to which different elements correlate. Recall that the covariance matrix associated with $p(d_i, d_j)$ are defined as:

$$C_{ij} = \int_{-\infty}^{+\infty} \int_{-\infty}^{+\infty} (d_i - \bar{d}_i)(d_j - \bar{d}_j)\, p(d_i, d_j)\, \mathrm{d}d_i\, \mathrm{d}d_j \qquad (9.1)$$

Here $\bar{d}_i$ and $\bar{d}_j$ are the means of d_i and d_j, respectively. We can estimate C_{ij} from a data set by approximating the probability density function with a histogram constructed from the observed data. We first divide the (d_i, d_j) plane into many small bins, numbered by the index s. Each bin has area $\Delta d_i \Delta d_j$ and is centered at $(d_i^{(s)}, d_j^{(s)})$ (Figure 9.1). We now denote the number of data pairs in bin s by N_s. The probability, $p(d_i, d_j)\, \Delta d_i \Delta d_j \approx N_s/N$, where N is the total number of data pairs, so

Environmental Data Analysis with MATLAB®. http://dx.doi.org/10.1016/B978-0-12-804488-9.00009-4

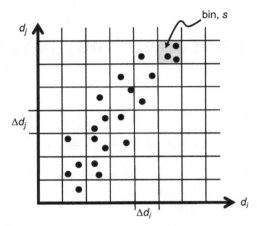

Figure 9.1 Scatter plot pairs of data (circles) are converted into an estimate of the covariance by binning the data in small patches of the (d_i, d_j) plane, and counting up the number of points in each bin. The bins are numbered with an index, s.

$$C_{ij} \approx \frac{1}{N} \sum_s [d_i^{(s)} - \bar{d}_i][d_j^{(s)} - \bar{d}_j]N_s \qquad (9.2)$$

We now shrink the size of the patches so that at most one data pair is in each bin. Then, N_s equals either zero or unity. Summation over the patches is equal to summation over the (d_i, d_j) pairs themselves:

$$C_{ij} \approx \frac{1}{N} \sum_{k=1}^{N} [d_i^{(k)} - \bar{d}_i][d_j^{(k)} - \bar{d}_j] \qquad (9.3)$$

The covariance is nonzero when the data exhibit some degree of correlation, but its actual numerical value depends on the overall range of the data. The range can be normalized to ±1 by scaling by the square root of the product of variances:

$$R_{ij} = \frac{C_{ij}}{\sqrt{C_{ii}C_{jj}}} \qquad (9.4)$$

The quantity **R** is called the *matrix of correlation coefficients*, and its elements are called *correlation coefficients* and are denoted by the lower-case letter, r. When, as above, they are estimated from the data (as contrasted to being computed from the probability density function), they are referred to as *sample* correlation coefficients. See Table 9.1 for a list of important quantities, such as **R**, that are introduced in this chapter. The covariance, **C**, and correlation coefficient matrix, **R**, can be estimated from a set of data, **D**, as follows:

```
C = cov(D); % covariance
R = corrcoef(D); % correlation coefficient (MatLab eda09_01)
```

Table 9.1 Important Quantities Used in Chapter 9.

Symbol	Name	Created from	Significance
C_d	Covariance matrix of the data, d	Probability density function of the data, $p(d)$	Diagonal elements, $[C_d]_{ij}$ with $i = j$: variance of the data, d_i; squared width of the univariate probability density function, $p(d_i)$ off-diagonal elements, $[C_d]_{ij}$ with $i \neq j$: degree of correlation between the pair of observations, d_i and d_j
R	Matrix of correlation coefficients	Probability density function of the data, $p(d)$	Normalized version of C_d with elements that vary between ± 1 elements of R given the symbol, r
$a = d \star d$	Autocorrelaton function	Time series, d	Element a_k: degree of correlation between two elements of d separated by a time lag, $\tau = (k-1)\Delta t$
$c = d^{(1)} \star d^{(2)}$	Cross-correlation function	Two time series, $d^{(1)}$ and $d^{(2)}$	Element c_k: degree of correlation between an element of $d^{(1)}$ and an element of $d^{(2)}$ separated by a time lag, $\tau = (k-1)\Delta t$
$f * d$	Convolution	Filter, f, and time series, d	Filters the times series, d, with the filter, f
$\tilde{d}(\omega)$	Fourier transform	Time series, $d(t)$	Amplitude of sines and cosines of frequency, ω, in the time series
$C^2(\omega_0, \Delta\omega)$	Coherence	Two time series, $d^{(1)}$ and $d^{(2)}$	Similarity between $d^{(1)}$ and $d^{(2)}$ at frequencies in the range, $\omega_0 \pm \Delta\omega$ varies between 0 and 1

Here, D, is an $N \times M$ matrix organized so that D_{ij} is the amount of element, j, in sample i (the same arrangement as in Equation 8.1). The matrix, R, is $M \times M$, so that R_{ij} expresses the degree of correlation of elements i and j. Figure 9.2A depicts the matrix of correlation coefficients for the Atlantic rock dataset, in which the elements are literal chemical elements. The diagonal elements are all unity, as a data type correlates perfectly with itself. Some pairs of chemical components, such as TiO_2 and NaO_2, strongly correlate with each other (Figure 9.2B). Other pairs, such as TiO_2 and Al_2O_3, are nearly uncorrelated.

The idea of correlation can also be applied to the elements of a time series. Neighboring samples in a time series are often highly correlated (and hence predictable), even though the time series as a whole may be random. Consider, for example, the stream flow of the Neuse River. On the one hand, a hydrologist, working a year ago, would not have been able to predict whether today's discharge is unusually high or low. It is just not possible to predict individual storms—the source of the river's water—a year in advance; they are best considered random phenomena. On the other hand, if today's discharge is high, the chances are excellent that tomorrow's discharge will be high as well. Stream flow persists for a few days, because the rain water takes time to drain away.

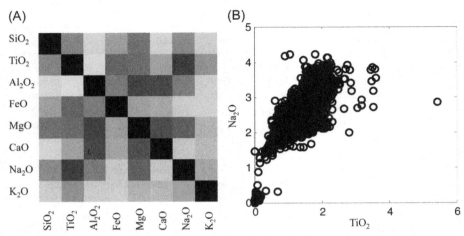

Figure 9.2 (A) Matrix of absolute values of correlation coefficients of chemical elements in the Atlantic rock dataset. (B) Scatter plot of TiO_2 and Na_2O, the most highly correlated elements ($r = 0.73$). *MatLab* script eda09_01.

The notion of short term correlation within the stream flow time series can also be described by a joint probability density function. If we denote the river's discharge at time t_i as d_i, and discharge at time t_j as d_j, then we can speak of the joint probability density function $p(d_i, d_j)$. In the case of stream flow, we expect that d_i and d_j will have a strong positive correlation when the time difference or *lag*, $\tau = t_i - t_j$, is small (Figure 9.3A). When the measurements are more widely separated in time, then we expect the correlation to be weaker (Figure 9.3B). We expect discharge to be uncorrelated at separations of, say, a month or so (Figure 9.3C). On the other hand, discharge will again be positively correlated, although maybe only weakly so, at separations of about a year, because patterns of stream flow have an annual cycle. Note that we must assume that the time series is stationary, meaning that its statistical properties do not change with time, or else the degree of correlation would depend on the measurement times, as well as the time difference between them.

We already have the methodology to quantify the degree of correlation of a joint probability density function: its covariance matrix, C_{ij}. In this case, we manipulate the formula to bring out the means, because in many cases we will be dealing with time series that fluctuate around zero:

$$C_{ij} = \int_{-\infty}^{+\infty} \int_{-\infty}^{+\infty} (d_i - \bar{d})(d_j - \bar{d}) \, p(d_i, d_j) \, dd_i \, dd_j$$

$$= \int_{-\infty}^{+\infty} \int_{-\infty}^{+\infty} d_i d_j \, p(d_i, d_j) \, dd_i \, dd_j - 2\bar{d}^2 + \bar{d}^2 = A_{ij} - \bar{d}^2$$

$$\text{with} \quad A_{ij} = \int_{-\infty}^{+\infty} \int_{-\infty}^{+\infty} d_i d_j \, p(d_i, d_j) \, dd_i \, dd_j \tag{9.5}$$

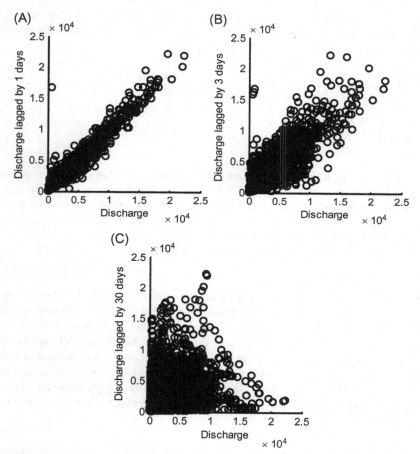

Figure 9.3 Scatter plots of the lagged Neuse River discharge. (A) Lag = 1 day, (B) 3 days, (C) 30 days. Note that the strength of the correlation decreases as lag is increased. *MatLab* script eda09_02.

Here, the mean, $\bar{d}$, of the time series is assumed to be independent of time (so it has no index). The matrix, **A**, is called the *autocorrelation matrix* of the time series. It is equal to the covariance matrix when the mean of the time series is zero.

Just as in the case of the covariance, the autocorrelation can be estimated from observations. The data are pairs of samples drawn from the time series, where one member of the pair is lagged by a fixed time interval, $\tau = (k - 1)\Delta t$, with respect to the other (with k an integer; note that $k = 1$ corresponds to $\tau = 0$). A time series of length N has $N - |k - 1|$ such pairs. We then form a histogram of the pairs, as we did in the case of covariance, so that the integral in Equation (9.5) can be approximated by a summation:

$$A_{i,j} = \int_{-\infty}^{+\infty} \int_{-\infty}^{+\infty} d_i d_j \, p(d_i, d_j) \, \mathrm{d}d_i \, \mathrm{d}d_j \approx \frac{1}{N - |k - 1|} \sum_s d_i^{(s)} d_j^{(s)} N_s. \tag{9.6}$$

Once again, we shrink the size of the bins so that at most one pair is in each bin and N_s equals either zero or unity, so summation over the bin is equal to summation over the data pairs themselves. For the $k > 0$ case, we have

$$A_{i,k+i-1} \approx \frac{1}{N - |k-1|} \sum_s d_i^{(s)} d_{k+i-1}^{(s)} N_s = \frac{1}{N - |k-1|} \sum_{i=1}^{N-k+1} d_i d_{k+i-1} = \frac{a_k}{N - |k-1|}$$

$$\text{with } a_k = \sum_{i=1}^{N-k+1} d_i d_{k+i-1} \quad \text{and} \quad k > 0 \tag{9.7}$$

The column vector, **a**, is called the *autocorrelation* of the time series. An element, a_k, is called the autocorrelation at time *lag*, $\tau = k - 1$. The autocorrelation at negative lags equals the autocorrelation at positive lags, as **A** is a symmetric matrix, that is, $A_{ij} = a_k$, with $k = |i - j| + 1$. As we have defined it above, a_k is *unnormalized*, in the sense that it omits the factor of $1/(N - |k - 1|)$.

In *MatLab*, the autocorrelation is calculated as follows:

```
a = xcorr(d);                    (MatLab Script eda09_03)
```

Here, d is a time series of length N. The xcorr() function returns a vector of length $2N - 1$ that includes both negative and positive lags so that the zero lag element is a(N).

The autocorrelation of the Neuse River hydrograph is shown in Figure 9.4. For small lags, say of less than a month, the autocorrelation falls off rapidly with lag, with a time scale that reflects the time that rain water needs to drain away after a storm. For larger lags, say of a few years, the autocorrelation oscillates around zero with a period of one year. This behavior reflects the seasonal cycle. Summer and winter discharges are *negatively* correlated, as one tends to be high when the other is low.

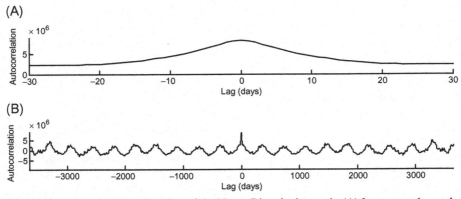

Figure 9.4 Autocorrelation function of the Neuse River hydrograph. (A) Lags up to 1 month. Note that the autocorrelation decreases with lag. (B) Lags up to 10 years. Note that the autocorrelation oscillates with a period of 1 year, reflecting the seasonal cycle. The autocorrelation function has been adjusted for the decrease in overlap at the larger lags. *MatLab* script eda09_03.

9.2 Computing autocorrelation by hand

The autocorrelation at zero lag ($k = 1$) can be calculated by hand by writing down two copies of the time series, one above the other, multiplying adjacent terms, and adding:

$$
\begin{array}{cccccc}
d_1 & d_2 & d_3 & \cdots & d_N & \\
d_1 & d_2 & d_3 & \cdots & d_N & \xrightarrow{\text{yields}} a_1 = d_1^2 + d_2^2 + d_3^2 + \cdots + d_N^2 \\
\times & & & & & \\
d_1^2 & d_2^2 & d_3^2 & d_N^2 & d_N^2 &
\end{array}
\tag{9.8}
$$

Note that a_1 is proportional to the power in the time series. Subsequent elements of a_k are calculated by progressively offsetting one copy of the time series with respect to the other, prior to multiplying and adding (and ignoring the elements with no overlap). The lag Δt ($k = 2$) element is as follows:

$$
\begin{array}{cccccc}
d_1 & d_2 & d_3 & \cdots & d_N & \\
 & d_1 & d_2 & \cdots & d_{N-1} & d_N \xrightarrow{\text{yields}} a_2 = d_2 d_1 + d_3 d_2 + d_4 d_3 + \cdots + d_N d_{N-1} \\
\times & & & & & \\
 & d_2 d_1 & d_3 d_2 & \cdots & d_N d_{N-1} &
\end{array}
\tag{9.9}
$$

and the lag $2\Delta t$ ($k = 3$) element is as follows:

$$
\begin{array}{cccccc}
d_1 & d_2 & d_3 & d_4 & \cdots & d_N \\
 & & d_1 & d_2 & \cdots & d_{N-2} \quad d_{N-1} \quad d_N \xrightarrow{\text{yields}} a_3 = d_1 d_3 + d_2 d_4 + \cdots + d_{N-2} d_N \\
\times & & & & & \\
 & & d_1 d_3 & d_2 d_4 & \cdots & d_{N-2} d_N
\end{array}
\tag{9.10}
$$

9.3 Relationship to convolution and power spectral density

The formula for the autocorrelation is very similar to the formula for the convolution (Equation 7.1):

$$
\begin{array}{cc}
\text{autocorrelation} & \text{convolution} \\[1em]
a_k = \sum_i d_i d_{k+i-1} & \theta_k = \sum_i g_i\, h_{k-i+1} \\[1em]
a(t) = \int_{-\infty}^{+\infty} d(\tau) d(t + \tau)\, d\tau & \theta(t) = \int_{-\infty}^{+\infty} g(\tau) h(t - \tau)\, d\tau \\[1em]
a = d \star d & \theta = g * h
\end{array}
\tag{9.11}
$$

Note that a five pointed star, $\star$, is used to indicate autocorrelation, in the same sense that an asterisk, $*$, is used to indicate convolution. The two formulas are very similar, except that in the case of the convolution, one of the two time series is backward in time, in contrast to the autocorrelation, where both are forward in time. The relationship between the two can be found by transforming the autocorrelation integral to a new variable, $\tau' = -\tau$,

$$a(t) = d(t) \star d(t) = \int_{-\infty}^{+\infty} d(\tau)d(t+\tau) \, d\tau = \int_{-\infty}^{+\infty} d(-\tau')d(t-\tau) \, d\tau'$$

$$= d(-t) * d(t) \qquad\qquad (9.12)$$

Thus, the autocorrelation is the convolution of a time-reversed time series with the original time series.

Two neighboring points on a time series will correlate strongly with each other if the time series varies slowly between them. A time series with an autocorrelation that declines slowly with lag is necessarily richer in low frequency energy than one that declines quickly with lag. This relationship can be explored by computing the Fourier transform of the autocorrelation. The calculation is simplified by recalling that the Fourier transform of a convolution is the product of the transforms. Thus,

$$\tilde{a}(\omega) = \mathcal{F}\{d(-t)\} \, \tilde{d}(\omega) \qquad\qquad (9.13)$$

where $\mathcal{F}\{-d(t)\}$ stands for the Fourier transform of $d(-t)$. We compute it as follows:

$$\mathcal{F}\{d(-t)\} = \int_{-\infty}^{+\infty} d(-t) \exp(i\omega t) \, dt \int_{-\infty}^{+\infty} d(t') \exp(i(-\omega)t') \, dt' = \tilde{d}(-\omega) = \tilde{d}^*(\omega)$$

$$\qquad\qquad (9.14)$$

Here, we have used the transformation of variables, $t' = -t$, together with the fact that, for real time series, $\tilde{d}(\omega)$ and $\tilde{d}(-\omega)$ are complex conjugates of each other. Thus,

$$\tilde{a}(\omega) = \tilde{d}^*(\omega) \, \tilde{d}(\omega) = |\tilde{d}(\omega)|^2 \qquad\qquad (9.15)$$

The Fourier transform of the autocorrelation is proportional to the power spectral density of the time series. As we have seen in Section 6.5, functions that are broad in time have Fourier transforms that are narrow in frequency. Hence, a time series with a broad autocorrelation function has most of its power at low frequencies.

9.4 Cross-correlation

The underlying idea behind the autocorrelation is that pairs of samples drawn from the same time series, and separated by a fixed time lag, τ, are correlated. This idea can be generalized to pairs of samples drawn from two *different* time series. As an example, consider time series of precipitation, **u**, and stream flow, **v**. At times when precipitation is high, we expect stream flow to be high, too. However, the time of peak stream flow will be delayed with respect to the time of maximum precipitation, as water takes time to drain from the land. Thus, the precipitation and stream flow time series will be most correlated when the former is lagged by a specific amount of time with respect to the latter.

We quantify this idea by defining the probability density function, $p(u_i, v_j)$, the joint probability for the i-th sample of time series, **u**, and the j-th sample of time series, **v**. The autocorrelation then generalizes to the *cross-correlation*, c_k (written side-by-size with the convolution, for comparison):

cross-correlation   convolution

$$c_k = \sum_i u_i\, v_{k+i-1} \qquad\qquad \theta_k = \sum_i g_i\, h_{k-i+1}$$

$$c(t) = \int_{-\infty}^{+\infty} u(\tau)v(t+\tau)\,\mathrm{d}\tau \quad \theta(t) = \int_{-\infty}^{+\infty} g(\tau)h(t-\tau)\,\mathrm{d}\tau \tag{9.16}$$

$$c = u \star v \qquad\qquad \theta = g * h$$

Note that the five pointed star is used to indicate cross-correlation, as well as autocorrelation, as the autocorrelation of a time series is its cross-correlation with itself. Here, $u(t_i)$ and $v(t_i)$ are two time series, each of length, N. The cross-correlation is related to the convolution by

$$c(t) = u(t) \star v(t) = u(-t) * v(t) \tag{9.17}$$

In *MatLab*, the cross-correlation is calculated with the function

```
c = xcorr(u,v);
```
(*MatLab* Script eda09_04)

Here, u and v are time series of length, N. The `xcorr()` function returns both negative and positive lags and is of length, 2N−1. The zero-lag element is `c(N)`. Unlike the autocorrelation, the cross-correlation is not symmetric in lag. Instead, the cross-correlation of v and u is the time-reversed version of the cross-correlation of u and v. Mistakes in ordering the arguments of the `xcorr()` function will lead to a result that is backwards in time; that is, if $u(t) \star v(t) = c(t)$, then $v(t) \star u(t) = c(-t)$.

We note here that the Fourier Transform of the cross-correlation is called the *cross-spectral density*:

$$\tilde{c}(\omega) = \tilde{u}^*(\omega)\, \tilde{v}(\omega) \tag{9.18}$$

However, we will put off discussion of its uses until Section 9.9.

9.5   Using the cross-correlation to align time series

The cross-correlation is useful in aligning two time series, one of which is delayed with respect to the other, as its peak occurs at the lag at which the two time series are best correlated, that is, the lag at which they best line up. In *MatLab*,

```
c = xcorr(u,v);
[cmax, icmax] = max(c);
tlag = -Dt * (icmax-N);                                    (MatLab eda09_04)
```

Here, Dt is the sampling interval of the time series and tlag is the time lag between the two time series. The lag is positive when features in v occur at later times than corresponding features in u. This technique is illustrated in Figure 9.5.

We apply this technique to an air quality dataset, in which the objective is to understand the diurnal fluctuations of ozone (O_3). Ozone is a highly reactive gas that

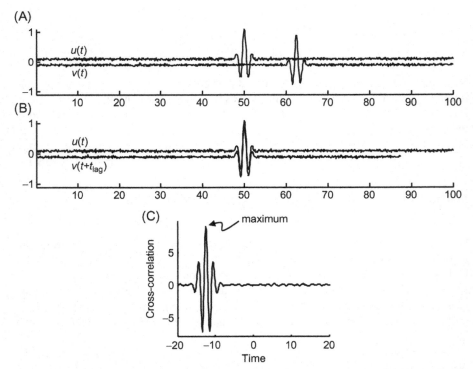

Figure 9.5 (A) Two time series, $u(t)$ and $v(t)$, with similar shapes but one shifted in time with respect to the other. (B) Time series aligned by lag determined through cross-correlation function. (C) Cross-correlation function. *MatLab* script eda09_04.

occurs in small (parts per billion) concentrations in the earth's atmosphere. Ozone in the stratosphere plays an important role in shielding the earth's surface from ultraviolet (UV) light from the sun, for it is a strong UV absorber. But its presence in the troposphere at ground level is problematical. It is a major ingredient in smog and a health risk, increasing susceptibility to respiratory diseases. Tropospheric ozone has several sources, including chemical reactions between oxides of nitrogen and volatile organic compounds in the presence of sunlight and high temperatures. We thus focus on the relationship between ozone concentration and the intensity of sunlight (that is, of solar radiation). Bill Menke provides the following information about the dataset:

> *A colleague gave me a text file of ozone data from the Weather Center at the United States Military Academy at West Point, NY. It contains tropospheric (ground level) ozone data for 15 days starting on August 1, 1993. Also included in the file are solar radiation, air temperature and several other environmental parameters. The original file is named* ozone_orig.txt *and has about a dozen columns of data. I used it to create a file* ozone_nohead.txt *that contains just 4 columns of data, time in days after 00:00 08/01/1993, ozone in parts per billion, solar radiation in W/m², and air temperature in °C.*

The solar radiation and ozone concentration data are shown in Figure 9.6. Both show a pronounced diurnal periodicity, but the peaks in ozone are delayed several hours behind the peaks in sunlight. The lag, determined by cross-correlating the two time series, is 3 h (Figure 9.7). Notice that excellent results are achieved, even though the two dataset do not exactly match.

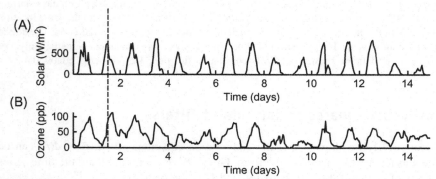

Figure 9.6 (A) Hourly solar radiation data, in W/m², from West Point, NY, for 15 days starting August 1, 1993. (B) Hourly tropospheric ozone data, in parts per billion, from the same location and time period. Note the strong diurnal periodicity in both time series. Peaks in the ozone lag peaks in solar radiation (see vertical line). *MatLab* script eda09_05.

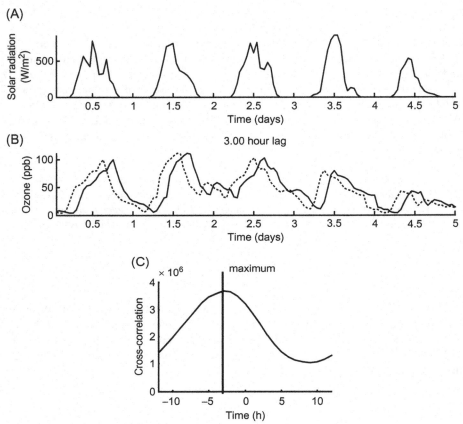

Figure 9.7 (A) Hourly solar radiation data, in W/m², from West Point, NY, for 5 days starting August 1, 1993. (B) Hourly tropospheric ozone data, in parts per billion, from the same location and time period. The solid curve is the original data. Note that it lags solar radiation. The dotted curve is ozone advanced by 3 h, an amount determined by cross-correlation. Note that only 5 of the 15 days of data are shown. (C) Cross-correlation function. *MatLab* script eda09_05.

9.6   Least squares estimation of filters

In Section 7.1, we showed that the convolution equation, $g(t)*m(t) = d(t)$, can be written as a matrix equation of the form, $\mathbf{Gm} = \mathbf{d}$, where $\mathbf{m}$ and $\mathbf{d}$ are the time series versions of $m(t)$ and $d(t)$, respectively, and $\mathbf{G}$ is the matrix:

$$\mathbf{G} = \begin{bmatrix} g_1 & 0 & 0 & \cdots & 0 \\ g_2 & g_1 & 0 & \cdots & 0 \\ g_3 & g_2 & g_1 & \cdots & 0 \\ \cdots & \cdots & \cdots & \cdots & 0 \\ g_N & g_{N-1} & g_{N-2} & \cdots & g_1 \end{bmatrix} \tag{9.19}$$

The least squares solution involves the matrix products, $\mathbf{G}^T\mathbf{G}$ and $\mathbf{G}^T\mathbf{d}$:

$$
\mathbf{G}^T\mathbf{G} =
\begin{bmatrix}
g_1 & g_2 & g_3 & \cdots & g_N \\
0 & g_1 & g_2 & \cdots & g_{N-1} \\
0 & 0 & g_1 & \cdots & g_{N-2} \\
\cdots & \cdots & \cdots & \cdots & 0 \\
0 & 0 & 0 & \cdots & g_1
\end{bmatrix}
\begin{bmatrix}
g_1 & 0 & 0 & \cdots & 0 \\
g_2 & g_1 & 0 & \cdots & 0 \\
g_3 & g_2 & g_1 & \cdots & 0 \\
\cdots & \cdots & \cdots & \cdots & 0 \\
g_N & g_{N-1} & g_{N-2} & \cdots & g_1
\end{bmatrix}
$$

$$
\approx
\begin{bmatrix}
a_1 & a_2 & a_3 & \cdots & a_N \\
a_2 & a_1 & a_2 & \cdots & \cdots \\
a_3 & a_2 & a_1 & \cdots & \cdots \\
\cdots & \cdots & \cdots & \cdots & \cdots \\
a_N & \cdots & \cdots & \cdots & a_1
\end{bmatrix}
\propto \mathbf{A}
\tag{9.20}
$$

$$
\mathbf{G}^T\mathbf{d} =
\begin{bmatrix}
g_1 & g_2 & g_3 & \cdots & g_N \\
0 & g_1 & g_2 & \cdots & g_{N-1} \\
0 & 0 & g_1 & \cdots & g_{N-2} \\
\cdots & \cdots & \cdots & \cdots & 0 \\
0 & 0 & 0 & \cdots & g_1
\end{bmatrix}
\begin{bmatrix}
d_1 \\ d_2 \\ d_3 \\ \cdots \\ d_N
\end{bmatrix}
=
\begin{bmatrix}
c_1 \\ c_2 \\ c_3 \\ \cdots \\ c_N
\end{bmatrix}
= \mathbf{c}
$$

Thus, the elements of $\mathbf{G}^T\mathbf{d}$ are the cross-correlation, $\mathbf{c}$, of the time series $\mathbf{d}$ and $\mathbf{g}$ and the elements of $\mathbf{G}^T\mathbf{G}$ are approximately the autocorrelation matrix, $\mathbf{A}$, of the time series, $\mathbf{g}$. The matrix, $\mathbf{G}^T\mathbf{G}$, is approximately Toeplitz, with elements $[\mathbf{G}^T\mathbf{G}]_{ij} = a_k$, where $k = |i - j| + 1$. This result is only approximate, because on close examination, elements that appear to refer to the same autocorrelation are actually different from one another. Thus, for example, $[\mathbf{G}^T\mathbf{G}]_{11}$ is exactly a_1, but $[\mathbf{G}^T\mathbf{G}]_{22}$ is not, as it is the autocorrelation of the first $N - 1$ elements of $\mathbf{g}$, not of all of $\mathbf{g}$. The difference grows towards the bottom-right of the matrix.

This technique is sometimes used to solve the filter estimation problem, that is, solve $\boldsymbol{\theta} = \mathbf{g} * \mathbf{h}$ for an estimate of $\mathbf{h}$. We examined this problem previously in Section 7.3, using *MatLab* script eda07_03. We provide here an alternate version of this script. A major modification is made to the function called by the biconjugate gradient solver, bicg(). It now uses the autocorrelation to perform the multiplication $\mathbf{F}^T\mathbf{F}\mathbf{v}$. The function was previously called filterfun() but is renamed here to autofun():

```
function y = autofun(v,transp_flag)
global a H;
N = length(v);
% FT F v = GT G v + HT H v
GTGv=zeros(N,1);
for i = [1:N]
    GTGv(i) = [fliplr(a(1:i)'), a(2:N-i+1)'] * v;
    end
Hv = H*v;
HTHv = H'*Hv;
y = GTGv + HTHv;
return
```
(*MatLab* autofun)

The global variable, a, contains the autocorrelation of g. It is computed only once, in the main script. The main script also performs the cross-correlation prior to the call to bicg():

```
clear a H;
global a H;
------
al = xcorr(g);
Na = length(al);
a = al((Na+1)/2: Na);
------
cl = xcorr(qobs2, g);
Nc = length(cl);
c = cl((Nc+1)/2: Nc);
------
% set up F'f = GT qobs + HT h
% GT qobs is c=qobs2*g
HTh = H'* h;
FTf = c + HTh;
% solve
hest3 = bicg(@autofun, FTf, 1e-10, 3*L);    (MatLab eda09_06)
```

The results, shown in Figure 9.8, can be compared to those in Figure 7.7. The method does a good job recovering the two peaks in **h**, but suffers from "edge effects," that is, putting spurious oscillations at the beginning and end of the time series.

9.7 The effect of smoothing on time series

As was discussed in Section 4.5, the smoothing of data is a linear process of the form, $\mathbf{d}^{\text{smooth}} = \mathbf{G}\mathbf{d}^{\text{obs}}$. Smoothing is also a type of filtering, as can be seen by examining the form of data kernel, **G** (Equation 4.16), which is Toeplitz. The columns of **G** define a *smoothing filter*, **s**. Usually, we will want the smoothing to be symmetric, so that the smoothed data, d_i^{smooth}, is calculated through a weighted average of the observed data, d_j^{obs}, both to its left and right of i (where $j > i$ corresponds to the future and $j < i$ corresponds to the past). The filter, s_i, is, therefore, noncausal with coefficients that are symmetric about the present value ($i = 1$). The coefficients need to sum to unity, to preserve the overall amplitude of the data. These traits are exemplified in the three-point smoothing filter (see Equation 4.15):

$$\mathbf{s} = [s_0, s_1, s_2]^{\text{T}} = [\tfrac{1}{4}, \tfrac{1}{2}, \tfrac{1}{4}]^{\text{T}} \tag{9.21}$$

It uses the present (element, i), the past (element, $i - 1$) and the future (element, $i + 1$) of $\mathbf{d}^{\text{obs}}$ to calculate d_i^{smooth}:

smoothed data = weighted average of observed data

or

$$d_i^{\text{smooth}} = \tfrac{1}{4}d_{i-1}^{\text{obs}} + \tfrac{1}{2}d_i^{\text{obs}} + \tfrac{1}{4}d_{i+1}^{\text{obs}} \tag{9.22}$$

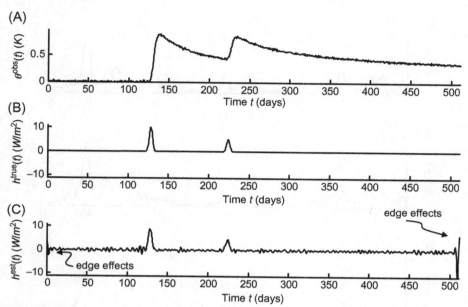

Figure 9.8 (A) Synthetic temperature data, $\theta^{obs}(t)$, constructed from the true temperature plus the same level of random noise as in Figure 7.6. (B) True heat production, $h^{true}(t)$. (C) Estimated heat production, $h^{est}(t)$, calculated with generalized least squares using prior information of smoothness. Note edge effects. *MatLab* script eda09_06.

As long as the filter is of finite length, L, we can view the output as delayed with respect to the input, and the filtering operation itself to be causal:

$$d_i^{\text{smoothed and delayed}} = \tfrac{1}{4}d_i^{obs} + \tfrac{1}{2}d_{i-1}^{obs} + \tfrac{1}{4}d_{i-2}^{obs} \qquad (9.23)$$

In this case, the delay is one sample. In general, the delay is $(L-1)/2$ samples. The length, L, controls the smoothness of the filter, with large Ls corresponding to large degrees of smoothing (Figure 9.9).

The above filter is triangular in shape, as it ramps up linearly to its central value and then linearly ramps down. It weights the central datum more than its neighbors. This is in contrast to the *uniform* filter, which has L constant coefficients, each of amplitude, L^{-1}. It weights all L data equally. Many other shapes are possible, too. An important issue is the best shape for the smoothing filter, **s**.

One way of understanding the choice of the filter is to examine its effect on the autocorrelation function of the smoothed time series. Intuitively, we expect that smoothing broadens the autocorrelation, because it makes the time series vary less between samples. This behavior can be verified by computing the autocorrelation of the smoothed time series

$$\{s(t) * d(t)\} \star \{s(t) * d(t)\} = s(-t) * d(-t) * s(t) * d(t) = \{s(t) \star s(t)\} * \{d(t) \star d(t)\} \qquad (9.24)$$

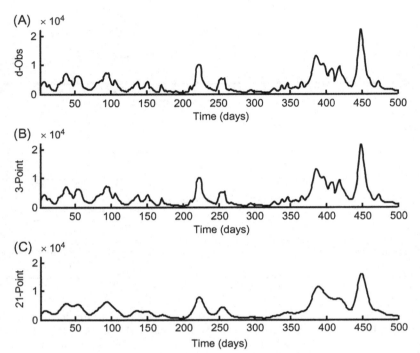

Figure 9.9 Smoothing of Neuse River hydrograph. (A) Observed data. (B) Observed data smoothed with symmetric three-point triangular filter. (C) Observed smoothed data with symmetric 21-point triangular filter. For clarity, only the first 500 days are plotted. *MatLab* script eda09_07.

Thus, the autocorrelation of the smoothed time series is the autocorrelation of the original time series convolved with the autocorrelation of the smoothing filter. The autocorrelation function of the smoothing filter is a broad function. When convolved with the autocorrelation function of the data, it smoothes and broadens it. Filters of different shapes have autocorrelation functions with different degrees of broadness. Each results in the smoothed data having a somewhat differently shaped autocorrelation function.

Another way of understanding the effect of the filter is to examine its effect on the power spectral density of the smoothed time series. The idea behind smoothing is to suppress high frequency fluctuations in the data while leaving the low frequencies unchanged. One measure of the quality of a filter is the evenness by which the suppression occurs. From this perspective, filters that evenly damp out high frequencies are better than filters that suppress them unevenly.

The behavior of the filter can be understood via the convolution theorem (Section 6.11), which states that the Fourier transform of a convolution is the product of the transforms. Thus, the Fourier transform of the smoothed data is just

$$\tilde{d}^{\text{smoothed}}(\omega) = \tilde{s}(\omega)\tilde{d}^{\text{obs}}(\omega) \tag{9.25}$$

That is, the transform of the smoothed data is the transform of the observed data multiplied by the transform of the filter. Thus, the effect of the filter can be understood by examining its amplitude spectral density, $|\tilde{s}(\omega)|$.

The uniform, or *boxcar*, filter with width, T, and amplitude, T^{-1} is the easiest to analyze:

$$\tilde{s}(\omega) = \frac{1}{T}\int_{-T/2}^{T/2} \exp(-i\omega t)\, dt = \frac{2}{T}\int_{0}^{T/2} \cos(\omega t)\, dt = \frac{2}{T}\frac{\sin(\omega t)}{\omega}\bigg|_{0}^{T/2} = \text{sinc}\left(\frac{\omega T}{2\pi}\right) \tag{9.26}$$

Here, we have used the rule, $exp(-i\omega t) = \cos(\omega t) + i\sin(\omega t)$ and the definition, $\text{sinc}(x) = \sin(\pi x)/(\pi x)$. The cosine function is symmetric about the origin, so its integral on the $(-\frac{1}{2}T, +\frac{1}{2}T)$ interval is twice that on $(0, +\frac{1}{2}T)$ interval. The sine function is anti-symmetric, so its integral on the $(-\frac{1}{2}T, 0)$ interval cancels its integral on the $(0, +\frac{1}{2}T)$ interval. While the sinc function (Figure 9.10) declines with frequency, it does so unevenly, with many *sidelobes* along the frequency axis. It does not smoothly damp out high frequencies and so is a poor filter, from this perspective.

A filter based on a Normal curve will have no sidelobes (Figure 9.11), as the Fourier transform of a Normal curve with variance, σ_t^2, in time is a Normal curve with variance, $\sigma_\omega^2 = \sigma_t^{-2}$, in frequency (Equation 6.27). It is a better filter, from the

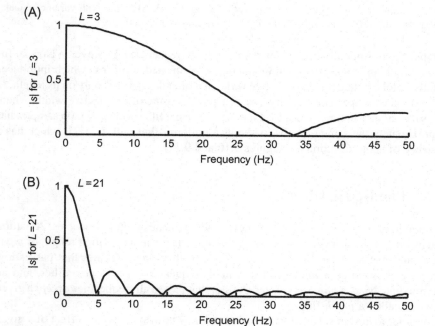

Figure 9.10 Amplitude spectral density of uniform smoothing filters (A) Filter of length, $L = 3$. (B) Filter of length, $L = 21$. *MatLab* script eda09_08.

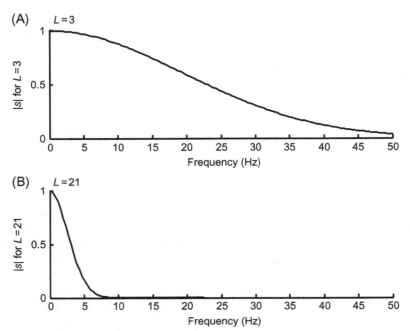

Figure 9.11 Amplitude spectral density of Normal smoothing filters. (A) Filter with variance equal to that of a uniform filter with, length, $L = 3$. (B) Filter with variance equal to that of a uniform filter with length, $L = 21$. *MatLab* script eda09_09.

perspective of smoothly and evenly damping high frequencies. However, a Normal filter is infinite in length and must, in practice, be truncated, a process which introduces small sidelobes. Note that the effective width of a filter depends not only on its length, L, but also on its shape. The quantity, $2\sigma_t$, is a good measure of its effective width, where σ_t^2 is its variance in time. Thus, for example, a Normal filter with $\sigma_t = 6.05$ samples has approximately the same effective width as a uniform filter with $L = 21$, which has a variance of about 6^2 (compare Figures 9.10 and 9.11).

9.8 Band-pass filters

A smoothing filter passes low frequencies and attenuates high frequencies. A natural extension of this idea is a filter that passes frequencies in a specified range, or *pass-band*, and that attenuates frequencies outside of this range. A filter that passes low frequencies is called a *low-pass* filter, high frequencies, a *high-pass* filter, and an intermediate band, a *band-pass* filter. A filter that passes all frequencies *except* a given range is called a *notch* filter.

In order to design such filters, we need to know how to assess the effect of a given set of filter coefficients on the power spectral density of the filter. We start with the definition of an Infinite Impulse Response (IIR) filter (Equation 7.21), $\mathbf{f} = \mathbf{v}^{\text{inv}} * \mathbf{u}$, where $\mathbf{u}$ and $\mathbf{v}$ are short filters of lengths, N_u and N_v, respectively, and $\mathbf{v}^{inv}$ is the inverse filter of $\mathbf{v}$. The z-transform of the filter, $\mathbf{f}$, is

$$\mathbf{f} = \mathbf{v}^{inv} * \mathbf{u} \rightarrow f(z) = \frac{u(z)}{v(z)} = c\frac{\prod_{j=1}^{N_u-1}(z - z_j^u)}{\prod_{k=1}^{N_v-1}(z - z_k^v)} \tag{9.27}$$

Here, z_j^u and z_k^v are the roots of $u(z)$ and $v(z)$, respectively and c is a normalization constant. As our goal involves spectral properties, we need to understand the connection between the z-transform and the Fourier transform. The Discrete Fourier Transform is defined as

$$\tilde{f}_k = \sum_{n=1}^{N} f_k \exp(-i\omega_k t_n) = \sum_{n=1}^{N} f_k \exp(-i(k-1)\Delta\omega(n-1)\Delta t) \tag{9.28}$$

as $\omega_k = (k-1)\Delta\omega$ and $t_n = (n-1)\Delta t$. Note that the factor of $(n-1)$ within the exponential can be interpreted as raising the exponential to the $(n-1)$ power. Thus,

$$\tilde{f}_k = \sum_{n=1}^{N} f_k z^{n-1} \quad \text{with } z = \exp(-i(k-1)\Delta\omega\Delta t) = \exp\left(-\frac{2\pi i(k-1)}{N}\right) \tag{9.29}$$

Here, we have used the relationship, $\Delta\omega\Delta t = 2\pi/N$. Thus, the Fourier transform is just the z-transform evaluated at a specific set of z's. There are N of these z's and they are equally spaced on the unit circle (that is, the circle $|z|^2 = 1$ in the complex z-plane, Figure 9.12). A point on the unit circle can be represented as, $z = \exp(-i\theta)$,

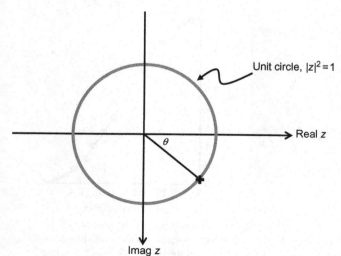

Unit circle, $|z|^2 = 1$

Real z

θ

Imag z

Figure 9.12 Complex z-plane. showing the unit circle, $|z|^2 = 1$. A point (+ sign) on the unit circle makes an angle, θ, with respect to the positive z-axis. It corresponds to a frequency, $\omega = \theta/\Delta t$, in the Fourier transform.

where θ is angle with respect to the real axis. Frequency, ω, is proportional to angle, θ, via $\theta = \omega \Delta t = (k-1)\Delta\omega\Delta t = 2\pi(k-1)/N$. As the points in a Fourier transform are evenly spaced in frequency, they are evenly spaced in angle around the unit circle. Zero frequency corresponds to $\theta = 0$ and the Nyquist frequency corresponds to $\theta = \pi$; that is, $180°$).

Now we are in a position to analyze the effect of the filters, $\mathbf{u}$ and $\mathbf{v}$ on the spectrum of the composite filter, $\mathbf{f} = \mathbf{v}^{inv}*\mathbf{u}$. The polynomial, $u(z)$, has $N_u - 1$ roots (or "*zeros*"), each of which creates a region of low amplitude in a patch of the z-plane near that zero. If the unit circle intersects this patch, then frequencies on that segment of the unit circle are attenuated. Thus, for example, zeros near $\theta = 0$ attenuate low frequencies (Figure 9.13A) and zeros near $\theta = \pi$ (the Nyquist frequency) attenuate high frequencies (Figure 9.13B).

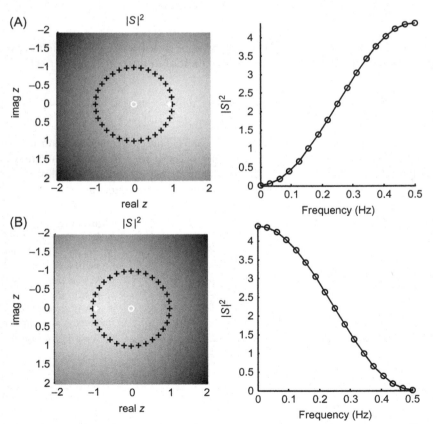

Figure 9.13 (A) Complex z-plane representation of the high-pass filter, $\mathbf{u} = [1, -1.1]^{\mathrm{T}}$ along with power spectral density of the filter. (B) Corresponding plots for the low-pass filter, $\mathbf{u} = [1, 1.1]^{\mathrm{T}}$. Origin (circle), Fourier transform points on the unit circle (black +), and zero (white *) are shown. *MatLab* script eda09_10 and eda09_11.

The polynomial, $v(z)$, has $N_u - 1$ roots, so that its reciprocal, $1/v(z)$, has $N_u - 1$ singularities (or *poles*), each of which creates a region of high amplitude in a patch of the z-plane near that pole. If the unit circle intersects this patch, then frequencies on that segment of the unit circle are amplified (Figure 9.14A). Thus, for example, poles near $\theta = 0$ amplify low frequencies and zeros near $\theta = \pi$ (the Nyquist frequency) amplify high frequencies. As was discussed in Section 7.6, the poles must lie outside the unit circle for the inverse filter, $\mathbf{v}^{inv}$, to exist. In order for the filter to be real, the poles and zeros either must be on the real z-axis or occur in complex-conjugate pairs (that is, at angles, θ and $-\theta$).

Filter design then becomes a problem of cleverly placing poles and zeros in the complex z-plane to achieve whatever attenuation or amplification of frequencies is desired. Often, just a few poles and zeros are needed to achieve the desired effect. For instance, two poles nearly collocated with two zeros suffice to create a notch filter

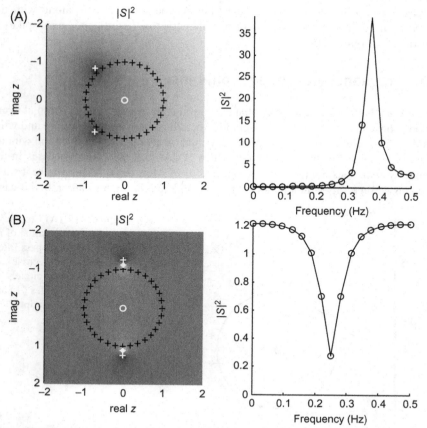

Figure 9.14 (A) Complex z-plane representation of a band-pass filter with $\mathbf{u} = [1, 0.60 + 0.66i]^T * [1, 0.60 - 0.66i]^T$ along with the power spectral density of the filter. (B) Corresponding plots for the notch filter, $\mathbf{u} = [1, 0.9i]^T * [1, - 0.9i]^T$ and $\mathbf{v} = [1, 0.8i]^T * [1, - 0.8i]^T$. Origin (circle), Fourier transform points on the unit circle (black +), zeros (white *), and poles (white +) are shown. *MatLab* script eda09_12 and eda09_13.

(Figure 9.14B), that is, one that attenuates just a narrow range of frequencies. With just a handful of poles and zeros—corresponding to filters **u** and **v** with just a handful of coefficients—one can create extremely effective and efficient filters.

As an example, we provide a *MatLab* function for a *Chebyshev* band-pass filter, chebyshevfilt.m. It passes frequencies in a specific frequency interval and attenuates frequencies outside that interval. It uses **u** and **v** each of length 5, corresponding to four zeros and four poles. The zeros are paired up, two at $\theta = 0$ and two at $\theta = \pi$, so that frequencies near zero and near the Nyquist frequency are strongly attenuated. The two conjugate pairs of poles are near θs corresponding to the ends of the pass-band interval (Figure 9.15). The function is called as follows:

```
[dout, u, v] = chebyshevfilt(din, Dt, flow, fhigh);
```
                                                        (*MatLab* eda09_14)

Here, din is the input time series, Dt is the sampling interval and flow, and fhigh the pass-band. The function returns the filtered time series, dout, along with the filters, u and v. The input response of the filter (that is, its influence on a spike) is illustrated in Figure 9.16.

9.9   Frequency-dependent coherence

Time series that track one another, that is, exhibit *coherence*, need not do so at every period. Consider, for instance, a geographic location where air temperature and wind speed both have annual cycles. Summer days are, on average, both hotter and windier than winter days. But this correlation, which is due to large scale processes in the climate system, does not hold for shorter periods of a few days. A summer heat wave is not, on average, any windier than in times of moderate summer weather. In this case,

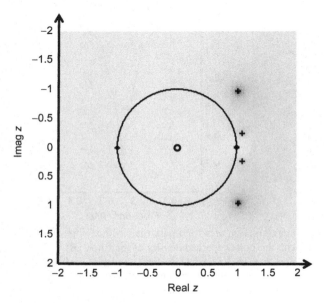

Figure 9.15 (A) Complex z-plane representation of a Chebychev band-pass filter. The origin (small circle), unit circle (large circle), zeros (*), and poles (+) are shown. *MatLab* script eda09_14.

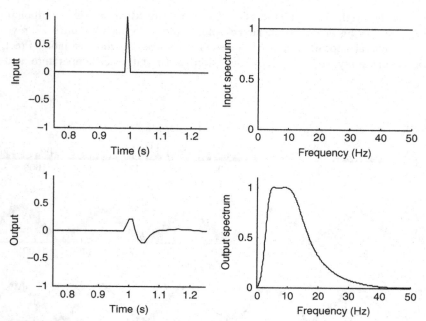

Figure 9.16 Impulse response and spectrum of a Chevyshev band-pass filter, for a 5-10 Hz pass-band. *MatLab* script eda09_14.

temperature and wind are correlated at long periods, but not at short ones. In another example, plant growth in a given biome might correlate with precipitation over periods of a few weeks, but this does not necessarily imply that plant growth is faster in winter than in summer, even when winter tends to be wetter, on average, than summer. In this case, growth and precipitation are correlated at short periods, but not at long ones.

We introduce here a new dataset that illustrates this behavior, water quality data from the Reynolds Channel, part of the Middle Bay estuary on the south shore of Long Island, NY. Bill Menke, who provided the data, says the following about it:

*I downloaded this Reynolds Channel Water Quality dataset from the US Geological Survey's National Water Information System. It consists of daily average values of a variety of environmental parameters for a period of about five years, starting on January 1, 2006. The original data was in one long text file, but I broke it into two pieces, the header (*reynolds_header.txt*) and the data (*reynolds_data. txt*). The data file has very many columns, and has time in a year-month-day format. In order to make the data more manageable, I created another file, *reynolds_ uninterpolated.txt*, that has time reformatted into days starting on January 1, 2006 and that retains only six of the original data columns: precipitation in inches, air temperature in °C, water temperature in°C, salinity in practical salinity units, turbidity in formazin nephelometric units and chlorophyll in micrograms per liter. Not every parameter had a data value for every time, so I set the missing values to the placeholder, −999. Finally I created a file, *reynolds_interpolated. txt*, in which missing data are filled in using linear interpolation. The MatLab script that I used is called *interpolate_reynolds.m*.*

Note that the original data had missing data that were filled in using interpolation. We will discuss this process in the next chapter. A plot of the data (Figure 9.17) reveals that the general appearance of the different data types is quite variable. Precipitation is very spiky, reflecting individual storms. Air and water temperature, and to

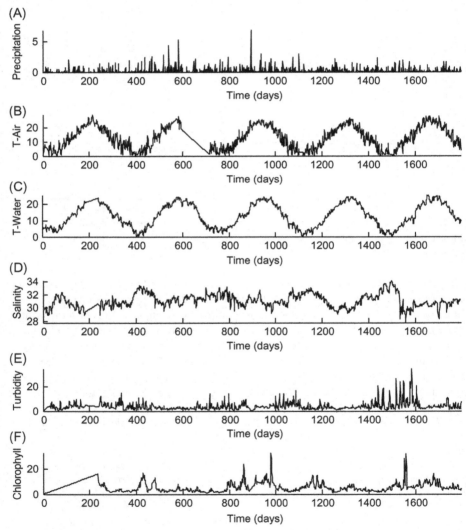

Figure 9.17 Daily water quality measurements from Reynolds Channel (New York) for several years starting January 1, 2006. Six environmental parameters are shown: (A) precipitation in inches; (B) air temperature in °C; (C) water temperature in °C; (D) salinity in practical salinity units; (E) turbidity; and (F) chlorophyll in micrograms per liter. While these data have been linearly interpolated to fill in gaps, an alternative (and maybe better) strategy would be to leave the gaps and compare only portions of pairs of timeseries with no gaps. *MatLab* script eda09_15.

a lesser degree, salinity, are dominated by the annual cycle. Moreover, turbidity (cloudiness of the water) and chlorophyll (a proxy for the concentration of algae and other photosynthetic plankton) have both long period oscillations and short intense spikes.

We can look for correlations at different periods by band-pass filtering the data using different pass bands, for example periods of about 1 year and periods of about 5 days (Figure 9.18). All six time series appear to have some coherence at periods of 1 year, with air and water temperature tracking each other the best and turbidity tracking nothing very well. The situation at periods of about 5 days is more complicated. The most coherent pair seems to be salinity and precipitation, which are anti-correlated (as one might expect, as rain dilutes the salt in the bay). Air and water temperature do not track each other nearly as well in this period band than at periods of 1 year, but they do seem to show some coherence. Chlorophyll does not seem correlated with any of the other parameters at these shorter periods.

Our goal is to quantify the degree of similarity between two time series, $u(t)$ and $v(t)$, at frequencies near a specified frequency, ω_0. We start by band-pass filtering the time series to produce filtered versions, $f(t) * u(t)$ and $f(t) * v(t)$. The band-pass filter, $f(t, \omega_0, \Delta\omega)$, is chosen to have a center frequency, ω_0, and a bandwidth, $2\Delta\omega$ (meaning that it passes frequencies in the range $\omega_0 \pm \Delta\omega$). We now compare these two filtered time series by cross-correlating them:

$$
\begin{aligned}
c(t, \omega_0, \Delta\omega) &= \{f(t, \omega_0, \Delta\omega) * u(t)\} \star \{f(t, \omega_0, \Delta\omega) * v(t)\} \\
&= f(-t, \omega_0, \Delta\omega) * f(t, \omega_0, \Delta\omega) * u(-t) * v(t)
\end{aligned}
\tag{9.30}
$$

If the two time series are similar in shape (and if they are aligned in time), then the zero-lag value of the cross-correlation, $c(t = 0, \omega_0, \Delta\omega)$ will have a large absolute value. Its value will be large and positive when the time series are nearly the same, and large and negative if they have nearly the same shape but are flipped in sign with respect to each other. It will be near-zero when the two time series are dissimilar.

Two undesirable aspects of Equation (9.30) are that a different band-pass filtered version of the time series is required for every frequency at which we want to evaluate similarity and the whole cross-correlation is calculated, whereas only its zero-lag value is needed. As we show below, these time-consuming calculations are unnecessary. We can substantially improve on Equation (9.30) by utilizing the fact that the value of a function, $c(t)$, at time, $t = 0$, is proportional to the integral of its Fourier transform over frequency:

$$
c(t = 0) = \frac{1}{2\pi} \int_{-\infty}^{+\infty} \tilde{c}(\omega) \exp(0) \, d\omega = \frac{1}{2\pi} \int_{-\infty}^{+\infty} \tilde{c}(\omega) \, d\omega
\tag{9.31}
$$

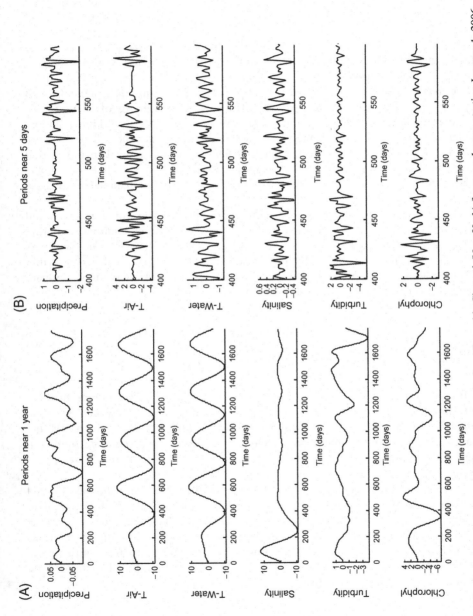

Figure 9.18 Band-pass filtered water quality measurements from Reynolds Channel (New York) for several years starting January 1, 2006. (A) Periods near 1 year; and (B) periods near 5 days. *MatLab* script eda09_16.

Applying this relationship to the cross-correlation at zero lag, $c(t = 0)$, and using the rule that the Fourier transform of a convolution is the product of the transforms, yields

$$c(t = 0, \omega_0, \Delta\omega) = \frac{1}{2\pi} \int_{-\infty}^{+\infty} \tilde{f}^*(\omega, \omega_0, \Delta\omega) \, \tilde{f}(\omega, \omega_0, \Delta\omega) \, \tilde{u}^*(\omega) \tilde{v}(\omega) \, d\omega$$

$$\approx \frac{1}{2\pi} \int_{-\omega_0-\Delta\omega}^{-\omega_0+\Delta\omega} \tilde{u}^*(\omega) \tilde{v}(\omega) \, d\omega + \frac{1}{2\pi} \int_{+\omega_0-\Delta\omega}^{+\omega_0+\Delta\omega} \tilde{u}^*(\omega) \tilde{v}(\omega) \, d\omega$$

$$= \frac{1}{\pi} \int_{\omega_0-\Delta\omega}^{\omega_0+\Delta\omega} \text{Re}\{\tilde{u}^*(\omega) \tilde{v}(\omega)\} \, d\omega = \frac{2\Delta\omega}{\pi} \overline{\text{Re}\{\tilde{u}^*(\omega_0) \tilde{v}(\omega_0)\}} \qquad (9.32)$$

Note that this formula involves the cross-spectral density, $\tilde{u}^*(\omega)\tilde{v}(\omega)$. Here, we assume that the band-pass filter can be approximated by two boxcar functions, one centered at $+\omega_0$ and the other at $-\omega_0$, so the integration limits, $\pm\infty$, can be replaced with integration over the positive and negative pass-bands. The cross-correlation is a real function, so the real part of its Fourier transform is symmetric in frequency and the imaginary part is anti-symmetric. Thus, only the real part of the integrand contributes. Except for a scaling factor of $1/(2\Delta\omega)$, the integral is just the average value of the integrand within the pass band, so we replace it with the average, defined as

$$\overline{\tilde{z}(\omega_0)} = \frac{1}{2\Delta\omega} \int_{\omega_0-\Delta\omega}^{\omega_0+\Delta\omega} \tilde{z}(\omega) \, d\omega \qquad (9.33)$$

The zero-lag cross-correlation can be normalized into a quantity that varies between ± 1 by dividing each time series by the square root of its power. Power is just the autocorrelation, $a(t)$, of the time series at zero lag, and the autocorrelation is just the cross-correlation of a time series with itself, so power satisfies an equation similar to the one above:

$$P_u = a_u(t = 0, \omega_0, \Delta\omega) = \frac{2\Delta\omega}{\pi} \overline{\tilde{u}^*(\omega_0)\tilde{u}(\omega_0)} = \frac{2\Delta\omega}{\pi} \overline{|\tilde{u}(\omega_0)|^2}$$

$$P_v = a_v(t = 0, \omega_0, \Delta\omega) = \frac{2\Delta\omega}{\pi} \overline{\tilde{v}^*(\omega_0)\tilde{v}(\omega_0)} = \frac{2\Delta\omega}{\pi} \overline{|\tilde{v}(\omega_0)|^2} \qquad (9.34)$$

Here, P_u and P_v, are the power in the band-passed versions of $u(t)$ and $v(t)$, respectively. Note that we can omit taking the real parts, for they are purely real. The quantity

$$\mathcal{C} = \frac{c(t = 0, \omega_0, \Delta\omega)}{P_u^{1/2} P_v^{1/2}} = \frac{\overline{\text{Re}\{\tilde{u}^*(\omega_0) \tilde{v}(\omega_0)\}}}{\left\{ \overline{|\tilde{u}(\omega_0)|^2} \, \overline{|\tilde{v}(\omega_0)|^2} \right\}^{1/2}} \qquad (9.35)$$

which varies between $+1$ and -1, is a measure of the degree of similarity of the time series, $u(t)$ and $v(t)$. However, the quantity

$$C_{uv}^2(\omega_0, \Delta\omega) = \frac{\left| \overline{\tilde{u}^*(\omega_0) \tilde{v}(\omega_0)} \right|^2}{\overline{|\tilde{u}(\omega_0)|^2} \, \overline{|\tilde{v}(\omega_0)|^2}} \qquad (9.36)$$

is more commonly encountered in the literature. It is called the *coherence* of time series $u(t)$ and $v(t)$. It is nearly the square of $\mathcal{C}$, except that it omits the taking of the real part, so that it does not have exactly the interpretation of the normalized zero-lag cross-correlation of the band-passed time series. It does, however, behave similarly (see Note 9.1). It varies between zero and unity, being small when the time series are very dissimilar and large when they are nearly identical. These formulas are demonstrated in *MatLab* script eda09_17.

The averaging in Equation (9.36) is over neighboring frequencies in a densely-sampled frequency series that is the result of taking the Fourier transform of long time series. However, a very similar result can be obtained by subdividing the long time series into shorter segments, taking the Fourier transform of each, and averaging the results. This correspondence follows from the frequency-spacing of the Fourier transform depending upon on the length of the time series. When a long time series is subdivided into K shorter segments of equal length, the number of frequencies decreases by a factor of K and the number of estimates at a given frequency increases by the same factor. Averaging the K estimates, all for the same frequency, from the short time series gives a result similar to averaging K adjacent values from the long one. This result, due to Welch (1967) is the basis for the MatLab coherence function mscohere() (see *MatLab* script eda09_18 for an example).

We return now to the Reynolds Channel water quality dataset, and compute the coherence of each pair of time series (several of which are shown in Figure 9.19).

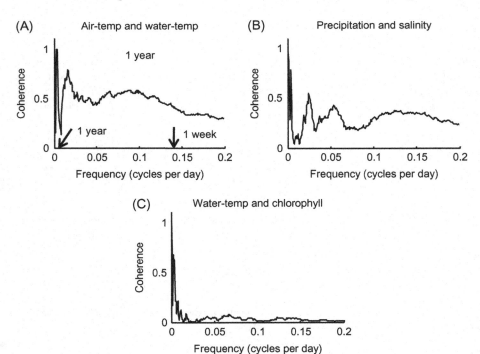

Figure 9.19 Coherence of water quality measurements from Reynolds Channel (New York). (A) Air temperature and water temperature; (B) precipitation and salinity; and (C) water temperature and chlorophyll. *MatLab* script eda09_18.

Air temperature and water temperature are the most highly coherent time series. They are coherent both at low frequencies (periods of a year or more) and high frequencies (periods of a few days). Precipitation and salinity are also coherent over most of frequency range, although less strongly than air and water temperature. Chlorophyll correlates with the other time series only at the longest periods, indicating that, while it is sensitive to the seasonal cycle, it is not sensitive to short time scale fluctuations in these parameters.

9.10   Windowing before computing Fourier transforms

When computing the power spectral density of continuous time series, we are faced with a decision of how long a segment of the time series to use. Longer is better, of course, both because a long segment is more likely to have properties representative of the time series as a whole, and because long segments provide greater resolution (recall that frequency sampling, $\Delta\omega$, scales with N^{-1}). Actually, as data are often scarce, more often the question is how to make do with a *short* segment.

A short segment of a time series can be created by multiplying an indefinitely long time series, $d(t)$, by a *window function*, $W(t)$; that is, a function that is zero everywhere outside the segment. The simplest window function is the *boxcar* function, which is unity within the interval and zero outside it. The key question is what effect windowing has on the Fourier transform of a time series; that is, how the Fourier transform of $W(t)d(t)$ differs from the Fourier transform of $d(t)$. This question can be analyzed using the convolution theorem. As discussed in Section 6.11, the convolution of two time series has a Fourier transform that is the product of the two individual Fourier transforms. But time and frequency play symmetric roles in the Fourier transform. Thus, the product of two time series has a Fourier transform that is the convolution of the two individual transforms. Windowing has the effect of convolving the Fourier transform of the time series with the Fourier transform of the window function.

From this perspective, a window function with a spiky Fourier transform is the best, because convolving a function with a spike leaves the function unchanged. As we have seen in Section 9.7, the Fourier transform of a boxcar is a sinc function. It has a central spike, which is good, but it also has sidelobes, which are bad. The sidelobes create peaks in the spectrum of the windowed time series, $W(t)d(t)$, that are not present in the spectrum of the original time series, $d(t)$ (Figure 9.20). These *artifacts* can easily be mistaken for real periodicities in the data.

The solution is a better window function, one that does not have a Fourier transform with such strong sidelobes. It must be zero outside the interval, but we have complete flexibility in choosing its shape within the interval. Many such functions (or *tapers*) have been proposed. A popular one is the *Hamming* window function (or Hamming taper)

$$W(t_k) = 0.54 - 0.46 \cos\left(\frac{2\pi(k-1)}{N_w - 1}\right)$$

(9.37)

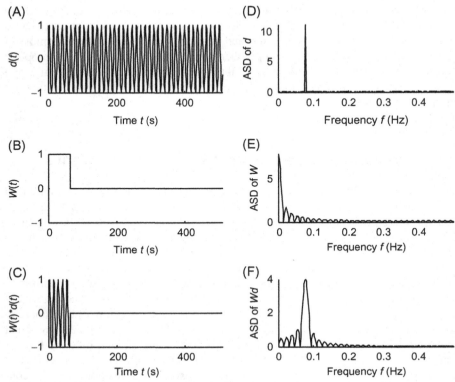

Figure 9.20 Effect of windowing a sinusoidal time series, $d(t) = \cos(\omega_0 t)$, with a boxcar window function, $W(t)$, prior to computing its amplitude spectral density. (A) Time series, $d(t)$. (B) Boxcar windowing function, $W(t)$. (C) Product, $d(t)W(t)$. (D-F) Amplitude spectral density of $d(t)$, $W(t)$ and $d(t)W(t)$. *MatLab* script eda09_19.

where N_w is the length of the window. Its Fourier transform (Figure 9.21) has significantly lower-amplitude sidelobes than the boxcar window function. Its central spike is wider, however (compare Figure 9.20E with Figure 9.21E), implying that it smoothes the spectrum of $d(t)$ more than does a boxcar. Smoothing is bad in this context, because it blurs features in the spectrum that might be important. Two narrow and closely-spaced spectral peaks, for instance, will appear as a single broad peak. Unfortunately, the width of the central peak and the amplitude of sidelobes trade off in window functions. The end result is always a compromise between the two.

Notwithstanding this fact, one can nevertheless do substantially better than the Hamming taper, as we will see in the next section.

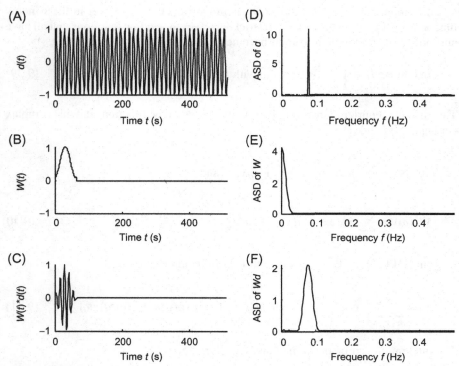

Figure 9.21 Effect of windowing a sinusoidal time series, $d(t) = \cos(\omega_0 t)$, with a Hamming window function, $W(t)$, prior to computing its amplitude spectral density. (A) Time series, $d(t)$. (B) Hamming windowing function, $W(t)$. (C) Product, $d(t)W(t)$. (D-F) Amplitude spectral density of $d(t)$, $W(t)$, and $d(t)W(t)$. *MatLab* script eda09_20.

9.11 Optimal window functions

A good window function is one that has a spiky power spectral density. It should have large amplitudes in a narrow range of frequencies, say $\pm\omega_0$, straddling the origin and have small amplitudes at higher frequencies. One way to quantify spikiness is through the ratio

$$R = \frac{\int_{-\omega_0}^{+\omega_0} \left| \tilde{W}(\omega) \right|^2 \, d\omega}{\int_{-\omega_{ny}}^{+\omega_{ny}} \left| \tilde{W}(\omega) \right|^2 \, d\omega} \tag{9.38}$$

Here, $\tilde{W}(\omega)$ is the Fourier transform of the window function and ω_{ny} is the Nyquist frequency. From this point of view, the best window function is the one that maximizes the ratio, R.

The denominator of Equation (9.38) is proportional to the power in the window function (see Equation 6.41). If we restrict ourselves to window functions that all have unit power, then the maximization becomes as follows:

$$\text{maximize } F = \int_{-\omega_0}^{+\omega_0} \left| \tilde{W}(\omega) \right|^2 d\omega \text{ with the constraint } \int \left| W(t) \right|^2 dt = 1 \qquad (9.39)$$

The discrete Fourier transform, $\tilde{W}(\omega)$, of the window function and its complex conjugate, $\tilde{W}^*(\omega)$, are

$$\tilde{W}(\omega) = \sum_{n=1}^{N} w_n \exp(-i(n-1)\omega \Delta t) \quad \text{and}$$

$$\tilde{W}^*(\omega) = \sum_{m=1}^{N} w_m \exp(+i(m-1)\omega \Delta t) \qquad (9.40)$$

Inserting $\left| \tilde{W}(\omega) \right|^2 = \tilde{W}^*(\omega)\tilde{W}(\omega)$ into F in Equation (9.39) yields

$$F = \sum_{m=1}^{N}\sum_{n=1}^{N} w_n w_m M_{nm} \quad \text{with } M_{nm} = \int_{-\omega_0}^{+\omega_0} \exp(i(m-n)\omega \Delta t) \, d\omega \qquad (9.41)$$

The integration can be performed analytically:

$$M_{nm} = \int_{-\omega_0}^{+\omega_0} \exp(i(m-n)\omega \Delta t) \, d\omega = 2\int_{0}^{+\omega_0} \cos((m-n)\omega \Delta t) \, d\omega$$

$$= \frac{2\sin((m-n)\omega_0 \Delta t)}{(m-n)\Delta t} = 2\omega_0 \, \text{sinc}((m-n)\omega_0 \Delta t/\pi) \qquad (9.42)$$

Note that $\mathbf{M}$ is a symmetric $N \times N$ matrix. The window function, $\mathbf{w}$, satisfies

$$\text{maximize } F = \sum_{m=1}^{N}\sum_{n=1}^{N} w_n w_m M_{nm} \text{ with the constraint } C = \sum_{n=1}^{N} w_n^2 - 1 = 0$$

or, equivalently

$$\text{maximize } F = \mathbf{w}^T\mathbf{M}\mathbf{w} \quad \text{with the constraint } C = \mathbf{w}^T\mathbf{w} - 1 = 0 \qquad (9.43)$$

The *Method of Lagrange Multipliers* (see Note 9.2) says that maximizing a function, F, with a constraint, $C = 0$, is equivalent to maximizing $F - \lambda C$ without a constraint, where λ is a new parameter that needs to be determined. Differentiating $\mathbf{w}^T\mathbf{M}\mathbf{w} - \lambda(\mathbf{w}^T\mathbf{w} - 1)$ with respect to $\mathbf{w}$ and setting the result to zero leads to the equation

$$\mathbf{M}\mathbf{w} = \lambda\mathbf{w} \qquad (9.44)$$

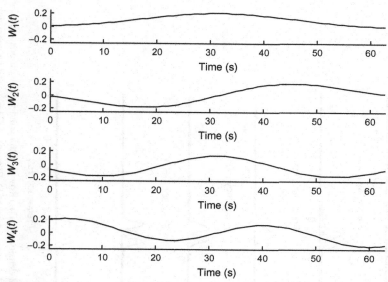

Figure 9.22 First four window functions, $W_i(t)$, for the case $N = 64$, $\omega_0 = 2\Delta f$. *MatLab* script eda09_21.

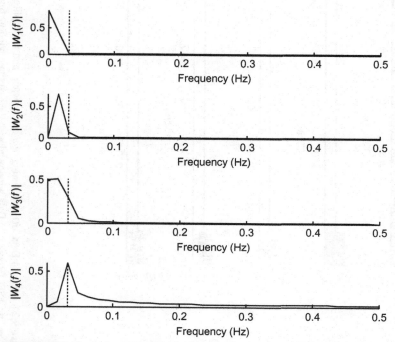

Figure 9.23 Amplitude spectral density, $|W_i(f)|$, of the first four window functions, for the case $N = 64$, $\omega_0 = 2\Delta\omega$. The dotted vertical line marks frequency, $2\Delta f$. *MatLab* script eda09_21.

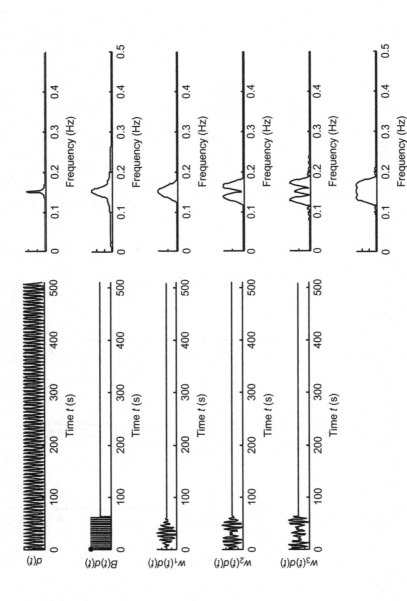

Figure 9.24 (Row 1) Data, $d(t)$, consisting of a cosine wave, and its amplitude spectral density (ASD). (Row 2) Data windowed with boxcar, $B(t)$, and its ASD. (Rows 3–5) Data windowed with the first three window functions, and corresponding ASD. (Row 6) ASD obtained by averaging the results of the first three window functions. *MatLab* script eda09_21.

This is just the algebraic eigenvalue problem (see Equation 8.6). Recall that this equation has N solutions, each with an eigenvalue, λ_i, and a corresponding eigenvector, $\mathbf{w}^{(i)}$. The eigenvalues, λ_i, satisfy $\lambda_i = \mathbf{w}^{(i)T}\mathbf{M}\mathbf{w}^{(i)}$, as can be seen by pre-multiplying Equation (9.44) by $\mathbf{w}^T$ and recalling that the eigenvectors have unit length, $\mathbf{w}^T\mathbf{w} = 1$. But $\mathbf{w}^T\mathbf{M}\mathbf{w}$ is the quantity, F, being maximized in Equation (9.43). Thus, the eigenvalues are a direct measure of the spikiness of the window functions. The best window function is equal to the eigenvector with the largest eigenvalue.

We illustrate the case of a 64-point window function with a width of $\omega_0 = 2\Delta\omega$ (Figures 9.22, 9.23 and 9.24). The six largest eigenvalues are 6.28, 6.27, 6.03, 4.54, 1.72, and 0.2. The first three eigenvalues are approximately equal in size, indicating that three different tapers come close to achieving the design goal of maximizing the spectral power in the $\pm\omega_0$ frequency range. The first of these, $W_1(t)$, is similar in shape to a Normal curve, with high amplitudes in the center of the interval that taper off towards its ends. One possibility is to consider $W_1(t)$ the best window function and to use it to compute power spectral density.

However, $W_2(t)$ and $W_3(t)$ are potentially useful, because they weight the data differently than does $W_1(t)$. In particular, they leave intact data near the ends of the interval that $W_1(t)$ strongly attenuates. Instead of using just the single window, $W_1(t)$, in the computation of power spectral density, alternatively we could use several to compute several different estimates of power spectral density, and then average the results (Figure 9.24). This idea was put forward by Thomson (1982) and is called the *multitaper* method.

Problems

9.1 The ozone dataset also contains atmospheric temperature, a parameter, which like ozone, might be expected to lag solar radiation. Modify the eda09_05 script to estimate its lag. Does it have the same lag as ozone?

9.2 Suppose that the time series $\mathbf{f}$ and $\mathbf{h}$ are related by the convolution with the filter, $\mathbf{s}$; that is, $\mathbf{f} = \mathbf{s}*\mathbf{h}$. As the autocorrelation represents the covariance of a probability density function, the autocorrelation of $\mathbf{f}$ should be related to the autocorrelation of $\mathbf{h}$ by the normal rules of error propagation. Verify that this is the case by writing the convolution in matrix form, $\mathbf{f} = \mathbf{Sh}$, and using the rule $\mathbf{C}_f = \mathbf{S}\mathbf{C}_h\mathbf{S}^T$, where the $\mathbf{C}$s are covariance matrices.

9.3 Modify *MatLab* script eda09_03 to estimate the autocorrelation of the Reynolds Channel chlorophyll dataset. How quickly does the autocorrelation fall off with lag (for small lags)?

9.4 Taper the Neuse River Hydrograph data using a Hamming window function before computing its power spectral density. Compare your results to the untapered results, commenting on whether any conclusions about periodicities might change. (Note: before tapering, you should subtract the mean from the time series, so that it oscillates around zero).

9.5 Band-pass filter the Black Rock Forest temperature dataset to highlight diurnal variations of temperature. Provide a new answer to Question 2.3 that uses these results.

References

Thomson, D.J., 1982. Spectrum estimation and harmonic analysis. Proc. IEEE 70, 1055–1096.
Welch, P.D., 1967. The Use of Fast Fourier Transform for the Estimation of Power 15 Spectra: A Method Based on Time Averaging Over Short, Modified Periodograms, IEEE 16 Transactions on Audio and Electroacoustics, Vol. AU-15, 70–73.

10 Filling in missing data

10.1   Interpolation requires prior information

Two situations in which data points need to be *filled in* are common:

> *The data are collected at irregularly spaced times or positions, but the data analysis method requires that they be evenly spaced. Spectral analysis is one such example, because it requires time to increase with constant increments, Δt.*
>
> *Two sets of data need to be compared with each other, but they have not been observed at common values of time or position. The making of scatter plots is one such example, because the pairs of data that are plotted need to have been made at the same time or position.*

In both cases, the times or positions at which the data have been observed are inconvenient. A solution to this dilemma is to *interpolate* the data; that is, to use the available data to estimate the data values at a more useful set of times or positions. We encountered the interpolation problem previously, in Chapter 5 (see Figure 10.1). Prior information about how the data behaved *between* the data points was a key factor in achieving the result.

The generalized least-squares methodology that we developed in Chapter 5 utilized both observations and the prior information to create an estimate of the data at all times or positions. Observations and prior information are both treated probabilistically. The solution is dependent on the observations and prior information, but does not satisfy either, anywhere. This is not seen as a problem, because both observations and prior information are viewed as subject to uncertainty. There is really no need to satisfy either of them exactly; we only need to find a solution for which the level of error is acceptable.

Environmental Data Analysis with MATLAB®. http://dx.doi.org/10.1016/B978-0-12-804488-9.00010-0

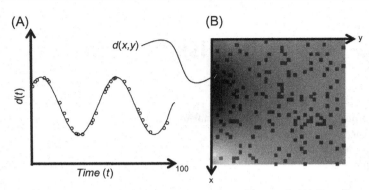

Figure 10.1 Examples of filling in data gaps drawn from Chapter 5. (A) One-dimensional data, $d(t)$; (B) Two-dimensional data, $d(x, y)$. In these examples, generalized least squares is used to find a solution that approximately fits the data and that approximately obeys prior information.

An alternative, deterministic approach is to find a solution that passes exactly through the observations and that exactly satisfies the prior information *between* them. Superficially, this approach seems superior—both observations and prior information are being satisfied exactly. However, this approach singles out the observation points as special. The solution will inevitably behave somewhat differently at the observation points than between them, which is *not* a desirable property.

Whether or not this approach is tractable—or even possible—depends on the type of prior information involved. Information about smoothness turns out to be especially easy to implement and leads to a set of techniques that might be called the *traditional* approach to interpolation. As we will see, these techniques are a straightforward extension of the *linear model* techniques that we have been developing in this book.

A purist might argue that interpolation of any kind is *never* the best approach to data analysis, because it will invariably introduce features not present in the original data, leading to wrong conclusions. A better approach is to generalize the analysis technique so that it can handle irregularly spaced data. This argument has merit, because interpolation adds additional—and often unquantifiable—error to the data. Nevertheless, if used sensibly, it is a valid data analysis tool that can simplify many data processing projects.

The basic idea behind interpolation is to construct an *interpolant*, a function, $d(t)$, that goes through all the data points, d_i, and does something sensible in between them. We can then evaluate the interpolant, $d(t)$, at whatever values of time, t, that we want—evenly spaced times, or times that match those of another dataset, to name two examples. Finding functions that go through a set of data points is easy. Finding functions that do something sensible in between the data points is more difficult as well as more problematical, because our notion of what is sensible will depend on prior knowledge, which varies from dataset to dataset.

Some obvious ideas do not work at all. A polynomial of degree $N - 1$, for instance, can easily be constructed to go through N data points (Figure 10.2). Unfortunately,

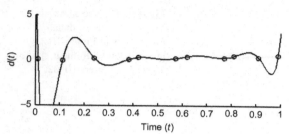

Figure 10.2 N irregularly spaced data, d_i (circles), fit by a polynomial (solid curve) or order, $(N - 1)$. *MatLab* script eda10_01.

the polynomial often takes wilds swings between the points. A high-degree polynomial does not embody the prior information that the function, $d(t)$, does not stray too far above or below the values of nearby data points.

10.2 Linear interpolation

A low-order polynomial is less wiggly than a high-order one and is better able to capture the prior information of smoothness. Unfortunately, a low-order polynomial can pass exactly through only a few data. The solution is to construct the function, $d(t)$, out of a sequence of polynomials, each valid for a short time interval. Such functions are called *splines*.

The simplest splines are line segments connecting the data points. The second derivative of a line is zero, so it embodies the prior information that the function is very smooth between the data points. The estimated datum at time, t, depends only on the values of the two *bracketing* observations, the one made immediately before time, t, and the one immediately after time, t. The interpolation formula is as follows:

estimated datum = weighted average of bracketing observations

or

$$d(t) = \frac{(t_{i+1} - t)d_i}{h_i} + \frac{(t - t_i)d_{i+1}}{h_i} \quad \text{with} \quad h_i = t_{i+1} - t_i \tag{10.1}$$

Here, the time, t, is bracketed by the two observation times, t_i and t_{i+1}. Note the use of the *local* quantities, $t_{i+1} - t$ and $t - t_i$; that is, time measured with respect to a nearby sample. In *MatLab*, linear interpolation is performed as follows:

```
dp=interp1(t,d,tp);                    (MatLab eda10_02)
```

Here, d is a column vector of the original data, measured at time, t, and dp is the interpolated data at time, tp.

Linear interpolation has the virtue of being simple. The interpolated data always lie between the observed data; they never deviate above or below them (Figure 10.3). Its major defect is that the function, $d(t)$, has kinks (discontinuities in slope) at the data points. The kinks are undesirable because they are an *artifact* not present in the original data; they arise from the prior information. Kinks will, for instance, add high frequencies to the power spectral density of the time series.

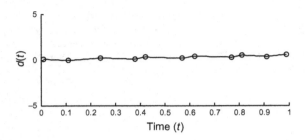

Figure 10.3 The same N irregularly spaced data, d_i, (circles) as in Figure 10.2, interpolated with linear splines. *MatLab* script eda10_02.

10.3 Cubic interpolation

A relatively simple modification is to use cubic polynomials in each of the intervals between the data points, instead of linear ones. A cubic polynomial has four coefficients and can satisfy four constraints. The requirement that the function must pass through the observed data at the ends of the interval places two constraints on these coefficients, leaving two constraints that can represent prior information. We can require that the first and second derivatives are continuous across intervals, creating a smooth function, $d(t)$, that has no kinks across intervals. (Its first derivative has no kinks either, but its second derivative does).

The trick behind working out simple formula for cubic splines is properly organizing the knowns and the unknowns. We start by defining the i-th interval as the one between time, t_i, and time, t_{i+1}. Within this interval, the spline function is a cubic polynomial, $S_i(t)$. As the second derivative of a cubic is linear, it can be specified by its values, y_i and y_{i+1}, at the ends of the interval:

second derivative = weighted average of bracketing values

or

$$\frac{d^2}{dt^2} S_i(t) = \frac{y_i(t_{i+1} - t)}{h_i} + \frac{y_{i+1}(t - t_i)}{h_i} \quad \text{with} \quad h_i = t_{i+1} - t_i \qquad (10.2)$$

(Compare with Equation 10.1). We can make the second derivative continuous across adjacent intervals if we equate y_i of interval $i + 1$ with y_{i+1} of interval i. Thus, only one second derivative, y_i, is defined for each time, t_i, even though two cubic polynomials touch this point. These ys are the primary unknowns in this problem. The formula for $S_i(t)$ can now be found by integrating Equation 10.2 twice:

$$S_i(t) = \frac{y_i(t_{i+1} - t)^3}{6h_i} + \frac{y_{i+1}(t - t_i)^3}{6h_i} + a_i(t_{i+1} - t) + b_i(t - t_i) \qquad (10.3)$$

Here a_i and b_i are integration constants. We now choose these constants so that the cubic goes through the data points, that is $S_i(t_i) = d_i$ and $S_i(t_{i+1}) = d_{i+1}$. This requirement leads to

$$S_i(t_i) = d_i = \frac{y_i h_i^2}{6} + a_i h_i \quad \text{or} \quad a_i = \frac{d_i}{h_i} - \frac{y_i h_i}{6}$$

$$S_i(t_{i+1}) = d_{i+1} = \frac{y_{i+1} h_i^2}{6} + b_i h_i \quad \text{or} \quad b_i = \frac{d_{i+1}}{h_i} - \frac{y_{i+1} h_i}{6} \tag{10.4}$$

The cubic spline is then

$$S_i(t) = \frac{y_i(t_{i+1} - t)^3}{6h_i} + \frac{y_{i+1}(t - t_i)^3}{6h_i} + \left(\frac{d_i}{h_i} - \frac{y_i h_i}{6}\right)(t_{i+1} - t) + \left(\frac{d_{i+1}}{h_i} - \frac{y_{i+1} h_i}{6}\right)(t - t_i) \tag{10.5}$$

and its first derivative is

$$\frac{d}{dt} S_i(t) = -\frac{y_i(t_{i+1} - t)^2}{2h_i} + \frac{y_{i+1}(t - t_i)^2}{2h_i} - \left(\frac{d_i}{h_i} - \frac{y_i h_i}{6}\right) + \left(\frac{d_{i+1}}{h_i} - \frac{y_{i+1} h_i}{6}\right)$$

$$= -\frac{y_i(t_{i+1} - t)^2}{2h_i} + \frac{y_{i+1}(t - t_i)^2}{2h_i} + \frac{(d_{i+1} - d_i)}{h_i} - \frac{(y_{i+1} - y_i)h_i}{6} \tag{10.6}$$

Finally, we determine the ys by requiring that two neighboring splines have first derivatives that are continuous across the interval:

$$\frac{d}{dt} S_{i-1}(t_i) = \frac{d}{dt} S_i(t_i)$$

or

$$\frac{y_i h_{i-1}}{2} + \frac{(d_i - d_{i-1})}{h_{i-1}} - \frac{(y_i - y_{i-1})h_{i-1}}{6} = -\frac{y_i h_i}{2} + \frac{(d_{i+1} - d_i)}{h_i} - \frac{(y_{i+1} - y_i)h_i}{6}$$

or

$$h_{i-1} y_{i-1} + 2(h_{i-1} + h_i)y_i + h_i y_{i+1} = \frac{6(d_{i+1} - d_i)}{h_i} - \frac{6(d_i - d_{i-1})}{h_{i-1}} \tag{10.7}$$

The y's satisfy a linear equation that can be solved by standard matrix methods, which we have discussed in Chapter 1. Note that the $i = 1$ and $i = N$ equations involve the quantities, y_0 and y_{N+1}, which represent the second derivatives at two undefined points, one to the left of the first data point and the other to the right of the last data point. They can be set to any value and moved to the right hand side of the equation. The choice $y_0 = y_{N+1} = 0$ leads to a solution called *natural* cubic splines.

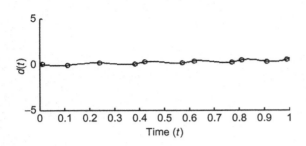

Figure 10.4 The same N irregularly spaced data, d_i, (circles) as in Figure 10.2, interpolated with cubic splines. *MatLab* script eda09_03.

Because of its smoothness (Figure 10.4), cubic spline interpolation is usually preferable to linear interpolation.

In *MatLab*, cubic spline interpolation is performed as follows:

```
dp=spline(t,d,tp);
```
(*MatLab* eda10_03)

Here, d is a column vector of the original data, measured at time, t, and dp is the interpolated data at time, tp.

10.4 Kriging

The generalized least-squared methodology that we developed in Chapter 5 to fill in data gaps was based on prior information, represented by the linear equation, $\mathbf{Hm} = \bar{\mathbf{h}}$. We quantified roughness, say R, of a vector, $\mathbf{m}$, by choosing $\mathbf{H}$ to be a matrix of second derivatives and by setting $\bar{\mathbf{h}}$ to zero. The minimization of the total roughness,

$$R = (\mathbf{Hm})^{\mathrm{T}}(\mathbf{Hm}) \tag{10.8}$$

in Equation (5.8) leads to a solution that is smooth. Note that we can rewrite Equation (10.8) as

$$R = (\mathbf{Hm})^{\mathrm{T}}(\mathbf{Hm}) = \mathbf{m}^{\mathrm{T}}\,\mathbf{H}^{\mathrm{T}}\mathbf{Hm} = \mathbf{m}^{\mathrm{T}}[\mathbf{C}_m]^{-1}\mathbf{m} \quad \text{with } \mathbf{C}_m = [\mathbf{H}^{\mathrm{T}}\mathbf{H}]^{-1} \tag{10.9}$$

We have encountered the quantity, $\mathbf{m}^{\mathrm{T}}[\mathbf{C}_m]^{-1}\mathbf{m}$, before in Equation (5.1), where we interpreted it as the error, $E_p(\mathbf{m})$, in the prior information. The quantity, $\mathbf{C_m} = [\mathbf{H}^{\mathrm{T}}\mathbf{H}]^{-1}$, can be interpreted as the covariance, $\mathbf{C}_m$, of the prior information. In the case of smoothness, the prior covariance matrix, $\mathbf{C}_m$, has nonzero off-diagonal elements. Neighboring model parameters are highly correlated, as a smooth curve is one for which neighboring points have similar values. Thus, the prior covariance, $\mathbf{C}_m$, of the model is intrinsically linked to the measure of smoothness. Actually we have already encountered this link in the context of time series. The autocorrelation function of a time series is closely related to *both* its covariance and its smoothness (see Section 9.1). This analysis suggests that prior information about a function's autocorrelation function can be usefully applied to the interpolation problem.

Kriging (named after its inventor, Danie G. Krige) is an interpolation method based on this idea. It provides an estimate of a datum, d_0^{est}, at an arbitrary time, t_0^{est}, based on a set of N observations, say $(t_i^{\text{obs}}, d_i^{\text{obs}})$, when prior information about its autocorrelation function, $a(t)$, is available. The method assumes a linear model in which d_0^{est} is a weighted average of *all* the observed data (as contrasted to just the two bracketing data) (Krige, 1951):

$$\text{estimated datum} \quad = \text{weighted average of all observations}$$
$$d_0^{\text{est}} = \sum_{i=1}^{N} w_i d_i^{\text{obs}} = (\mathbf{d}^{\text{obs}})^{\text{T}} \mathbf{w} \tag{10.10}$$

Here, $\mathbf{d}^{\text{obs}}$ is a column vector of the observed data and $\mathbf{w}$ is an unknown column-vector of weights. As we shall see, the weights can be determined using prior knowledge of the autocorrelation function. The approach is to minimize the variance of $d_0^{\text{est}} - d_0^{\text{true}}$, the difference between the estimated and true value of the time series. As we will see, we will not actually need to know the true value of the time series, but only its true autocorrelation function. The formula for variance is

$$\sigma_{\Delta d}^2 = \int [(d_0^{\text{est}} - d_0^{\text{true}}) - (\bar{d}_0^{\text{est}} - \bar{d}_0^{\text{true}})]^2 p(\mathbf{d}) \mathrm{d}^N d = \int (d_0^{\text{est}} - d_0^{\text{true}})^2 p(\mathbf{d}) \mathrm{d}^N d$$

$$= \int (d_0^{\text{est}})^2 p(\mathbf{d}) \mathrm{d}^N d + \int (d_0^{\text{true}})^2 p(\mathbf{d}) \mathrm{d}^N d - 2 \int d_0^{\text{est}} d_0^{\text{true}} p(\mathbf{d}) \mathrm{d}^N d$$

$$= \sum_{i=1}^{N} \sum_{j=1}^{N} w_i w_j \int d_i^{\text{obs}} d_i^{\text{obs}} p(\mathbf{d}) \mathrm{d}^N d + \int (d_0^{\text{true}})^2 p(\mathbf{d}) \mathrm{d}^N d - 2 \sum_{i=1}^{N} w_i \int d_0^{\text{est}} d_j^{\text{true}} p(\mathbf{d}) \mathrm{d}^N d \tag{10.11}$$

Here, $p(\mathbf{d})$ is the probability density function of the data. We have assumed that the estimated and true data have the same mean, that is, $\bar{d}_0^{\text{est}} = \bar{d}_0^{\text{true}}$. We now set $d_i^{\text{obs}} \approx d_i^{\text{true}}$ so that

$$\sigma_{\Delta d}^2 \propto \sum_{i=1}^{N} \sum_{j=1}^{N} w_i w_j a(|t_j^{\text{obs}} - t_i^{\text{obs}}|) + a(0) - 2 \sum_{i=1}^{N} w_i a(|t_i^{\text{obs}} - t_0|) \tag{10.12}$$

Note that we have used the definition of the autocorrelation function, $a(t)$ (see Equation 9.6) and that we have ignored a normalization factor (hence the $\propto$ sign), which appears in all three terms of the equation. Finally, we differentiate the variance with respect to w_k and set the result to zero:

$$\frac{\mathrm{d}\sigma_{\Delta d}^2}{\mathrm{d}w_k} = 0 \propto 2 \sum_{j=1}^{N} w_i a(|t_k^{\text{obs}} - t_i^{\text{obs}}|) - 2a(|t_k^{\text{obs}} - t_0|) \tag{10.13}$$

which yields the matrix equation

$$\mathbf{M}\mathbf{w} = \mathbf{v} \quad \text{where } M_{ij} = a(|t_i^{\text{obs}} - t_j^{\text{obs}}|) \quad \text{and} \quad v_i = a(|t_i^{\text{obs}} - t_0^{\text{est}}|) \tag{10.14}$$

Thus,

$$\mathbf{w} = \mathbf{M}^{-1}\mathbf{v} \quad \text{and} \quad d_0^{\text{est}} = (\mathbf{d}^{\text{obs}})^{\mathrm{T}}\mathbf{w} = (\mathbf{d}^{\text{obs}})^{\mathrm{T}}\mathbf{M}^{-1}\mathbf{v} \tag{10.15}$$

Note that the matrix, $\mathbf{M}$, and the column-vector, $\mathbf{v}$, depend only on the autocorrelation of the data. Normally, data needs to be interpolated at many different times, not just one as in the equation above. In such case, we define a column vector, $\mathbf{t}_0^{\text{est}}$, of all the times at which it needs to be evaluated, along with a corresponding vector of interpolated data, $\mathbf{d}_0^{\text{est}}$. The solution becomes

$$\mathbf{d}_0^{\text{est}} = (\mathbf{d}^{\text{obs}})^{\mathrm{T}}\mathbf{M}^{-1}\mathbf{V} \quad \text{with} \quad [\mathbf{V}]_{ij} = a(|t_i^{\text{obs}} - [\mathbf{t}_0^{\text{est}}]_j|) \tag{10.16}$$

Note that the results are insensitive to the overall amplitude of the autocorrelation function, which cancels from the equation; only its shape matters. Prior notions of smoothness are implemented by specifying a particular autocorrelation function. A wide autocorrelation function will lead to a smooth estimate of $d(t)$ and a narrow function to a rough one. For instance, if we use a Normal function,

$$a(|t_i - t_j|) = \exp\left\{ -\frac{(t_i - t_j)^2}{2L^2} \right\} \tag{10.17}$$

then its variance, L, will control the degree of smoothness of the estimate.

In *MatLab*, this approach is very simply implemented:

```
A=exp(-abs(tobs*ones(N,1)'-ones(N,1)*tobs').^2/(2*L2));
V=exp(-abs(tobs*ones(M,1)'-ones(N,1)*test').^2/(2*L2));
dest=dobs'*((A+1e-6*eye(N))\V);          (MatLab eda10_04)
```

Note that we have damped the matrix, $\mathbf{A}$, by addition of a small amount of the identity matrix. This modification guards against the possibility that the matrix, $\mathbf{A}$, is near-singular. It has little effect on the solution when it is not near-singular, but drives the solution towards zero when it is. An example is shown in Figure 10.5.

Finally, we note that the power spectral density of a time series is uniquely determined by its autocorrelation function, as the former is the Fourier transform of the latter (see Section 9.3). Thus, while we have focused here on prior information in the form of the autocorrelation function of the time series, we could alternatively have used prior information about its power spectral density.

Crib Sheet 10.1 When and how to interpolate

Rule of thumb

If you can't already guess what the interpolated values should be, don't interpolate; limit your analysis to the uninterpolated data.

Good reasons to interpolate

To convert an irregularly-sampled independent variable into an evenly-sampled one, in order to use an algorithm that requires the latter.

Continued

Crib Sheet 10.1—cont'd

To align the independent variable of two different data sets, so that they can be compared.

Poor reasons to interpolate
To increase the number N of data and to fill in large data gaps.
(You are just fooling yourself when you do this).

Linear interpolation, advantages
The results are *local* (depend only upon bracketing values)
The overall range of the data is preserved.

disadvantage
The interpolated data may contain sharp kinks
(Don't use when computing derivatives or power spectra)

Cubic interpolation, advantage
Extremely smooth
(Use when computing derivatives and power spectra)

disadvantages
The overall range of the data is not preserved.
The results are *non-local* (an outlier can lead to *ringing*).

Caveats when computing power spectra of interpolated data
While decreasing Δt appears to increase f_{max}, the results are bogus.
Interpolating large data gaps creates strong correlations between spectral peaks that complicate tests of significance.

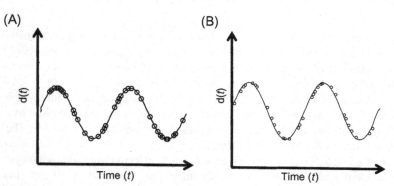

Figure 10.5 Example of filling in data gaps using Kriging. (A) Kriging using prescribed normal autocorrelation function. (B) Generalized least-squares result from Chapter 5. *MatLab* script eda10_04.

10.5   Interpolation in two-dimensions

Interpolation is not limited to one dimension. Equally common is the case where data are collected on an irregular two-dimensional grid but need to be interpolated onto a regular, two-dimensional grid. The basic idea behind two-dimensional interpolation is the same as in one dimension: construct a function, $d(x_1, x_2)$, that goes through all the data points and does something sensible in between them, and use it to estimate the data at whatever points, (x_1, x_2), are of interest.

A key part of spline interpolation is defining intervals over which the spline functions are valid. In one-dimensional interpolation, the intervals are both conceptually and computationally simple. They are just the segments of the time axis bracketed by neighboring observations. As long as the data are in the order of increasing time, each interval is between a data point and its successor. In two dimensions, intervals become two-dimensional patches (or *tiles*) in (x_1, x_2). Creating and manipulating these tiles is complicated.

One commonly used tiling is based on connecting the observation points together with straight lines, to form a mesh of triangles. The idea is to ensure that the vertices of every triangle coincide with the data points and that the (x_1, x_2) plane is completely covered by triangles with no holes or overlap. A spline function is then defined within each triangle. Such triangular meshes are nonunique as the data points can be connected in many alternative ways. *Delaunay triangulation* is a method for constructing a mesh that favors equilateral triangles over elongated ones. This is a desirable property as then the maximum distance over which a spline function acts is small. *MatLab* provides two-dimensional spline functions that rely on Delaunay triangulation, but perform it behind-the-scenes, so that normally you do not need to be concerned with it. Linear interpolation is performed using

```
dp=griddata(xobs,yobs,dobs,xp,yp,'linear');
```
(*MatLab* eda09_05)

and cubic spline interpolation by

```
dp=griddata(xobs,yobs,dobs,xp,yp,'cubic');
```
(*MatLab* eda10_05)

In both cases, xobs and yobs are column vectors of the (x_1, x_2) coordinates of the data, dobs is a column vector of the data, xp and yp are column vectors of the (x_1, x_2) coordinates at which interpolated data are desired, and dp is a column vector of the corresponding interpolated data. An example is given in Figure 10.6.

The *MatLab* documentation indicates that the griddata() function is being *depreciated* in favor of another, similar function, TriScatteredInterp(), meaning that it may not be supported in future releases of the software. The use of this new function is illustrated in *MatLab* script eda10_06. As of the time of publication, TriScatteredInterp() does not perform cubic interpolation.

Occasionally, the need arises to examine the triangular mesh that underpins the interpolation (Figure 10.6B). Furthermore, triangular meshes have many useful

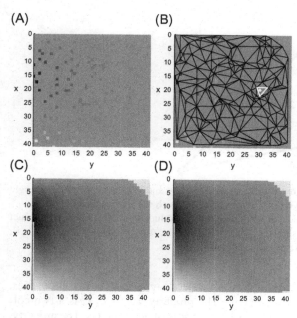

Figure 10.6 The pressure data, d_i (squares), are a function of two spatial variables, (x, y). (B) A Delaunay triangular mesh with vertices at the locations of the data. The triangle containing the point, (20, 30) (circle), is highlighted. (C) Linear interpolation of the pressure data. (D) Cubic spline interpolation of the pressure data. *MatLab* script eda10_05.

applications in addition to interpolation. In *MatLab*, a triangular grid is created as follows:

```
mytri = DelaunayTri(xobs,yobs);
XY=mytri.X;
[NXY, i] = size(XY);
TRI = mytri.Triangulation;
[NTRI, i] = size(TRI);                    (MatLab eda10_05)
```

The DelaunayTri() function takes column-vectors of the (x_1, x_2) coordinates of the data (called xobs and yobs in the script) and returns information about the triangles in the *object*, mytri. Until now, we have not encountered *MatLab objects*. They are a type of variable that shares some similarity with a vector, as can be understood by the following comparison:

A *vector*, v, contains *elements*, that are referred to using a numerical index, for example, v(1), v(2), and so forth. The parentheses are used to separate the index from the name of the vector, so that there is no confusion between the two. Each element is a variable that can hold a scalar value. Thus, for example, v(1)=10.5.

An *object*, o, contains *properties*, that are referred to using symbolic names, for example, o.a, o.b, and so forth. The period is used to separate the name of the property from the name of the object so that there is no confusion between the two. Each property is a variable that can hold *anything*. Thus, for instance, the o.a property could hold a scalar value, o.a=10.5, but the o.b property could hold a 3000 × 100 matrix, o.b=zeros(3000,100).

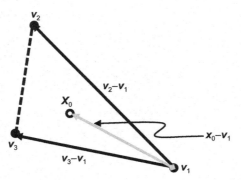

Figure 10.7 A triangle is defined by its three vertices, $\mathbf{v}_1$, $\mathbf{v}_2$, and $\mathbf{v}_3$. Each side of the triangle consists of a line connecting two vertices (e.g., $\mathbf{v}_1$ and $\mathbf{v}_2$) with the third vertex excluded (e.g., $\mathbf{v}_3$). A point, $\mathbf{x}_0$, is on the same side of the line as the excluded vertex if the cross product $(\mathbf{v}_3 - \mathbf{v}_1) \times (\mathbf{v}_2 - \mathbf{v}_1)$ has the same sign as $(\mathbf{x}_0 - \mathbf{v}_1) \times (\mathbf{v}_2 - \mathbf{v}_1)$. A point is within the triangle if, for each side, it is on the same side as the excluded vertex.

The `DelaunayTri()` function returns an object, `mytri`, with two properties, `mytri.X` and `mytri.Triangulation`. The property, `mytri.X`, is a two-column wide array of the (x_1, x_2) coordinates of the vertices of the triangles. The property, `mytri.Triangulation`, is a three-column wide array of the indices, (`t1, t2, t3`), in the `mytri.X` vertex array, of the three vertices of each triangle.

In the script above, the `mytri.X` property is copied to a normal array, `XY`, of size `NXY×2`, and the `mytri.Triangulation` property is copied to a normal array, `TRI`, of size `NTRI×3`. Thus, the i-th triangle has one vertex at `XY(v1,1)`, `XY(v1,2)`, another at `XY(v2,1)`, `XY(v2,2)`, and a third at `XY(v3,1)`, `XY(v3,2)`, where

```
v1=TRI(i,1); v2=TRI(i,2); v3=TRI(i,3);      (MatLab eda10_05)
```

A commonly encountered problem is to determine within which triangle a given point, x_0, is. *MatLab* provides a function that performs this calculation (see Figure 10.7 for a description of the theory behind this calculation):

```
tri0 = pointLocation(mytri, x0);            (MatLab eda10_05)
```

Here, `x0` is a two-column wide array of (x_1, x_2) coordinates and `tri0` is a column vector of indices to triangles enclosing these points (i.e., indices into the `TRI` array).

10.6   Fourier transforms in two dimensions

Periodicities can occur in two dimensions, as for instance in an aerial photograph of storm waves on the surface of the ocean. In order to analyze for these periodicities, we must Fourier transform over both spatial dimensions. A function, $f(x, y)$, of two spatial variables, x and y, becomes a function of two spatial frequencies, or *wavenumbers*, k_x and k_y. The integral transform and its inverse are

$$\tilde{\tilde{f}}(k_x, k_y) = \int_{-\infty}^{+\infty} \int_{-\infty}^{+\infty} f(x, y) \exp(-ik_x x - ik_y y) dy dx$$

and

$$f(x, y) = \frac{1}{(2\pi)^2} \int_{-\infty}^{+\infty} \int_{-\infty}^{+\infty} \tilde{\tilde{f}}(k_x, k_y) \exp(+ik_x x + ik_y y) dk_y dk_x \quad (10.18)$$

The two-dimensional transform can be thought of as two transforms applied successively. The first one transforms $f(x, y)$ to $\tilde{f}(x, k_y)$ and the second transforms $\tilde{f}(x, k_y)$ to $\tilde{\tilde{f}}(k_x, k_y)$. *MatLab* provides a function, $\texttt{fft2()}$, that compute the two-dimensional discrete Fourier transform.

In analyzing the discrete case, we will assume that x increases with row number and y increases with column number, so that the data are in a matrix, $\mathbf{F}$, with $F_{ij} = f(x_i, y_j)$. The transforms are first performed on each row of, $\mathbf{F}$, producing a new matrix, $\tilde{\mathbf{F}}$. As with the one-dimensional transform, the positive k_ys are in the left-hand side of $\tilde{\mathbf{F}}$ and the negative k_ys are in its right-hand side. The transform is then performed on each column of $\tilde{\mathbf{F}}$ to produce $\tilde{\tilde{\mathbf{F}}}$. The positive k_xs are in the top of $\tilde{\tilde{\mathbf{F}}}$ and the negative k_xs are in the bottom of $\tilde{\tilde{\mathbf{F}}}$. The output of the *MatLab* function, $\texttt{Ftt=fft2(F)}$, in the $N_x \times N_y = 8 \times 8$ case, looks like

$$\tilde{\tilde{\mathbf{F}}} = \begin{bmatrix}
\tilde{\tilde{f}}(0,0) & \tilde{\tilde{f}}(0,1) & \tilde{\tilde{f}}(0,2) & \tilde{\tilde{f}}(0,3) & \tilde{\tilde{f}}(0,4) & \tilde{\tilde{f}}(0,5) & \tilde{\tilde{f}}(0,-3) & \tilde{\tilde{f}}(0,-2) & \tilde{\tilde{f}}(0,-1) \\
\tilde{\tilde{f}}(1,0) & \tilde{\tilde{f}}(1,1) & \tilde{\tilde{f}}(1,2) & \tilde{\tilde{f}}(1,3) & \tilde{\tilde{f}}(1,4) & \tilde{\tilde{f}}(1,5) & \tilde{\tilde{f}}(1,-3) & \tilde{\tilde{f}}(1,-2) & \tilde{\tilde{f}}(1,-1) \\
\tilde{\tilde{f}}(2,0) & \tilde{\tilde{f}}(2,1) & \tilde{\tilde{f}}(2,2) & \tilde{\tilde{f}}(2,3) & \tilde{\tilde{f}}(2,4) & \tilde{\tilde{f}}(2,5) & \tilde{\tilde{f}}(2,-3) & \tilde{\tilde{f}}(2,-2) & \tilde{\tilde{f}}(2,-1) \\
\tilde{\tilde{f}}(3,0) & \tilde{\tilde{f}}(3,1) & \tilde{\tilde{f}}(3,2) & \tilde{\tilde{f}}(3,3) & \tilde{\tilde{f}}(3,4) & \tilde{\tilde{f}}(3,5) & \tilde{\tilde{f}}(3,-3) & \tilde{\tilde{f}}(3,-2) & \tilde{\tilde{f}}(3,-1) \\
\tilde{\tilde{f}}(4,0) & \tilde{\tilde{f}}(4,1) & \tilde{\tilde{f}}(4,2) & \tilde{\tilde{f}}(4,3) & \tilde{\tilde{f}}(4,4) & \tilde{\tilde{f}}(4,5) & \tilde{\tilde{f}}(4,-3) & \tilde{\tilde{f}}(4,-2) & \tilde{\tilde{f}}(4,-1) \\
\tilde{\tilde{f}}(5,0) & \tilde{\tilde{f}}(5,1) & \tilde{\tilde{f}}(5,2) & \tilde{\tilde{f}}(5,3) & \tilde{\tilde{f}}(5,4) & \tilde{\tilde{f}}(5,5) & \tilde{\tilde{f}}(5,-3) & \tilde{\tilde{f}}(5,-2) & \tilde{\tilde{f}}(5,-1) \\
\tilde{\tilde{f}}(-3,0) & \tilde{\tilde{f}}(-3,1) & \tilde{\tilde{f}}(-3,2) & \tilde{\tilde{f}}(-3,3) & \tilde{\tilde{f}}(-3,4) & \tilde{\tilde{f}}(-3,5) & \tilde{\tilde{f}}(-3,-3) & \tilde{\tilde{f}}(-3,-2) & \tilde{\tilde{f}}(-3,-1) \\
\tilde{\tilde{f}}(-2,0) & \tilde{\tilde{f}}(-2,1) & \tilde{\tilde{f}}(-2,2) & \tilde{\tilde{f}}(-2,3) & \tilde{\tilde{f}}(-2,4) & \tilde{\tilde{f}}(-2,5) & \tilde{\tilde{f}}(-2,-3) & \tilde{\tilde{f}}(-2,-2) & \tilde{\tilde{f}}(-2,-1) \\
\tilde{\tilde{f}}(-1,0) & \tilde{\tilde{f}}(-1,1) & \tilde{\tilde{f}}(-1,2) & \tilde{\tilde{f}}(-1,3) & \tilde{\tilde{f}}(-1,4) & \tilde{\tilde{f}}(-1,5) & \tilde{\tilde{f}}(-1,-3) & \tilde{\tilde{f}}(-1,-2) & \tilde{\tilde{f}}(-1,-1)
\end{bmatrix}$$

$$(10.19)$$

For real data, $\tilde{\tilde{f}}(k_x, k_y) = \tilde{\tilde{f}}^*(-k_x, -k_y)$, so that only the left-most $N_y/2 + 1$ columns are independent. Note, however, that reconstructing a right-hand column from the corresponding left-hand column requires reordering its elements, in addition to complex conjugation:

```
for m = [2:Ny/2]
    mp=Ny-m+2;
    Ftt2(1,mp) = conj(Ftt(1,m));
    Ftt2(2:Nx,mp) = flipud(conj(Ftt(2:Nx,m)));
end
```
(*MatLab* eda10_07)

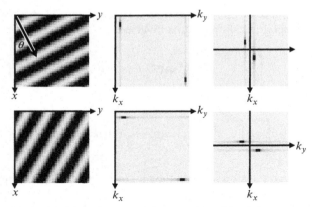

Figure 10.8 Amplitude spectral density of a two-dimensional function, $d(x, t)$. (Top row) (Left) Cosine wave, $d(x, t)$, inclined $\theta = 30°$ from x-axis. (Middle) Amplitude spectral density, as a function of wavenumbers, k_x and k_y, in order returned by *MatLab* function, `fft2()`. (Right) Amplitude spectral density after rearranging the order to put the origin in the middle of the plot. (Bottom row) Same as for 60° cosine wave. *MatLab* script eda10_07

Here, `Ftt2` is reconstructed from the left-size of `Ftt`, the two-dimensional Fourier transform of the `Nx` by `Ny` matrix, `F` (calculated, for example, as `Ftt=fft2(F)`).

We illustrate the two-dimensional Fourier transform with a cosine wave with wavenumber, k_r, oriented so that a line perpendicular to its wavefronts makes an angle, θ, with the x-axis (Figure 10.8):

$$f(x,y) = \cos(n_x k_r x + n_y k_r y) \quad \text{with} \quad [n_x, n_y]^T = [\sin(\theta), \cos(\theta)]^T \quad (10.20)$$

The Fourier transform, $\tilde{\tilde{f}}(k_x, k_y)$, has two spectral peaks, one at $(k_x, k_y) = (k_r\cos\theta, k_r\sin\theta)$ and the other at $(k_x, k_y) = (-k_r\cos\theta, -k_r\sin\theta)$. Sidelobes, which appear as vertical and horizontal streaks extending from the spectral peaks in Figure 10.8, are as much of a problem in two dimensions are they are in one.

Problems

10.1 What is the covariance matrix of the estimated data in linear interpolation? Assume that the observed data have uniform variance, σ_d^2. Do the interpolated data have constant variance? Are they uncorrelated?

10.2 Modify the `interpolate_reynolds` script of Chapter 9 to interpolate the Reynolds Water Quality dataset with cubic splines. Compare your results to the original linear interpolation.

10.3 Modify the eda10_04 script to computing the kriging estimate for a variety of different widths, L^2, of the autocorrelation function. Comment on the results.

10.4 Modify the eda10_07 script to taper the data in both the x and y directions before computing the Fourier transform. Use a two-dimensional taper that is the product of a Hamming window function in x and a Hamming window function in y.

10.5 Estimate the power spectral density in selected monthly images of the CAC Sea Surface Temperature dataset (Chapter 8). Which direction has the longer-wavelength features, latitude or longitude? Explain your results.

References

Krige, D.G., 1951. A statistical approach to some basic mine valuation problems on the Witwatersrand. J. Chem. Metal. Mining Soc. South Africa 52, 119–139.

11 "Approximate" is not a pejorative word

11.1 The value of approximation

As schoolchildren, we were taught to admire the exactness of mathematics. The length of the hypotenuse of a right triangle with legs of 3 and 4 m is 5 m exactly, not a little less or a little more. Archimedes's 250 B.C.E. value of 22/7 for π was merely an upper bound for the ratio of a circle's circumference to its diameter. The exact value, known to one million digits by 1973, is something different and worth striving for.

Later in life, we learned that transcribing all 10 digits off the screen of a calculator is not really necessary; 3 or 4 is sufficient for most purposes. Yet we know that we can recompute all 10, should the need arise. We also came to appreciate the value of a back-of-the-envelope calculation, where the goal is merely to estimate the correct order of magnitude of the answer. Yet we view such calculations as imperfect—a tool to guide the development of a more accurate solution.

Environmental Data Analysis with MATLAB®. http://dx.doi.org/10.1016/B978-0-12-804488-9.00011-2

The situation in data analysis is fundamentally different. The accuracy of a result is a combination of methodological error, meaning error associated with the underlying theory and its computation, and observational error. Once methodological error is reduced to a level well below observation error, further improvement accomplishes little. The appropriate goal is to develop a methodology with acceptably small error, not one with none.

Of course, one might decide to err on the safe side; that is, to use only exact theories and to carry out all calculations to as many significant digits as possible. Superficially, such a strategy would not seem to be a bad one and, indeed, solutions achieved this way are certainly no worse than those that employ approximations. Yet the approach has serious limitations, over and above the obvious one that exact theories are unavailable for many imperfectly understood real-world phenomena. Approximations can bring simplicity, speed of computation, and adaptability to a solution which, in its exact form, is inscrutable, slow, and inflexible.

A *simple* method is one whose structure is sufficiently well-understood that its general behavior can be anticipated and its pitfalls identified and avoided. The grid search procedure (Section 4.6) for finding the minimum error solution is simple in this sense. This exhaustive search is easy to understand (and to code) and, although not completely foolproof, fails only when the search limits have been so poorly chosen that they fail to enclose the solution or when the solution varies over scale lengths smaller than the grid spacing. Linear theories of the form $\mathbf{Gm} = \mathbf{d}$ are simple, too. The matrix $\mathbf{G}$ may be large and complicated, yet it is straightforward to create and interpret. Furthermore, its least squares solution (Section 4.7) will succeed except in cases that can be easily detected (e.g., when the matrix inverse is near-singular) and corrected (e.g., by adding prior information). As we will see below, approximations that linearize an exact nonlinear theory have wide application.

Approximations can also improve the speed of computation. Speed is an extremely important issue in very large problems that challenge the capabilities of the current generation of computers and in real-time applications, where a solution is needed immediately after the data becomes available. A Global Positioning System (GPS) receiver would be of limited utility in an automobile if it took 10 min to determine a location!

Finally, some data analysis scenarios need to adapt to evolving conditions that affect the character of the data. The character of a river hydrograph, for example, may evolve significantly over the years as its watershed becomes urbanized. The methodology needed to predict discharge from precipitation may need to be correspondingly updated. Some approximate methods are particularly well-suited for incorporating *machine learning*.

11.2   Polynomial approximations and Taylor series

A polynomial can always be designed to behave similarly to a smooth function $y(t)$ in the vicinity of a specified point t_0. This assertion can be demonstrated by constructing a polynomial $y_p(t)$ that has the same value, first derivative, second derivative, etc. at this point. The polynomial becomes a better and better approximation to the function

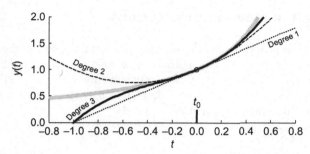

Figure 11.1 The function $y(t) = 1/(1-t)$ (gray curve) approximated by a sequence of polynomials of increasing degree. Note that the approximations become more accurate, especially near the point t_0, as the degree is increased. *MatLab* script eda11_01.

as derivatives of higher and higher order are matched (Figure 11.1). A polynomial with unknown coefficients c_i is

$$y_p(t) = c_0 + c_1(t - t_0) + c_2(t - t_0)^2 + c_3(t - t_0)^3 + \dots \tag{11.1}$$

and its derivates are

$$\frac{dy_p}{dt} = c_1 + 2c_2(t - t_0) + 3c_3(t - t_0)^2 + \dots$$

$$\frac{d^2 y_p}{dt^2} = 2c_2 + 6c_3(t - t_0) + \dots \tag{11.2}$$

etc.

When evaluated at the point $t - t_0$, these functions each depends upon a single constant:

$$y_p(t_0) = c_0 \quad \text{and} \quad \left.\frac{dy_p}{dt}\right|_{t_0} = c_1 \quad \text{and} \quad \left.\frac{d^2 y_p}{dt^2}\right|_{t_0} = 2c_2 \quad \text{etc.} \tag{11.3}$$

After solving for the constants and inserting them into Equation (11.1), a polynomial $y_p(t)$ is obtained, whose value and derivatives match the smooth function $y(t)$ at the point t_0:

$$y_p(t) = y(t_0) + \left.\frac{dy}{dt}\right|_{t_0}(t - t_0) + \frac{1}{2}\left.\frac{d^2 y}{dt^2}\right|_{t_0}(t - t_0)^2 + \dots \tag{11.4}$$

This polynomial can be shown to converge to the smooth function when the number of terms becomes indefinitely large; that is, $y_p(t) \to y(t)$ (at least for a finite interval of the t axis containing t_0). This result is known as *Taylor theorem* and Equation (11.4) is known as the *Taylor series* representation of the function $y(t)$. As we will show below, Taylor series play a critical role in many approximation methods.

11.3   Small number approximations

An important class of approximations uses just the first two terms of the Taylor series to represent a function in the *neighborhood* of a point $t = t_0$, since the higher order terms are typically negligible there. Consider the function $y(t) = (1 \pm t)^n$. At the point $t_0 = 0$, it has value $y(0) = 1$ and first derivative $dy/dt|_0 = \pm n(1 \pm t)^{n-1}|_0 = \pm n$. Inserting these values into Equation (11.4) yields

$$(1 \pm t)^n \approx 1 \pm nt \tag{11.5}$$

This approximation is surprisingly accurate. For instance, $(1 + 0.04)^{1/2} = 1.0198$, whereas $(1 + \frac{1}{2} \times 0.04) = 1.02$; the difference is only 0.02%.

Two additional examples are the functions $\sin t$ and $\cos t$ in the neighborhood of the point $t_0 = 0$. The sine has a value of 0, a first derivative of 1, and a second derivative of 0 at this point, while the cosine has a value of 1, a first derivative of 0, and a second derivative of -1 there. Inserting these values into Equation (11.4) yields

$$\sin t \approx t \quad \text{and} \quad \cos t \approx 1 - \frac{1}{2}t^2 \tag{11.6}$$

These and other *small number approximations* can be used to simplify and speed up data analysis problems (see Crib Sheet 11.1).

Crib Sheet 11.1 Small number approximations

$$(\text{valid when } x \ll 1)$$

$$(1 \pm x)^n \approx 1 \pm nx$$

$$\sin x \approx x$$

$$\cos x \approx 1 - \frac{1}{2}x^2$$

$$\tan x \approx x$$

$$\exp x \approx 1 + x$$

$$\ln(1 \pm x) \approx \pm x$$

$$a^x \approx 1 + x \ln a$$

$$\sinh x \approx x$$

$$\cosh x \approx 1 + \frac{1}{2}x^2$$

$$\tanh x \approx x$$

11.4   Small number approximation applied to distance on a sphere

A common problem encountered when working with geographical data is to deter-mine the *great circle* distance r between two points on a sphere (such as the Earth). Spherical trigonometry can be used to show that this distance, measured in degrees of arc, is given by

$$\cos r = \sin \lambda_1 \sin \lambda_2 + \cos \lambda_1 \cos \lambda_2 \cos \Delta L \qquad (11.7)$$

Here, λ_1 and λ_2 are the latitudes of the two points, respectively; L_1 and L_2 are their longitudes; and $\Delta L = L_2 - L_1$. Six distinct trigonometric functions must be evaluated in order to determine r—a time-consuming process. A small r approximation can re-duce the count of functions. The first step is to define the difference in latitude as $\Delta \lambda = \lambda_2 - \lambda_1$ and the mean latitude as $\bar{\lambda} = \frac{1}{2}(\lambda_2 + \lambda_1)$. Standard trigonometric identities are then used to rewrite Equation (11.7) in terms of these quantities:

$$\cos r = \frac{1}{2} \cos \Delta \lambda - \frac{1}{2} \cos \bar{\lambda} + \frac{1}{2} + \frac{1}{2} \cos \Delta \lambda \cos \Delta L + \cos^2 \bar{\lambda} \cos \Delta L$$
$$- \frac{1}{2} \cos \Delta L \qquad (11.8)$$

By assumption, ΔL, $\Delta \lambda$, and r are all small. After applying the small number approx-imations of Equation (11.6) and after considerable algebra, the approximation is found to be

$$r^2 \approx (\Delta \lambda)^2 + (\Delta L)^2 \cos^2 \bar{\lambda} - (\Delta \lambda)^2 (\Delta L)^2 \left[\frac{1}{4} - \frac{\cos^2 \bar{\lambda}}{6} \right] \qquad (11.9)$$

The first two terms on the right-hand side represent the Euclidian distance between the points; that is, distance as measured on a plane. The last term corrects this *flat-earth* distance for the curvature of the surface. Only two functions, a cosine and a square root, need to be evaluated in order to determine r, which is a factor-of-three reduction in effort compared with the exact formula. Furthermore, the approximation is very accurate, at least out to distances of a few degrees (Figure 11.2).

11.5   Small number approximation applied to variance

A commonly encountered problem is that of estimating the variance of a nonlinear function of the model parameters. As an example, suppose that a least squares proce-dure estimates an angular frequency m^{est} with variance σ_m^2, but we would rather state confidence bounds for the period $T = 2\pi/m$. As discussed in Chapter 3, this problem can be solved exactly by assuming that m is normally distributed, working out the

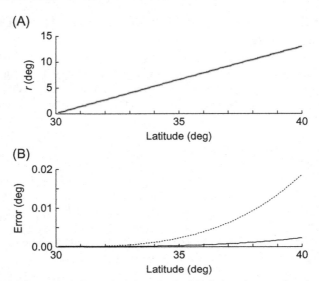

Figure 11.2 (A) Linear (dotted curve) and quadratic (black curve) approximation for distance r on a sphere compares well with the exact value (gray curve). (B) The corresponding error. This calculation is for the case $L = \lambda$. *MatLab* script eda11_02.

probability density function $p(T)$, and then calculating its mean and variance. However, the exact procedure is difficult and time-consuming; the approximation that is developed here is accurate when $m^{\text{est}} \gg \sigma_m$.

The idea is to examine how a small fluctuation Δm in frequency causes a small fluctuation ΔT in period. Writing $m = m^{\text{est}} + \Delta m$ and $T = T^{\text{est}} + \Delta T$ and inserting them into the formula relating frequency and period yield

$$
T = \frac{2\pi}{m} = 2\pi[m^{\text{est}} + \Delta m]^{-1} = \frac{2\pi}{m^{\text{est}}} \left[1 + \frac{\Delta m}{m^{\text{est}}} \right]^{-1}
$$

$$
\approx \frac{2\pi}{m^{\text{est}}} \left[1 - \frac{\Delta m}{m^{\text{est}}} \right] = \frac{2\pi}{m^{\text{est}}} - \frac{2\pi}{(m^{\text{est}})^2} \Delta m
$$

(11.10)

After equating T^{est} with $2\pi/m^{\text{est}}$, small fluctuations in period and frequency are found to be related through

$$
\Delta T \approx \frac{2\pi}{(m^{\text{est}})^2} \Delta m
$$

(11.11)

If the fluctuations in Δm have a variance of σ_m^2, then according to the usual rules of error propagation, the fluctuations in ΔT will have variance:

$$
\sigma_T^2 \approx \left[\frac{2\pi}{(m^{\text{est}})^2} \right]^2 \sigma_m^2
$$

(11.12)

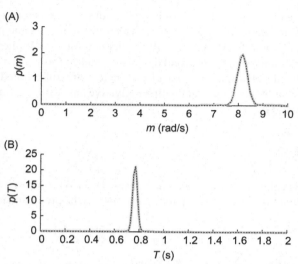

Figure 11.3 Normal probability density function $p(m)$ for angular frequency m with mean $\bar{m}$ and variance $\sigma_m^2 \ll \bar{m}$ (gray curve). An empirical probability density function (dashed curve) based on a histogram of 10,000 realizations of m_i closely matches this function. (B) Normal probability density function $p(T)$ for period $T = 2\pi/m$, with mean $\bar{T}$ and variance σ_T^2 calculated using the linear approximation discussed in the text (gray curve). An empirical probability density function (dashed curve) based on a histogram of $T_i = 2\pi/m_i$ closely matches this function. *MatLab* script eda11_03.

The period T has the same variance, since it differs from ΔT only by an additive constant. As an example, we consider the case $m^{est} = 2\pi \times 1.3$ rad/s, $\sigma_\omega = 0.2$ rad/s, which leads to $T^{est} = 0.769$ s and $\sigma_T = 0.019$ s, estimates that agree well with a numerical simulation (Figure 11.3).

11.6 Taylor series in multiple dimensions

In a typical data analysis problem, the predicted data $\mathbf{d}^{pre}(\mathbf{m})$ is a function of model parameters $\mathbf{m}$. The Taylor series for $\mathbf{d}^{pre}(\mathbf{m})$ is a straightforward generalization of the one-dimensional series and can be shown to be

$$d_i^{pre}(\mathbf{m}) = d_i^{pre}(\mathbf{m}_0) + \sum_{j=1}^{M} \frac{\partial d_i^{pre}}{\partial m_j}\bigg|_{\mathbf{m}_0} (m_j - m_{0j})$$

$$+ \frac{1}{2}\sum_{j=1}^{M}\sum_{k=1}^{M} \frac{\partial^2 d_i^{pre}}{\partial m_j \partial m_k}\bigg|_{\mathbf{m}_0} (m_j - m_{0j})(m_k - m_{0k}) + \dots$$

(11.13)

The sequence of terms is the same as in the one-dimensional case: a constant term; a linear term, a quadratic term, and so forth. Derivatives of the same order appear in the same places in both formulas, as well: the linear term is multiplied by a first derivative; the quadratic term by a second derivative; and so forth. The main complication is that both the data and the model parameters are subscripted quantities, so summations are involved. Note that the first derivative has two subscripts and is therefore a matrix, say $\mathbf{G}^{(0)}$:

$$
G_{ij}^{(0)} = \left. \frac{\partial d_i^{\text{pre}}}{\partial m_j} \right|_{\mathbf{m}_0}
\tag{11.14}
$$

The matrix $\mathbf{G}$ is called the *linearized data kernel*. The Taylor series is a little simpler when a scalar function of the model parameters, such as the prediction error E, is being considered:

$$
\begin{aligned}
E(\mathbf{m}) = {}& E(\mathbf{m}_0) + \sum_{j=1}^{M} \left. \frac{\partial E}{\partial m_j} \right|_{\mathbf{m}_0} \left(m_j - m_{0j} \right) \\
& + \frac{1}{2} \sum_{j=1}^{M} \sum_{k=1}^{M} \left. \frac{\partial^2 E}{\partial m_j \partial m_k} \right|_{\mathbf{m}_0} \left(m_j - m_{0j} \right)\left(m_k - m_{0k} \right) + \dots
\end{aligned}
\tag{11.15}
$$

The first derivative has one subscript and so is a vector, say $\mathbf{b}^{(0)}$, and the second derivative has two subscripts and so is a matrix, say $\mathbf{B}^{(0)}$:

$$
b_j^{(0)} = \left. \frac{\partial E}{\partial m_j} \right|_{\mathbf{m}_0} \quad \text{and} \quad B_{ij}^{(0)} = \left. \frac{\partial^2 E}{\partial m_j \partial m_k} \right|_{\mathbf{m}_0}
\tag{11.16}
$$

The quantities $\mathbf{b}$ and $\mathbf{B}$ are called the *gradient* vector and the *curvature* matrix, respectively. Equations (11.13) and (11.15) can then be written in vector notation as

$$
\mathbf{d}^{\text{pre}} \approx \mathbf{d}_0^{\text{pre}} + \mathbf{G}^{(0)} \Delta \mathbf{m}
$$

$$
E \approx E_0 + \mathbf{b}^{(0)\text{T}} \Delta \mathbf{m} + \frac{1}{2} \left(\Delta \mathbf{m} \right)^{\text{T}} \mathbf{B}^{(0)} \Delta \mathbf{m}
\tag{11.17}
$$

with $\quad \mathbf{d}_0^{\text{pre}} = \mathbf{d}^{\text{pre}}(\mathbf{m}_0) \quad$ and $\quad E_0 = E(\mathbf{m}_0) \quad$ and $\quad \Delta \mathbf{m} = \mathbf{m} - \mathbf{m}_0$

11.7   Small number approximation applied to covariance

The procedure for estimating variance (Section 11.5) can be generalized to the nonlinear vector function $\mathbf{T}(\mathbf{m})$, providing a way to estimate the covariance matrix $\mathbf{C}_T$. The first step is to expand the function $\mathbf{T}(\mathbf{m})$ in a Taylor series around the point $\mathbf{m}^{\text{est}}$ and keep only the first two terms. The resulting linear approximation relates small fluctuations in $\mathbf{m}$ to small fluctuations in $\mathbf{T}$:

$$\Delta T = M \Delta m$$

with $\quad \Delta m = m - m^{est} \quad$ and $\quad \Delta T = T - T(m^{est}) \quad$ and $\quad M_{ij} = \left. \dfrac{\partial T_i}{\partial m_j} \right|_{m^{est}}$

$$(11.18)$$

The standard error propagation rule (Section 3.12) then gives $C_T = MC_mM^T$.

11.8   Solving nonlinear problems with iterative least squares

The error $E(m)$ and its derivatives can be computed as long as the theory $d^{pre}(m)$ is known, irrespective of whether it is linear or nonlinear. Consequently, we can apply the principle of least square to nonlinear problems; the solution m^{est} is the one that minimizes the error $E(m)$, or equivalently, the one that solves $\partial E / \partial m_i = 0$. Unfortunately, this set of nonlinear equations is usually difficult to solve exactly. However, if we can guess a *trial solution* m_0 that is *close* to the minimum error solution, then we can use Taylor's theorem to approximate $E(m)$ as a low order polynomial centered about this point. If m_0 is close enough to the minimum that a quadratic approximation will suffice, then the process of solving $\partial E / \partial m_i = 0$ is very simple, because the derivative of a quadratic function is a linear function. Differentiating the expression for E in Equation (11.17) and setting the result to zero yield

$$0 \approx b^{(0)} + B^{(0)}\Delta m \quad \text{or} \quad \Delta m \approx -\left[B^{(0)}\right]^{-1} b^{(0)} \qquad (11.19)$$

This equation enables us to find an *updated* solution, $m = m_0 + \Delta m$, that is closer to the minimum than is m_0. This process can be iterated indefinitely, to provide a sequence of ms that, under favorable circumstances, will converge to the minimum error solution. It is sometimes referred to as *Newton's method*. A limitation is that the curvature matrix B must be calculated, in addition to the gradient vector b. Calculating this matrix may be difficult.

In the special case of the linear theory $d^{pre} = Gm$, the error is

$$E = \left[d^{obs} - Gm\right]^T \left[d^{obs} - Gm\right] = d^{obsT}d^{obs} - 2d^{obsT}Gm + m^TG^TGm$$

$$(11.20)$$

The unknown quantities m_0, b, and B are found by comparison with Equation (11.17):

$$m_0 = 0 \quad \text{and} \quad b = -2G^Td^{obs} \quad \text{and} \quad B = 2G^TG \qquad (11.21)$$

Inserting these values into Equation (11.19) yields the familiar least squares solution: $m = \left[G^TG\right]^{-1}G^Td^{obs}$. This correspondence suggests the strategy of approximating the curvature matrix as $\mathbf{B}^{(0)} \approx \mathbf{G}^{(0)T}\mathbf{G}^{(0)}$, in which case the procedure for updating a trial solution $\mathbf{m}^{(k)}$ is (Tarantola and Valette, 1982; see also Menke, 2014)

$$
\begin{aligned}
\Delta\mathbf{d} &= \mathbf{d}^{\text{obs}} - \mathbf{d}^{\text{pre}}\left(\mathbf{m}^{(k)}\right) \\
G_{ij}^{(k)} &= \left.\frac{\partial d_i^{\text{pre}}}{\partial m_j}\right|_{\mathbf{m}^{(k)}} \\
\Delta\mathbf{m} &= \left[\mathbf{G}^{(k)T}\mathbf{G}^{(k)}\right]^{-1}\mathbf{G}^{(k)T}\Delta\mathbf{d} \\
\mathbf{m}^{(k+1)} &= \mathbf{m}^{(k)} + \Delta\mathbf{m}
\end{aligned}
\tag{11.22}
$$

The covariance of the solution is approximately

$$
\mathbf{C}_m = \sigma_d^2\left[\mathbf{G}^{(k_{\max})T}\mathbf{G}^{(k_{\max})}\right]^{-1}
\tag{11.23}
$$

where $k_{\max}$ signifies the final iteration.

This procedure is known as *iterative least squares*. An important issue is when to stop iterating. Possibilities include when the error of the fit, as quantified by its posterior variance, declines below a specified value that represents the fit being acceptably good; or when the error no longer decreases significantly between iterations; that is, when $\Delta E = E\left(m^{(k)}\right) - E\left(m^{(k+1)}\right)$ declines below a specified value; or when the solution no longer changes significantly between iterations; that is $\{[\Delta\mathbf{m}]^T[\Delta\mathbf{m}]\}^{1/2}$ declines below a specified value.

Prior information, represented by the linear equation $\mathbf{Hm} = \mathbf{h}^{\text{pri}}$, can be added to the procedure in order to generalize it. The only nuance is that the prior information is satisfied by the model parameters $\mathbf{m}^{(k+1)}$ and not by the model perturbation $\Delta\mathbf{m}$. Inserting $\mathbf{m}^{(k+1)} = \mathbf{m}^{(k)} + \Delta\mathbf{m}$ into the prior information equation yields

$$
\mathbf{H}\Delta\mathbf{m} = \mathbf{h} - \mathbf{Hm}^{(k)}
\tag{11.24}
$$

See Crib Sheet 11.2 for the complete formulation.

Crib Sheet 11.2 Iterative generalized least squares

data equation $\mathbf{g}(\mathbf{m}) = \mathbf{d}^{\text{obs}}$ and prior information $\mathbf{Hm} = \mathbf{h}^{\text{pri}}$

trial solution $\mathbf{m}^{(k)}$

linearized equations $\mathbf{G}^{(k)}\Delta\mathbf{m} = \Delta\mathbf{d}^{(k)}$ and $\mathbf{H}\Delta\mathbf{m}^{(k)} = \Delta\mathbf{h}^{(k)}$

with

$$
G_{ij}^{(k)} = \left.\frac{\partial g_i}{\partial m_j}\right|_{\mathbf{m}^{(k)}}
$$

Continued

Crib Sheet 11.2—cont'd

and with

$$\Delta \mathbf{m}^{(k)} = \mathbf{m} - \mathbf{m}^{(k)} \quad \text{and} \quad \Delta \mathbf{d}^{(k)} = \mathbf{d} - \mathbf{g}\left(\mathbf{m}^{(k)}\right) \quad \text{and} \quad \Delta \mathbf{h}^{(k)} = \mathbf{h} - \mathbf{H}\mathbf{m}^{(k)}$$

combined equations

$$\mathbf{F}^{(k)} \Delta \mathbf{m}^{(k)} = \mathbf{f}^{(k)} \quad \text{with}$$

$$\mathbf{F}^{(p)} = \begin{bmatrix} \mathbf{C}_d^{-1/2} \mathbf{G}^{(k)} \\ \mathbf{C}_h^{-1/2} \mathbf{H} \end{bmatrix} \quad \text{and} \quad \mathbf{f}^{(p)} = \begin{bmatrix} \mathbf{C}_d^{-1/2} \Delta \mathbf{d}^{(k)} \\ \mathbf{C}_h^{-1/2} \Delta \mathbf{h}^{(k)} \end{bmatrix}$$

solution $\Delta \mathbf{m}^{(k)} = \left[\mathbf{F}^{(k)\mathrm{T}} \mathbf{F}^{(k)}\right]^{-1} \mathbf{F}^{(k)\mathrm{T}} \mathbf{f}^{(k)}$

update rule $\mathbf{m}^{(k+1)} = \mathbf{m}^{(k)} + \Delta \mathbf{m}^{(k)}$

covariance $\mathbf{C}_m = \left[\mathbf{F}^{(k)\mathrm{T}} \mathbf{F}^{(k)}\right]^{-1}$

11.9 Fitting a sinusoid of unknown frequency

Suppose that the data consists of a single sinusoid superimposed on a background level, and that the amplitude, phase, and frequency of the sinusoid, as well as the background level, is unknown. This scenario is described by an equation that is nonlinear in the model parameters:

$$d_i^{\mathrm{pre}} = m_1 \bar{d} + m_2 A \sin(m_4 \omega_a t_i) + m_3 A \cos(m_4 \omega_a t_i) \tag{11.25}$$

The constants $\bar{d}$, A, and ω_a have been introduced in order to normalize the size of the model parameters, which will be in the ± 1 range (or close to it) when these constants are properly chosen. The matrix $\mathbf{G}^{(k)}$ has elements:

$$
\begin{aligned}
G_{i1}^{(k)} &= \bar{d} \\
G_{i2}^{(k)} &= A \sin\left(m_4^{(k)} \omega_a t_i\right) \\
G_{i3}^{(k)} &= A \cos\left(m_4^{(k)} \omega_a t_i\right) \\
G_{i3}^{(k)} &= m_2^{(k)} A \omega_a t_i \cos\left(m_4^{(k)} \omega_a t_i\right) - m_3^{(k)} A \omega_a t_i \sin\left(m_4^{(k)} \omega_a t_i\right)
\end{aligned}
\tag{11.26}
$$

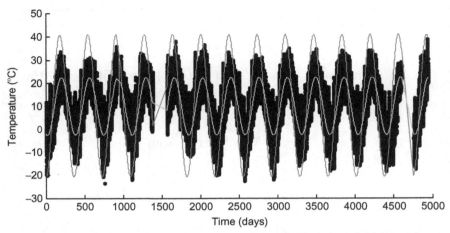

Figure 11.4 Black Rock Forest temperature data: observed (black dots); initial model (dark gray curve); final model (light gray curve). See text for further discussion. *MatLab* script eda11_04.

This model can be applied to the Black Rock Forest temperature data, which has a pronounced annual cycle. We set $\bar{d}$ to the mean of the data, A to half its range, and ω_a to the frequency $2\pi/365$ rad/day. A reasonable trial solution is $\mathbf{m} = [1.0, 0.1, -0.9, 1.0]^T$. (The choice $m_2 = 0.1$, $m_3 = -0.9$ implies that the highest temperature is in summer.) The iterative procedure of Equation (11.22) is then used to refine this estimate. The iteration process is terminated when the change in model parameters, from one iteration to the next, drops below a predetermined value (in this example, 10^{-4}).

This method produces an estimated period of 365.61 ± 0.06 days (95%), where the confidence interval was computed by the method of Section 11.5. Whether the non-linear fit (Figure 11.4) provides a better estimate of the period that can be determined from standard amplitude spectral density analysis is debatable. The annual peak in the amplitude spectral density is broad, owing to the relatively few annual oscillations in the temperature time series (only 11). Nevertheless, its center is very close to the value determined by the nonlinear analysis (Figure 11.5).

11.10 The gradient method

The gradient vector $b_i^{(k)} = \partial E^{(k)}/\partial m_i$ represents the slope of the error function $E(\mathbf{m})$ in the vicinity of the trial solution $\mathbf{m}^{(k)}$. It points in the direction in which the error most rapidly increases, which is to say, directly away from the direction in which $\mathbf{m}^{(k)}$ must be perturbed to lower the error. For this reason, its appearance in the rule for updating a trial solution $\mathbf{m}^{(k)}$ (Equation 11.19) is unsurprising. This relationship is further examined by rewriting the update rule as

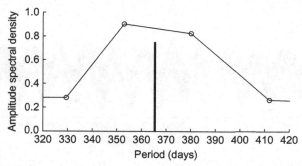

Figure 11.5 Amplitude spectral density (curve with circles) of the Black Rock Forest temperature dataset, for a restricted period range centered on the annual period of 365 days. Note that this curve has a broad peak centered on the annual period. The period (vertical bar) determined by the nonlinear least squares fit is close to the center of this peak. See text for further discussion. *MatLab* script eda11_04.

$$\Delta \mathbf{m} = -\left[\mathbf{B}^{(k)}\right]^{-1}\mathbf{b}^{(k)} = \mathbf{M}^{(k)}\,\boldsymbol{v}^{(k)} \quad \text{with} \quad \mathbf{M} = \left|\mathbf{b}^{(k)}\right|\left[\mathbf{B}^{(k)}\right]^{-1}$$
$$\text{and} \quad \boldsymbol{v} = -\mathbf{b}^{(k)}/\left|\mathbf{b}^{(k)}\right| \tag{11.27}$$

Here $|\mathbf{b}|$ is the length of $\mathbf{b}$. This rule has a simple interpretation: the vector $\boldsymbol{v}^{(k)}$ is the *downslope direction*; that is, the direction in which $\mathbf{m}^{(k)}$ should be perturbed to most reduce the error. The matrix $\mathbf{M}^{(k)}$ encodes information about the *size* of the perturbation needed to reach the minimum in the error $E(\mathbf{m})$, and although it also can change the direction of the perturbation, this effect is usually secondary.

The gradient vector $\mathbf{b}^{(k)}$ can be computed easily from the linearized data kernel $\mathbf{G}^{(k)}$:

$$b_i^{(k)} = \left.\frac{\partial E}{\partial m_i}\right|_{\mathbf{m}^{(k)}} = \frac{\partial}{\partial m_i}\sum_{j=1}^{N}\left(d_j^{\text{obs}} - d_j^{\text{pre}}\right)^2 = -2G_{ji}^{(k)}\left(d_j^{\text{obs}} - d_j^{\text{pre}}\right)$$
$$\text{or} \quad \mathbf{b}^{(k)} = -2\mathbf{G}^{(k)\text{T}}\left(\mathbf{d}^{\text{obs}} - \mathbf{d}^{\text{pre}}\right) \tag{11.28}$$

The calculation of the matrix $\mathbf{M}^{(k)}$ is more difficult because it involves taking a matrix inverse. A trial solution would be easier to update if it could be omitted from the formula. The *gradient method* approximates $\mathbf{M}^{(k)}$ as $\alpha\mathbf{I}$, where α is an unknown *step size*. An initial guess for α is made at the start of the error minimization process and is used to tentatively update the trial solution according to the rule $\mathbf{m}^{(k+1)} = \mathbf{m}^{(k)} + \alpha\boldsymbol{v}^{(k)}$. The iteration process continues if the error $E^{(k+1)}$ has decreased. However, if the error has increased, the new solution has *overshot* the minimum because the step size is too big. The step size is decreased, say by replacing α with $\alpha/2$, and another tentative solution is tried.

The main limitation of the gradient method, compared to iterative least squares, is that more iteration is needed to achieve an acceptable solution. While each step moves the solution *towards* the minimum, the step size has not been correctly set to move it *to*

(A)

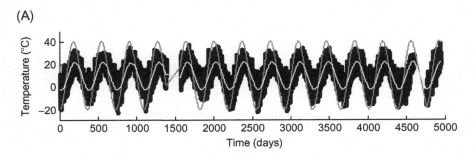

(B)

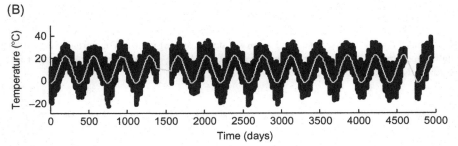

(C)

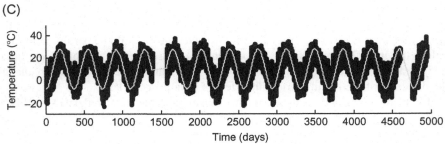

Figure 11.6 Black Rock Forest temperature dataset (black dots) modeled as a single sinusoid of unknown period. (A) Iterative least squares fit. (B) Gradient method. (C) Stochastic gradient method. See text for further discussion. *MatLab* scripts eda11_05 and eda11_06.

the minimum, so convergence is much slower. When applied to Black Rock Forest temperature problem described in the previous section, about 3500 iterations of the gradient method are needed to provide an acceptable solution (Figure 11.6) whereas fewer than 10 iterations of the iterative least squares method are needed for a solution with similar accuracy.

The least squares error $E(\mathbf{m})$ is the sum of many individual errors:

$$E(\mathbf{m}) = \sum_{i=1}^{N} e_i^2 \quad \text{where } e_i = d_i^{\text{obs}} - d_i^{\text{pre}}(\mathbf{m}) \tag{11.29}$$

In many practical cases, the number of individual errors is very large ($N \approx 100,000$ in the Black Rock Forest case). The updating rule $\mathbf{m}^{(k+1)} = \mathbf{m}^{(k)} + \alpha \boldsymbol{\nu}^{(k)}$ is only approximate. The omission of some of the individual errors from the formula for $\boldsymbol{v}^{(k)}$ makes it somewhat more approximate, but as long as $\boldsymbol{v}^{(k)}$ is not too inaccurate, each step will still move the solution towards the minimum. Furthermore, if a different random subset of individual errors is used in each step, the entire dataset can be *cycled through* during the course of the solution. This approximation is called the *stochastic gradient method*. In the case of the Black Rock Forest dataset, a reduction of the number of individual errors by a factor of 1000 still leads to an acceptable solution (Figure 11.6).

11.11   Precomputation of a function and table lookups

Many data analysis methods require that the same function be evaluated numerous times. If the function is complicated but has only one or two arguments of known range, then precalculating it and storing its values in a table may prove computationally efficient. As long as the time needed to perform a table lookup is less than the time needed to evaluate the function, computation time is reduced. Some accuracy is lost, too, since a table lookup provides only an approximate value of the function, but in many cases this inaccuracy is overwhelmed by other sources of error (and especially measurement error) that enter into the problem.

The lookup table for a one-dimensional function $d(x)$ is just the time series $d_k = d(x_k)$ with $x_k = k\Delta x + x_{\min} (1 < k < K)$. The range of x_k must encompass all values of x that need to be estimated, and the sampling must be small enough to faithfully represent fluctuations in the function. A simple estimate of the function $d(x)$ is the value of the corresponding table entry, d_n with $n = \text{floor}((z - z_{\min})/\Delta x) + 1$. Here the function floor(x) is the largest integer smaller than x. Alternatively, $d(x)$ could be estimated by linearly interpolating between the two bracketing d_n and d_{n+1}, though the interpolation process adds computation time.

As an example, we consider the problem of using a grid search to determine two model parameters (m_1, m_2) where the data $d_i(t_i)$ obeys

$$d_i(t_i) = \sin\left\{\pi \sin\left[\pi x_i \sin \pi (1 - x_i^2)\right]\right\} \quad \text{with} \quad x_i = m_1 + m_2 t_i \quad (11.30)$$

Note that $d_i(x_i)$ is a complicated function of just one variable $x_i(t_i)$. In order to set up the lookup table, we need to know the range of plausible values of the model parameters (m_1, m_2) and time t. Let us suppose that they are $0 \le m_1 \le 1$, $0 \le m_2 \le 1$, and $0 \le t \le 1$. These ranges allow us to compute the range for the xs in the lookup table as $0 \le x \le 2$. The version of the grid search that uses the lookup table executes in only about half the time of the version that directly computes $d_i(t_i)$ each time it is needed (Figure 11.7).

Lookup tables are applicable to functions of two or more variables, but tables of more than two dimensions are seldom useful, because the time needed to construct them is very large.

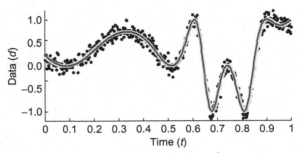

Figure 11.7 Results of grid search, showing observed data $\mathbf{d}^{obs}$ (dots), true data $\mathbf{d}^{true}$ (light gray curve), estimated data $\mathbf{d}^{obs}$ using a standard grid search (dark gray curve), and estimated data using a grid search with a table lookup (black curve). *MatLab* script eda11_07.

11.12   Artificial neural networks

Suppose that the function $d(x)$ is not accurately known. One strategy is to start with an initially poor estimate of it and improve it as new information becomes available. One advantage of using a lookup table to represent the function is that a table is easily updated. The values of its elements can be trivially changed to reflect new information. However, a table is also inflexible, in the sense that, once defined, the width Δx of its elements cannot easily be changed to reflect new information about the scale length over which $d(x)$ varies. Furthermore, the value of $d(x)$ jumps abruptly as an interval boundary is crossed, which may violate prior information about the function's smoothness. In this section, a new approximation, called the *artificial neural network* (or *neural net*, for short), is introduced. It shares some similarity with a lookup table but is more adaptable and smooth. (Notice that the attribute of the lookup table upon which we are focusing is updatability, not computation speed; neural nets are not especially speedy.)

An arbitrary function $d(x)$ can be represented schematically as a row of three boxes, with interconnecting lines that represent information flowing from left to right (Figure 11.8A). In this *network* representation, the left box represents input x, the right box represents output $d(x)$, and the middle box represents the workings of the function, which may be arbitrarily complicated. The middle box encapsulates (or *hides*) all the details of the function, and for this reason is said to be *hidden*. The table approximation replaces the workings in the middle box with a simple table lookup (Figure 11.8B). Since each element of the table specifies the value of the function $d(x)$ in a distinct interval of the x-axis, the process of interrogating the table lookup can be viewed as summing boxcar or *tower* functions, each of which is constant in the interval and zero outside of it (Figure 11.8C).

The left and right edges of the each table entry can be made separate by viewing a tower as the sum of two step functions:

$$d_c H(w_L x + b_L) - d_c H(w_R x + b_R) \tag{11.31}$$

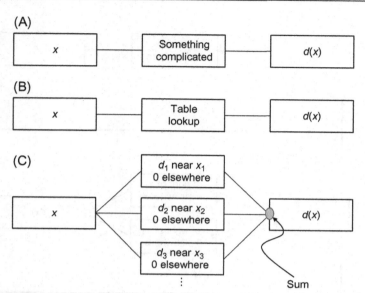

Figure 11.8 Schematic representation of a function, in which information flows from left to right. (A) The function $d(x)$. (B) The table lookup approximation to the function $d(x)$. (C) The middle box of the table lookup in (B) is subdivided into several distinct boxcars (or *towers*) each of which represents one element of the table. Note that information diverging out of a box is duplicated along the several connections and information from multiple connections converging on a box is summed.

Here $H(z)$ (with $z = wx + b$) is the Heaviside step function, defined as zero when $z < 0$ and unity when $z > 0$. The *weights* w and the *biases* b parameterize the location of the edges, with the left edge at $x_L = -b_L/w_L$ and the right edge at $x_R = -b_R/w_R$. The constant d_c specifies the height of the tower (Figure 11.9). In a lookup table, the edges of adjacent towers match one another; that is, $x_L^{(i)} = x_R^{(i-1)}$ and $x_R^{(i)} = x_L^{(i+1)}$ (where the superscript consecutively numbers the towers). However, in a neural net, this requirement is relaxed and step functions are allowed to overlap in any fashion.

The issue of smoothness is addressed by replacing the step function $H(z)$ by the *sigmoid* function:

$$\sigma(z) = \frac{1}{1 + \exp(-z)} \tag{11.32}$$

Like the step function, this function asymptotes to zero and unity, respectively, as $z \to \pm\infty$ and has a transition centered on $z = 0$ (or $x_0 = -b/w$). Unlike the step function, the sigmoid function is smooth. The parameter w controls the sharpness of the transition, with a maximum slope at x_0 of $s = d\sigma/dx = w/4$ (Figure 11.10).

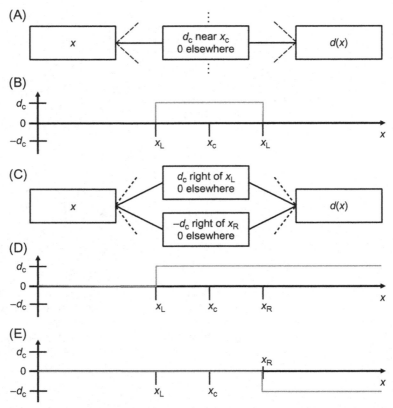

Figure 11.9 (A) Neural net, with one tower function singled out. (B) Graph of the tower function. (C) Neural net with the tower function replaced by two step functions. (D) and (E) Graphs of the two step functions.

11.13 Information flow in a neural net

A neural net consists of L columns (or *layers*) of boxes, with the kth layer containing $N^{(k)}$ boxes. Because the neural net approximation was motivated by studies of information flow in animal brains, the boxes are called *neurons*. Information flows from left to right along *connections*, denoted by line segments connecting the boxes in adjacent layers. Neuron i in layer k has one parameter associated with it, its bias $b_i^{(k)}$. The connection between neuron i in layer k and neuron j in layer $(k-1)$ has one parameter associated with it, its weight $w_{ij}^{(k)}$; the absence of a connection is indicated by a zero weight. Neuron i in layer k also has an input value $z_i^{(k)}$ and an output value (or *activity*) $a_i^{(k)}$. The activities of neurons in layer 1 represent the input to the function and the activities of the neurons in layer L represent the output of the function. In the one-dimensional case considered above, layer 1 has one neuron with activity x and layer L has one neuron with activity $d(x)$. However, neural nets can handle a

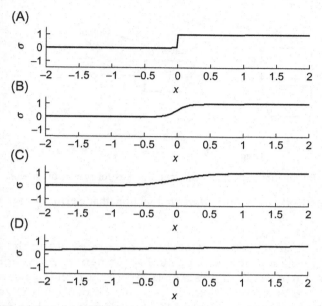

Figure 11.10 Exemplary sigmoid functions, $\sigma(wx + b)$. (A) Slope $s = w/4 = 100$, $b = 0$. (B) Slope $s = 3$, $b = 0$. (C) Slope $s = 1$, $b = 0$. (D) Slope $s = 0.1$, $b = 0$. Note that the width of the transition is sharper for the larger ss. *MatLab* script eda11_08.

multidimensional function $\mathbf{d}(\mathbf{x})$ simply by having multiple input and output neurons. The rule for propagating information between one layer and the next is

Input of	weighted sum of	bias of
neuron i	= activities of neurons	+ neuron i
on layer k	on layer $(k-1)$	on layer k

$$z_i^{(k)} = \sum_{j=1}^{N^{(k-1)}} w_{ij}^{(k)} a_j^{(k-1)} + b_i^{(k)}$$

Activity of	$k = L$: input of that neuron	(11.33)
neuron i	= $k < L$: sigmoid function applied to input
on layer k	of that neuron

$$a_i^{(k)} = \begin{cases} z_i^{(k)} & \text{when } k = L \\ \sigma\!\left(z_i^{(k)}\right) & \text{when } k < L \end{cases}$$

These relationships are depicted in Figure 11.11. Once the activities in layer 1 are specified, this rule can be used to propagate information layer by layer through

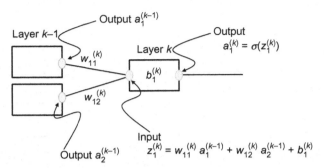

Figure 11.11 Information flow in a neural net. See text for further discussion.

the neural network. A simple *MatLab* function that performs this operation is provided:

```
a = eda_net_eval(N,w,b,a);
% set the activities of the layer 1 before calling this function
% N(i): column-vector with number of neurons in each layer;
% biases: b(1:N(i),i) is a column-vector that gives the
%          biases of the N(i) neurons on the i-th layer
% weights: w(1:N(i),1:N(i-1),i) is a matrix that gives the
%          weights between the N(i) neurons in the i-th layer
%          and the N(i-1) neurons on the (i-1)-th layer.
% activity: a(1:N(i),i) is a column-vector that gives the
%          activities of the N(i) neurons on the i-th layer
```
(MatLab eda_net_eval)

The neural nets for smoothed versions of a step function and a tower function are shown in Figure 11.12. Since the sigmoid function is omitted on the output layer, the activity of this layer is just a weighted sum of its inputs. This feature allows the easy amalgamation of networks. For instance, several single-tower networks can be amalgamated to yield a sum of towers (Figure 11.13).

A two-dimensional tower, say for input (x, y), can be created by amalgamating two one-dimensional tower networks, a tower in x and a tower in y (Figure 11.14A). However, their sum is not quite a two-dimensional tower function but rather consists of two intersecting ridges, one parallel to the x-axis and the other parallel to the y-axis, with a hump where the tower should be (Figure 11.14B). A fourth layer, implementing a step function with a transition higher than the ridges but lower than the hump, is needed to delete the ridges and form the tower (Figure 11.14C).

A function can be approximated to any degree of fidelity by summing together enough towers. Thus, a neural network is capable of approximating any continuous function. However, a network constructed in this fashion may not be the smallest capable of adequately approximating the function. In the next section we will develop a technique to *train* an arbitrary network to produce the best approximation of a function of which it is capable. This is a simple form of *machine learning*.

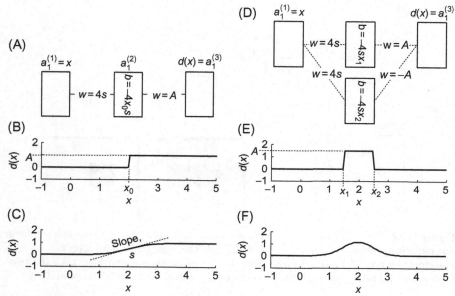

Figure 11.12 Simple neural nets. (A) Neural net for a smoothed version of a step function $d(x)$ with amplitude A, transition at x_0, and a slope s. All values not shown are zero. (B) Graph for the $A = 1, x_0 = 2, s = 250$ case. (C) Graph for the $A = 1, x_0 = 2, s = 1$ case. (D) Neural net for a smoothed version of a tower function $d(x)$ with amplitude A, transitions at x_1 and at x_2, and slope s. (E) Graph for the $A = 1.5, x_1 = 1.5, x_2 = 2.5, s = 250$ case. (F) Graph for the $A = 1.5$, $x_1 = 1.5, x_2 = 2.5, s = 1$ case. *MatLab* scripts eda11_09 and eda11_10.

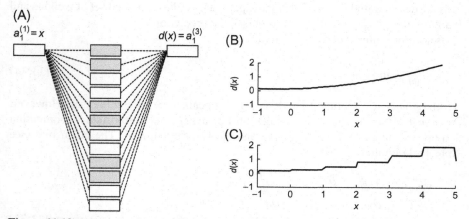

Figure 11.13 (A) A neural net with six towers (six pairs of sigmoid functions) is used to approximate a function $d(x)$. (B) The true function $d(x)$, a polynomial. (C) The neural net approximation. *MatLab* script eda11_11.

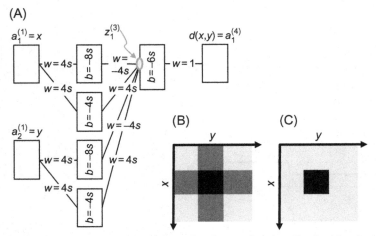

Figure 11.14 (A) A neural net for a two-dimensional tower function $d(x,y)$ with unit amplitude and nonzero between $1 < x < 2$ and $1 < y < 2$. (B) The input to the third layer has low-amplitude ridges (gray) that intersect to produce a high amplitude hump (black). (C) The fourth layer deletes the ridges but leaves the hump. Its output is the tower. *MatLab* script eda11_12.

11.14   Training a neural net

The goal in training a neural net is to match the output $d_i^{\text{pre}} = d^{\text{pre}}(x_i)$ of the neural net to some set of observed values (x_i, d_i^{obs}); that is, to find a set of weights and biases such that $d_i^{\text{pre}} \approx d_i^{\text{obs}}$. This problem can be solved by defining an error $e_i = d_i^{\text{obs}} - d_i^{\text{pre}}$ and finding the weights and biases that minimize the total error $E = \mathbf{e}^{\text{T}}\mathbf{e}$. Since the neural net is inherently nonlinear, the error minimization must be performed with the iterative least squares method (Section 11.8) or the gradient method (Section 11.10). The process refines an initial guess of the model parameters (the combined set of weights and biases) into a final estimate that minimizes the prediction error.

Both procedures require the derivatives:

$$\frac{\partial d_i^{\text{pre}}}{\partial b_n^{(k)}} \quad \text{and} \quad \frac{\partial d_i^{\text{pre}}}{\partial w_{nm}^{(k)}} \tag{11.34}$$

At first glance, computing these derivatives appears to be a daunting task. However, recall that d_i^{pre} is just the activity of neuron 1 in layer L, and the formula for computing activities (Equation 11.33) is simple and easily differentiated. The chain rule (see Note 11.1) implies

$$\frac{\partial d_i^{\text{pre}}}{\partial b_j^{(k)}} = \frac{\partial a_1^L}{\partial b_j^{(k)}} = \frac{\partial z_j^{(k)}}{\partial b_j^{(k)}} \frac{\partial a_j^{(k)}}{\partial z_j^{(k)}} \frac{\partial a_1^{(L)}}{\partial a_j^{(k)}}$$

and

$$\frac{\partial d_i^{\text{pre}}}{\partial w_{jm}^{(k)}} = \frac{\partial a_1^L}{\partial w_{jm}^{(k)}} = \frac{\partial z_j^{(k)}}{\partial w_{jm}^{(k)}} \frac{\partial a_j^{(k)}}{\partial z_j^{(k)}} \frac{\partial a_1^{(L)}}{\partial a_j^{(k)}} \tag{11.35}$$

While the derivatives of Equation (11.35) are of varying complexity, they are all tractable. The easiest is

$$\frac{\partial z_j^{(k)}}{\partial b_j^{(k)}} = 1 \tag{11.36}$$

The two next most complicated are

$$\frac{\partial z_j^{(k)}}{\partial w_{jm}^{(k)}} = a_m^{(k-1)} \quad \text{and} \quad \frac{\partial z_j^{(k)}}{\partial a_m^{(k-1)}} = w_{jm}^{(k)} \tag{11.37}$$

More complicated still is

$$\frac{\partial a_j^{(k)}}{\partial z_j^{(k)}} = \begin{cases} \dfrac{\partial \sigma}{\partial z_j^{(k)}} = \dfrac{\exp\left(-z_j^{(k)}\right)}{\left[1 + \exp\left(-z_j^{(k)}\right)\right]^2} = a_j^{(k)}\left[a_j^{(k)} - 1\right] & (k < L) \\ 1 & (k = L) \end{cases} \tag{11.38}$$

Finally, the $\partial a_i^{(L)}/\partial a_j^{(k)}$ derivative is built layer by layer, using another application of the chain rule. The derivative that connects changes in activities on *adjacent* layers is

$$\frac{\partial a_i^{(k)}}{\partial a_j^{(k-1)}} = \frac{\partial a_i^{(k)}}{\partial z_i^{(k)}} \frac{\partial z_i^{(k)}}{\partial a_j^{(k-1)}} \tag{11.39}$$

A succession of these derivatives can be used to connect the activities on the Lth layer to all the other layers, a process called *back-propagation*, since it starts with the Lth layer and works backward towards the first layer (Figure 11.15) (Werbos, 1974):

$$\frac{\partial a_i^{(L)}}{\partial a_j^{(L-1)}} = \frac{\partial a_i^{(L)}}{\partial z_i^{(L)}} \frac{\partial z_i^{(L)}}{\partial a_j^{(L-1)}} \quad \text{and}$$

$$\frac{\partial a_i^{(L)}}{\partial a_j^{(L-2)}} = \sum_{p=1}^{N^{(L-1)}} \frac{\partial a_i^{(L)}}{\partial a_p^{(L-1)}} \frac{\partial a_p^{(L-1)}}{\partial a_j^{(L-2)}} \quad \text{and} \tag{11.40}$$

$$\frac{\partial a_i^{(L)}}{\partial a_j^{(L-3)}} = \sum_{p=1}^{N^{(L-2)}} \frac{\partial a_i^{(L)}}{\partial a_p^{(L-2)}} \frac{\partial a_p^{(L-2)}}{\partial a_j^{(L-3)}} \quad \text{etc.}$$

The summation arises because a perturbation in the activity of a neuron on the kth layer is due to the combined effect of perturbations of the activities of all the neurons on the $(k-1)$th layer that connect with it (see Note 11.1).

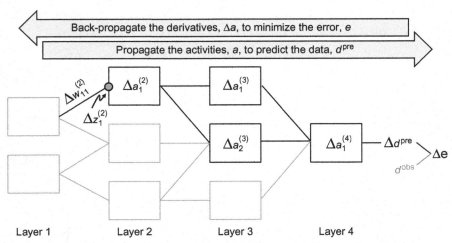

Figure 11.15 The back-propagation process follows the chain of events that lead to a change in the error Δe, backwards, from effect to cause. A perturbation in Δe is caused by a perturbation in the activity $a_1^{(4)} = d^{\mathrm{pre}}$, which in turn is caused by perturbations $\Delta a_1^{(3)}$ and $\Delta a_2^{(3)}$, which in turn are caused by perturbation $\Delta a_1^{(2)}$, which in turn is caused by input perturbation $\Delta z_1^{(2)}$, which in turn is caused by a perturbation $\Delta w_{11}^{(2)}$ in a specific weight.

A *MatLab* function that calculates the derivatives is provided:

```
[daLmaxdw, daLmaxdb] = eda_net_deriv(N,w,b,a);
% update the activities by calling a=eda_net_eval(N,w,b,a);
%       before calling this function
% daLmaxdb(i,j,k): the change in the activity of the i-th neuron
%       in layer Lmax due to a change in the bias of the of
%       the j-th neuron in the layer k
% daLmaxdw(i,j,k,l): the change in the activity of the i-th
%       neuron in layer Lmax due to a change in the weight
%       connecting the j-th neuron in the layer l with the k-th
%       neuron of layer l-1
```
                                          (*MatLab* eda_net_deriv)

The weights and biases together constitute the model parameters **m** of the iterative least squares problem (Section 11.8). The matrix $G_{ij}^{(p)}$ contains the derivatives, and has one row i for each observation and one column j for each model parameter. Careful bookkeeping is necessary to keep track of the correspondence between the jth model parameter and a particular weight or bias. Index tables can be used to translate back and forth between these two representations. We provide a *MatLab* function eda_net_map() that creates them but do not describe it further, here (see the comments in the script). An important issue is the starting values of the weights and activities. For simple networks that represent towers or other simple behaviors, it may be possible to choose the weights and biases so that initial behavior is somewhat

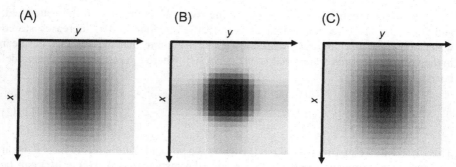

Figure 11.16 A single-tower network is trained to match an observed function $d^{obs}(x,y)$. (A) The observed function $d^{obs}(x,y)$, a Normal function. (B) The initial estimate $d^{obs}(x,y)$, based on a single-tower neural net with a slope $s = 1.25$. (C) The prediction $d^{pre}(x,y)$, after training. *MatLab* script eda11_13.

close to matching the observations. In this case, training merely refines the behavior and only a few iterations are necessary. In more complicated cases, the weights and biases may need to be initialized to random numbers, with the hope that the training will converge eventually to the best-fitting values. In both cases, imposing prior information that the correction $\Delta \mathbf{m}^{(k)}$ is small can help prevent overshoots and improve the rate of convergence.

As an example, a network consisting of a single tower is trained to match observations $d^{obs}(x,y)$ that have the shape of a two-dimensional Normal function (Figure 11.16A). The initial weights and biases are chosen to that the tower is centered on the peak of the Normal function but has edges that, while smooth, fall off more quickly than is observed (Figure 11.16B). The iterative least squares procedure converges rapidly and reduces the error to 6×10^{-5} of its starting value (Figure 11.16C).

11.15 Neural net for a nonlinear response

In Section 7.4, river discharge $d(t)$ is predicted from precipitation $x(t)$ by developing a filter $c(t)$ such that $d(t) = c(t) * x(t)$. A filter is appropriate when the response of the river to precipitation is linear. It assumes that the river responds similarly to a shower and a storm; the latter is merely a scaled up version of the former. In reality, few rivers are completely linear in this sense, because they behave differently in periods of low and high water. Storms cause a river to widen and, in extreme cases, to overflow its banks. Both effects lead to nonlinearities in the river's response to precipitation.

Consider the following very simple model of a river, in which the function $y(t)$ represents the volume of water stored in its watershed:

$$\frac{dy}{dx} = A_w x(t) - d(t) \quad \text{with} \quad d(t) = A_r v(t) \tag{11.41}$$

Here $v(t)$ is the mean velocity of water in the river, A_w is the area of the watershed, and A_r is the cross-sectional area of the river. Precipitation $x(t)$ increases stored water and river discharge $d(t)$ decreases it. Now assume that the velocity is linear in the volume of stored water; that is, $v(t) = c_1 y(t)$, where c_1 is a constant. The model becomes

$$\frac{dy}{dx} = A_w x(t) - c_1 A_r y(t) \tag{11.42}$$

This differential equation is linear in the unknown $y(t)$ and will therefore have a linear response to an impulsive precipitation event (the response to $A_w x(t) = \delta(t)$ can be shown to be $y(t) = \exp(-c_1 A_r t)$ for $t > 0$ and zero otherwise). Since discharge is proportional to stored water, it too has a linear response. A filter can completely capture the relationship between precipitation and discharge.

The model is now modified into a nonlinear one by assuming that the cross-sectional area of the river is proportional to stored water, so that $A_r = c_2 y(t)$, where c_2 is a constant. The model becomes

$$\frac{dy}{dx} = A_w g(t) - c_1 c_2 \, y^2(t) \tag{11.43}$$

This differential equation is nonlinear in the unknown $y(t)$ and consequently does not have a linear response. While an impulse in precipitation will cause a transient discharge signal, the signal for a storm will not have the same shape as the signal for a shower. Furthermore, the response to several closely spaced precipitation events will not linearly superimpose. As we will demonstrate below, a neural net is capable of predicting this nonlinear behavior.

Initially, a neural net is constructed that is capable of representing convolution by the filter $c(t)$. Later, this network is modified by adding connections that enhance its ability to represent nonlinearities. The filtering operation $d(t) = c(t) * x(t)$ is just a multidimensional generalization of the linear function $d(x) = cx + h$ (with $h = 0$):

$$d(t_i) = c(t) * x(t) = c_1 x(t_i) + c_2 x(t_{i-1}) + c_3 x(t_{i-2}) + \dots \tag{11.44}$$

Therefore, a *linear function-emulating network* is first created (Figure 11.17) and K of these networks are amalgamated into a *filter-emulating network* that implements convolution by a length-K filter. The linear function-emulating network exploits the property of the sigmoid function $\sigma(wx - b)$ that, for small weight w, its output is a linear function of its input, at least near its transition point $x_0 = -b/w$ (see Figure 11.10D). This linear behavior is demonstrated by expanding the sigmoid function in a Taylor series:

$$\sigma(wx - b) \approx \frac{1}{2} + \frac{1}{4} w(x - x_0) \quad \text{for} \quad x \approx x_0 \text{ and } w \ll 1 \tag{11.45}$$

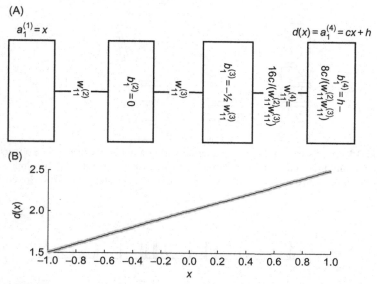

Figure 11.17 (A) Four layer neural net that implements the function $d(x) = h + cx$. (B) Graph of $d^{\text{true}}(x) = 2 + 1.5x$ (gray curve) and $d^{\text{pre}}(x)$ (black curve). *MatLab* script eda11_14.

The linear function-emulating network has four layers containing six parameters (three weights and three biases). These parameters can be initialized so that the network behaves linearly; subsequent training enables it to capture more complicated nonlinear behaviors.

This utility of the filter-emulating network is demonstrated using a precipitation-discharge test dataset. A synthetic precipitation time series $x(t)$, of length $N = 1000$, is created from a random patterns of spiky precipitation events. The corresponding discharge time series $d^{\text{obs}}(t)$ is determined by solving Equation (11.43) numerically. The nonlinear nature of the relationship between precipitation and discharge is apparent; large precipitation events produce disproportionately large peaks in discharge. A $K = 10$ filter-emulating network is initialized to the filter $c(t) = A \exp(-bt)$, where the parameters A and b are determined by least squares (see Section 7.4). Several additional connections are added to the network to allow it to better approximate a nonlinear response function (Figure 11.18) and these new connections are assigned small, randomly chosen weights. The final network contains $M = 69$ unknown weights and biases. When trained on the first half of the dataset, the network achieves an error reduction of a factor of about 10 when compared to the filter (Figure 11.19). Finally, without any retraining, the network is used to predict the second half of the test dataset. The network predicts it equally well, suggesting that it has successfully captured the nonlinear and, from the point of view of the test, unknown dynamics of the river system.

While not pursued here, the neural net could in principal be retrained on a daily basis, as more data arrives, using a rolling window of several years of data (see

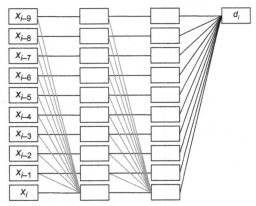

Figure 11.18 Exemplary neural net for approximating a nonlinear response. Connections in black are the minimum number needed to implement a filter of length $K = 10$. Connections in gray have been added to improve the net's accuracy of prediction. *MatLab* script eda11_15.

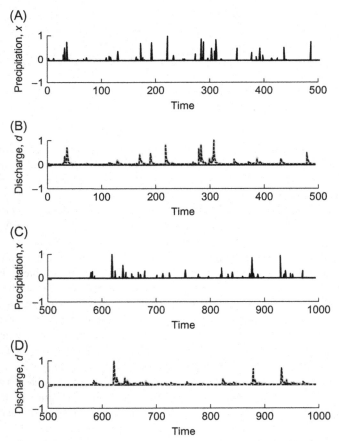

Figure 11.19 Application of a neural net to a nonlinear response problem. (A) First half of a synthetic observed precipitation time series $x(t)$. (B) Corresponding observed river discharge time series $d^{obs}(t)$, obtained by solving Equation (11.41) (gray) and its neural net prediction $d^{pre}(x)$ (solid) after training with these observations. (C) Second half of synthetic precipitation time series $x(t)$. (D) Corresponding observed river discharge time series $d^{obs}(x)$ (gray) and its prediction $d^{pre}(x)$ (solid) (without retraining). *MatLab* script eda11_15.

Problem 11.7). Such a strategy—a form of machine learning—would allow the network to adapt to unquantified factors that affects river dynamics, such as land-use changes. In the case we have been considering, the problem is small enough that full iterative least squares could be used for each daily update. In larger problems or ones in which very frequent updates are needed, the stochastic gradient method (Section 11.10) might prove preferable.

Problems

11.1. Derive the small number approximation for $\exp(x)$ near the point $x = 0$. Plot the function and its approximation. What is the range of x for which the error is less than 5%?

11.2. Suppose that uncorrelated normally distributed random variables x and y have means $\bar{x}$ and $\bar{y}$, respectively, and variances σ_x^2 and σ_y^2, respectively, with $\sigma_x \ll \bar{x}$ and $\sigma_y \ll \bar{y}$. Use small number approximations to estimate the covariance matrix $\mathbf{C}_{uv}$ of $u = xy$ and $v = x/y$.

11.3. Modify eda11_05 to include both annual and diurnal frequencies of oscillation. By how much is the error reduced?

11.4. Suppose that the time needed to evaluate a function of one variable is T_f, the time needed to lookup its value in a table is $T_t = T_f/10$, and the number of table entries is K. How many times must the function be evaluated in a grid search in order to achieve an 80% reduction in computation time? Be sure to include the time needed to create the table in your estimate.

11.5. Modify eda11_13.m to fit the function:

$$d^{\text{obs}}(x, y) = \sin(x)\cos(y) \quad \text{with} \quad 0 \leq x \leq \pi \text{ and } 0 \leq y \leq \pi$$

with a neural net containing two towers.

11.6. (A) Modify the neural net in eda11_11.m to model the parabola with just two towers of slope $s = 12.5$. (B) Add code, modeled on the iterative least squares procedure in eda11_13.m, to train the network. How good is the fit?

11.7. (This problem is especially difficult.) The test dataset in the nonlinear response example (Section 11.15) was generated by the script nonlinearresponse.m. This script solves Equation (11.43) with $c_1 c_2 = c_f = 0.2$. Modify this script to generate a test dataset of length 10,000 in which the coefficient c_f systematically increases from 0.2 at the start of the dataset to 0.5 at its end, emulating land-use changes in the watershed. Then modify the network training script eda11_15. Retain the section that uses iterative least squares to train on the first 500 points, but then predict the data in two different ways: by applying the neural net, as is, to the entire dataset; and by retraining the network each time step using a moving window of the previous 500 points. Use the same iterative least squares method, but with only one iteration, to retrain. Compare the results of the two methods. Can the retrained network keep up with changing river conditions?

References

Menke, W., 2014. Review of the generalized least squares method. Surv. Geophys. 36, 1–25.

Tarantola, A., Valette, B., 1982. Generalized non-linear inverse problems solved using the least squares criterion. Rev. Geophys. Space Phys. 20, 219–232.

Werbos, P.J., 1974. Beyond Regression: New Tools for Prediction and Analysis in the Behavioral Sciences. Ph.D. Thesis. Harvard University.

12 Are my results significant?

12.1   The difference is due to random variation!

This is a dreadful phrase to hear after spending hours or days distilling a huge data set down to a few meaningful numbers. You think that you have discovered a meaningful difference. Perhaps an important parameter is different in two geographical regions that you are studying. You begin to construct a narrative that explains why it is different and what the difference implies. And then you discover that the difference is caused by *noise in your data*. What a disappointment! Nonetheless, you are better off having found out earlier than later. Better to uncover the unpleasant reality by yourself in private than be criticized by other scientists, publicly.

On the other hand, if you can show that the difference is *not* due to observational noise, your colleagues will be more inclined to believe your results.

As noise is a random process, you can never be completely sure that any given pattern in your data is not due to observational noise. If the noise can really be anything, then there is a finite probability that it will mimic any difference, regardless of its magnitude. The best that one can do is assess the probability of a difference having been caused by noise. If the probability that the difference is caused by noise is small, then the probability of the difference being "real" is high.

This thinking leads to a formal strategy for testing significance. We state a *Null Hypothesis*, which is some variation on the following theme:

The difference is due to random processes. $\qquad\qquad$ (12.1)

Environmental Data Analysis with MATLAB®. http://dx.doi.org/10.1016/B978-0-12-804488-9.00012-4

The difference is taken to be significant if the Null Hypothesis can be *rejected* with high probability. How high is high will depend on the circumstances, but an exclusion probability of 95% is the minimum standard. While 95% may sound like a high number, it implies that a wrong conclusion about the significance of a result is made once in twenty times, which arguably does not sound all that low. A higher rejection probability is warranted in high-stakes situations.

We have encountered this kind of analysis before, in the discussion of a long-term trend in cooling of the Black Rock Forest temperature data set (Section 4.8). The estimated rate of change in temperature was -0.03 °C/year, with a 2σ error of $\pm 10^{-5}$ °C/year. In this case, a reasonable Null Hypothesis is that the rate differs from zero only because of observational noise. The Null Hypothesis can be rejected with better than 95% confidence because -0.03 is more than 2σ from zero. This analysis relies on the parameter being tested (distance from the mean) being Normally distributed and on our understanding of the Normal probability density function (that 95% of its probability is within $\pm 2\sigma$ of its mean).

Generically, a parameter computed from data is called a *statistic*. In the above example, the statistic being tested is the difference of a mean from zero, which, in this case, is Normally distributed. In order to be able to assess other kinds of Null Hypotheses, we will need to examine cases that involve statistics whose corresponding probability density functions are less familiar than the Normal probability density function.

12.2   The distribution of the total error

One important statistic is the total error, E. It is defined (see Section 5.6) as the sum of squares of the individual errors, weighted by their variance; that is, $E = \Sigma_i e_i^2$ with $e_i = (d_i^{\mathrm{obs}} - d_i^{\mathrm{pre}})/\sigma_{di}$. Each of the es are assumed to be Normally distributed with zero mean and, owing to being weighted by $1/\sigma_{di}$, unit variance. As the error, E, is derived from noisy data, it is a random variable with its own probability density function, $p(E)$. This probability density function is not Normal as the relationship between the es and E is nonlinear. We now turn to working out this probability density function.

We start on a simple basis and consider the special case of only one individual error, that is, $E = e^2$. We also use only the nonnegative values of the Normal probability density function of e, because the sign of e is irrelevant when we form its square. The Normal probability density function for a nonnegative e is

$$p(e) = (2/\pi)^{1/2}\exp(-\tfrac{1}{2}e^2) \tag{12.2}$$

Note that this function is a factor of two larger-than-usual Normal probability density functions, defined for both negative and positive values of e. This probability density function can be transformed into $p(E)$ using the rule $p(d) = p[d(E)]|de/dE|$ (Equation 3.8), where $e = E^{1/2}$ and $de/dE = \tfrac{1}{2}E^{-1/2}$:

$$p(E) = (2\pi E)^{-1/2}\exp(-\tfrac{1}{2}E) \tag{12.3}$$

This formula is reminiscent of the formula for a uniformly distributed random variable (Section 3.5). Both have a square-root singularity at the origin (Figure 12.1).

(A)   (B)

$p(e)$ $p(E)$

Figure 12.1 (A) Probability density function, $p(e)$, of a Normally distributed variable, e, with zero mean and unit variance. (B) Probability density function of $E = e^2$. *MatLab* script eda12_01.

Now let us consider the slightly more complicated case, $E = e_1^2 + e_2^2$, where the es are uncorrelated so that their joint probability density function is

$$p(e_1, e_2) = p(e_1)p(e_2) = \left(\frac{2}{\pi}\right) \exp\left(-\tfrac{1}{2}(e_1^2 + e_2^2)\right) \tag{12.4}$$

We compute the probability density function, $p(E)$, by first pairing E with another variable, say θ, to define a joint probability density function, $p(E,\theta)$, and then integrating over θ to reduce the joint probability density function to the univariate probability density function, $p(E)$. We have considerable flexibility in choosing the functional form of θ. Because of the similarity of the formula, $E = e_1^2 + e_2^2$, to the formula, $r^2 = x^2 + y^2$, of polar coordinates, we use $\theta = \tan^{-1}(e_1/e_2)$, which is analogous to the polar angle. Inverting these formulas for $e_1(E,\theta)$ and $e_2(E,\theta)$ yields

$$e_1 = E^{\frac{1}{2}} \sin\theta \quad \text{and} \quad e_2 = E^{\frac{1}{2}} \cos\theta \tag{12.5}$$

The Jacobian determinant, $J(E,\theta)$, is (see Equation 3.21)

$$J(E,\theta) = \begin{vmatrix} \dfrac{\partial e_1}{\partial E} & \dfrac{\partial e_2}{\partial E} \\ \dfrac{\partial e_1}{\partial \theta} & \dfrac{\partial e_2}{\partial \theta} \end{vmatrix} = \begin{vmatrix} -\tfrac{1}{2}E^{-\frac{1}{2}}\sin\theta & -\tfrac{1}{2}E^{-\frac{1}{2}}\sin\theta \\ E^{\frac{1}{2}}\cos\theta & -E^{\frac{1}{2}}\sin\theta \end{vmatrix} = \tfrac{1}{2}(\sin^2\theta + \cos^2\theta) = \tfrac{1}{2}$$

$$\tag{12.6}$$

The joint probability density function is therefore

$$p(E,\theta) = p(e_1(E,\theta), e_2(E,\theta))J(E,\theta) = \left(\frac{1}{\pi}\right) \exp(-\tfrac{1}{2}E) \tag{12.7}$$

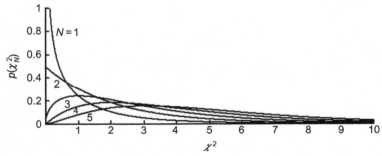

Figure 12.2 Chi-squared probability density function for $N = 1, 2, 3, 4$, and 5. *MatLab* script eda12_02.

Note that the probability density function is *uniform* in the polar angle, θ. Finally, the univariate probability density function, $P(E)$, is determined by integrating over polar angle, θ.

$$p(E) = \int_0^{\pi/2} p(E, \theta)d\theta = \tfrac{1}{2} \exp(-\tfrac{1}{2}E) \tag{12.8}$$

The polar integration is performed over only one quadrant of the (e_1, e_2) plane as all the es are nonnegative. As you can see, the calculation is tedious, but it is neither difficult nor mysterious. The general case corresponds to summing up N squares: $E_N = \Sigma_{i=1}^{N}e_i^2$. In the literature, the symbol χ^2_N is commonly used in place of E_N and the probability density function is called the *chi-squared* probability density function with N degrees of freedom (Figure 12.2). The general formula, valid for all N, can be shown to be

$$p(\chi^2_N) = \frac{1}{2^{N/2}((N/2) - 1)!} [\chi^2_N]^{\frac{N}{2}-1} \exp(-\tfrac{1}{2}\chi^2_N) \tag{12.9}$$

This probability density function can be shown to have mean, N, and variance $2N$. In *MatLab*, the probability density function is computed as

```
pX2 = chi2pdf(X2,N);                          (MatLab eda12_02)
```

where X2 is a vector of χ^2_N values. We will put off discussion of its applications until after we have examined several other probability density functions.

12.3   Four important probability density functions

A great deal (but not all) of hypothesis testing can be performed using just four probability density functions. Each corresponds to a different function of the error, **e**, which is presumed to be uncorrelated and Normally distributed with zero mean and unit variance:

(1) $p(Z)$ with $Z = e$

(2) $p(\chi_N^2)$ with $\chi_N^2 = \sum_{i=1}^{N} e_i^2$

(3) $p(t_N)$ with $t_N = \dfrac{e}{\left(\left(1/N\right)\sum_{i=1}^{N} e_i^2\right)^{\frac{1}{2}}}$

(4) $p(F_{N,M})$ with $F_{N,M} = \dfrac{N^{-1}\sum_{i=1}^{N} e_i^2}{M^{-1}\sum_{i=1}^{M} e_i^2}$

(12.10)

Probability density function 1 is just the Normal probability density function with zero mean and unit variance. Note that any Normally distributed variable, d, with mean, $\bar{d}$, and variance, σ_d^2, can be transformed into one with zero mean and unit variance with the transformation, $Z = (d - \bar{d})/\sigma_d$.

Probability density function 2 is the chi-squared probability density function, which we discussed in detail in the previous section.

Probability density function 3 is new and is called *Student's t-probability density function*. It is the ratio of a Normally distributed variable and the square root of the sum of squares of N Normally distributed variables (the e in the numerator is assumed to be different from the es in the denominator). It can be shown to have the functional form

$$p(t_N) = \frac{\left(\frac{N+1}{2} - 1\right)!}{(N\pi)^{\frac{1}{2}}\left(\frac{N}{2} - 1\right)!}\left[1 + \frac{t_N^2}{N}\right]^{-((N+1)/2)} \qquad (12.11)$$

The t-probability density function (Figure 12.3) has a mean of zero and, for $N > 2$, a variance of $N/(N-2)$ (its variance is undefined for $N \leq 2$). Superficially, it looks like a Normal probability density function, except that it is *longer-tailed* (i.e., it falls off with distance from its mean much more slowly than does a Normal probability density function. In *MatLab*, the t-probability density function is computed as

```
pt = tpdf(t,N);
```
(MatLab eda12_03)

where t is a vector of t values.

Probability density function 4 is also new and is called the *Fisher-Snedecor F-probability density function*. It is the ratio of the sum of squares of two different sets of random variables. Its functional form cannot be written in terms of elementary functions and is omitted here. Its mean and variance are

$$\bar{F} = \frac{M}{M-2} \quad \text{and} \quad \sigma_F^2 = \frac{2M^2(M+N-2)}{N(M-2)^2(M-4)} \qquad (12.12)$$

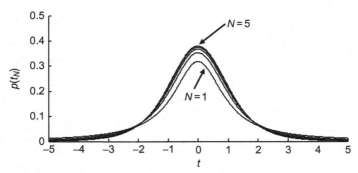

Figure 12.3 Student's t-probability density function *for $N = 1, 2, 3, 4$, and 5. MatLab script eda12_03.

Note that the mean of the F-probability density function approaches $\bar{F} = 1$ as $M \to \infty$. For small values of M and N, the F-probability density function is skewed towards low values of F. At large values of M and N, it is more symmetric around $F = 1$. In *MatLab*, the F-probability density function is computed as

        pF = fpdf(F,N,M);   (*MatLab* eda12_04)

where F is a vector of F values.

12.4   A hypothesis testing scenario

The general procedure for determining whether a result is significant is to first state a Null Hypothesis, and then identify and compute a statistic whose value will *probably be small* if the Null Hypothesis is true. If the value is large, then the Null Hypothesis is *unlikely to be true* and can be rejected. The probability density function of the statistic is used to assess just how unlikely, given any particular value. Four Null Hypotheses are common:

(1) *Null Hypothesis: The mean of a random variable differs from a prescribed value only because of random fluctuation.* This hypothesis is tested with the statistic, Z or t, depending on whether the variance of the data is known beforehand or estimated from the data. The tests of significance use the Z-probability density function or t-probability density function, and are called the Z-test and the t-test, respectively.

(2) *Null Hypothesis: The variance of a random variable differs from a prescribed value only because of random fluctuation.* This hypothesis is tested with the statistic, χ_N^2, and the corresponding test is called the chi-squared test.

(3) *Null Hypothesis: The means of two random variables differ from each other only because of random fluctuation.* The Z-test or t-test is used, depending on whether the variances are known beforehand or computed from the data.

(4) *Null Hypothesis: The variances of two random variables differ from each other only because of random fluctuation.* The F-test is used.

As an example of the use of these tests, consider the following scenario. Suppose that you are conducting research that requires measuring the sizes of particles (e.g., aerosols) in the 10-1000 nm range. You purchase a laboratory instrument capable of measuring particle diameters. Its manufacturer states that (1) the device is extremely well-calibrated, in the sense that particles diameters will exactly scatter about their true means, and (2) that the variance of any single measurement is $\sigma_d^2 = 1$ nm^2. You test the machine by measuring the diameter, d_i, of $N = 25$ specially purchased calibration particles, each exactly 100 nm in diameter. You then use these data to calculate a variety of useful statistics (Table 12.1) that you hope will give you a sense about how well the instrument performs. A few weeks later, you repeat the test, using another set of calibration particles.

The synthetic data for these tests that we analyze below were drawn from a Normal probability density function with a mean of 100 nm and a variance of 1 nm^2 (*MatLab* script eda12_05). Thus, the data do not violate the manufacturer's specifications and (hopefully) the statistical tests should corroborate this.

Question 1: Is the calibration correct? Because of measurement noise, the estimated mean diameter, $\bar{d}^{\text{est}}$, of the calibration particles will always depart slightly from the true value, $\bar{d}^{\text{true}}$, even if the calibration of the instrument is perfect. Thus, the Null Hypothesis is that the observed deviation of the average particle size from its true value is due to observational error (as contrasted to a bias in the calibration). If the data are Normally distributed, so is their mean, with the variance being smaller by a factor of $1/\sqrt{N}$. The quantity, Z^{est} (Table 12.1, row 7), which quantifies how different the observed mean is from the true mean, is Normally distributed with zero mean and unit variance. It has the value $Z^{\text{est}} = 0.278$ for the first test and $Z^{\text{est}} = 0.243$ for the second. The critical question is how frequently Zs of this size or larger occur. Only if they occur extremely infrequently can the Null Hypothesis be rejected. Note that a small Z is one that is close to zero, regardless of its sign, so the absolute value of Z is the quantity that is tested—a *two-sided* test. The Null Hypothesis can be rejected only when values greater than or equal to the observed Z are very uncommon; that is, when $P(|Z| \geq Z^{\text{est}})$ is small, say less than 0.05 (or 5%). We find (Table 12.1, row 8) that $P(|Z| \geq Z^{\text{est}}) = 0.78$ for test 1 and 0.81 for test 2, so the Null Hypothesis cannot be rejected in either case.

MatLab provides a function, normcdf(), that computes the cumulative probability of the Normal probability density function:

$$P(Z') = \int_{-\infty}^{Z'} p(Z)dZ = \int_{-\infty}^{Z'} \frac{1}{\sqrt{2\pi}} \exp(-\tfrac{1}{2}Z^2)dZ \tag{12.13}$$

The probability that Z is between $-Z^{\text{est}}$ and $+Z^{\text{est}}$ is $P(|Z^{\text{est}}|) - P(-|Z^{\text{est}}|)$. Thus, the probability that Z is outside this interval is $P(|Z| \geq Z^{\text{est}}) = 1 - [P(|Z^{\text{est}}|) - P(-|Z^{\text{est}}|)]$. In *MatLab*, this probability is computed as

```
PZA = 1 - (normcdf(ZA,0,1)-normcdf(-ZA,0,1));
```

(*MatLab* eda12_06)

where ZA is the absolute value of Z^{est}. The function, normcdf(), computes the cumulative Z-probability distribution (i.e., the cumulative Normal probability distribution).

Question 2: Is the variance within specs? Because of measurement noise, the estimated variance, $(\sigma_d^{\text{est}})^2$, of the diameters of the calibration particles will always

Table 12.1 Statistics arising from two calibration tests.

	Calibration Test 1	Calibration Test 2	Inter-Test Comparison
1 N	25	25	
2 $\overline{d}^{\text{true}}$	100	100	
3 $\overline{d}^{\text{est}} = \frac{1}{N}\sum_{i=1}^{N} d_i$	100.055	99.951	
4 $(\sigma_d^{\text{true}})^2$	1	1	
5 $(\sigma_d^{\text{est}})^2 = \frac{1}{N}\sum_{i=1}^{N}(d_i^{\text{obs}} - \overline{d}^{\text{true}})^2$	0.876	0.974	
6 $(\sigma_d^{\text{est}'})^2 = \frac{1}{N-1}\sum_{i=1}^{N}(d_i^{\text{obs}} - \overline{d}^{\text{est}})^2$	0.910	1.012	
7 $Z^{\text{est}} = \dfrac{\overline{d}^{\text{est}} - \overline{d}^{\text{true}}}{\sigma_d^{\text{true}}/\sqrt{N}}$	0.278	0.243	
8 $P(|Z| \geq Z^{\text{est}})$	0.780	0.807	
9 $\chi^2_{\text{est}} = \sum_{i=1}^{N}\dfrac{(d_i^{\text{obs}} - \overline{d}^{\text{true}})^2}{(\sigma_d^{\text{true}})^2}$	21.921	24.353	
10 $P(\chi^2 \geq \chi^2_{\text{est}})$	0.640	0.499	
11 $t^{\text{est}} = \dfrac{\overline{d}^{\text{est}} - \overline{d}^{\text{true}}}{\sigma_d^{\text{est}}/\sqrt{N}}$	0.297	0.247	
12 $P(|t_{25}| \geq t^{\text{est}})$	0.768	0.806	
13 $Z^{\text{est}} = \dfrac{(\overline{d}^{\text{est1}} - \overline{d}^{\text{est2}})}{\sqrt{((\sigma_{d1}^{\text{true}})^2/N_1) + ((\sigma_{d2}^{\text{true}})^2/N_2)}}$			0.368

14   $P(|Z| \geq Z^{est})$ 0.712

15   $t^{est} = \dfrac{(\overline{d}^{est1} - \overline{d}^{est2})}{\sqrt{((\sigma_{d1}^{est'})^2/N_1) + ((\sigma_{d2}^{est'})^2/N_2)}}$ 0.376

16   $M = \dfrac{[(\sigma_{d1}^{est'})^2/N_1 + (\sigma_{d1}^{est'})^2/N_2]^2}{((\sigma_{d1}^{est'})^2/N_1)^2/(N_1 - 1)) + ((\sigma_{d2}^{est'})^2/N_2)^2/(N_2 - 1)}$ 48

17   $P(|t_M| \geq t^{est})$ 0.707

18   $F^{est} = \dfrac{(\chi_1^2/N_1)}{(\chi_2^2/N_2)}$ 1.110

19   $P(F \leq 1/F^{est}$ or $F \geq F^{est})$ 0.794

depart slightly from the true value, $(\sigma_d^{\text{true}})^2$, even if the instrument is functioning correctly. Thus, the Null Hypothesis is that the observed deviation is due to random fluctuation (as contrasted to the instrument really being noisier than specified). The quantity, χ_{est}^2 (Table 12.1, row 9) is chi-squared distributed with 25 degrees of freedom. It has a value of, $\chi_{\text{est}}^2 = 21.9$ for the first test and 24.4 for the second. The critical question is whether these values occur with high probability; if so, the Null Hypothesis cannot be rejected. In this case, we really care only if the variance is *worse* than what the manufacturer stated, so a *one-sided* test is appropriate. That is, we want to know the probability that a value is greater than χ_{est}^2. We find that $P(\chi^2 \geq \chi_{\text{est}}^2) = 0.64$ for test 1 and 0.50 for test 2. Both of these numbers are much larger than 0.05, so the Null Hypothesis cannot be rejected in either case. In *MatLab*, the probability, $P(\chi^2 \geq \chi_{\text{est}}^2)$ is calculated as follows:

```
Pchi2A = 1 - chi2cdf(chi2A,NA);            (MatLab eda12_07)
```

Here, `chi2A` is χ_{est}^2 and `NA` stands for the degrees of freedom (25, in this case). The function, `chi2cdf()`, computes the cumulative chi-squared probability distribution.

Question 1, Revisited: Is the calibration correct? Suppose that the manufacturer had not stated a variance. We cannot form the quantity, Z, as it depends on the variance, σ_d^{true}, which is unknown. However, we can estimate the variance from the data, $(\sigma_d^{\text{est}})^2 = N^{-1} \sum_{i=1}^{N} (d_i - \bar{d}^{\text{true}})^2$. But because this estimate is a random variable, we cannot use it in the formula for Z, for Z would not be Normally distributed. Such a quantity would instead be *t*-distributed, as can be seen from the following:

$$t = \frac{(\bar{d}^{\text{est}} - \bar{d}^{\text{true}})}{N^{-1/2} \left(\dfrac{1}{N} \sum_{i=1}^{N} (d_i - \bar{d}^{\text{true}})^2 \right)^{1/2}} = \frac{(\bar{d}^{\text{est}} - \bar{d}^{\text{true}})}{\left(\dfrac{\sigma_d^{\text{true}}}{\sqrt{N}} \right) \left(\dfrac{1}{N} \sum_{i=1}^{N} \dfrac{(d_i - \bar{d}^{\text{true}})^2}{(\sigma_d^{\text{true}})^2} \right)^{1/2}} = \frac{e}{\left(\dfrac{1}{N} \sum_{i=1}^{N} e_i^2 \right)^{1/2}}$$

$$(12.14)$$

Note that we have inserted $\sigma_d^{\text{true}}/\sigma_d^{\text{true}}$ into the denominator of the third term in order to normalize d_i and $\bar{d}$ into random variables, e_i and e, that have unit variance. In our case, $t^{\text{est}} = 0.294$ for test 1 and 0.247 for test 2.

The Null Hypothesis is the same as before; that the observed deviation of the average particle size is due to observational error (as contrasted to a bias in the calibration). We again use a two-sided test and find that $P(|t| \geq t^{\text{est}}) = 0.77$ for test 1 and 0.81 for test 2. These probabilities are much higher than 0.05, so the Null Hypothesis cannot be rejected in either case. In *MatLab* we compute the probability as

```
PtA = 1 - (tcdf(tA,NA)-tcdf(-tA,NA));     (MatLab eda12_07)
```

Here, `tA` is $|t^{\text{est}}|$ and $NA = 25$ denotes the degrees of freedom (25 in this case). The function, `tcdf()`, computes the cumulative *t*-probability distribution.

Question 3: Has the calibration changed between the two tests? The Null Hypothesis is that any difference between the two means is due to random variation. The quantity, $(\bar{d}^{\text{est1}} - \bar{d}^{\text{est2}})$, is Normally distributed, as it is a linear function of two Normally distributed random variables. Its variance is just the sum of the variances of the two terms (see Table 12.1, row 13). We find $Z^{\text{est}} = 0.368$ and $P(|Z| \geq Z^{\text{est}}) = 0.712$.

Once again, the probability is much larger than 0.05, so the Null Hypothesis cannot be excluded.

Question 3, Revisited. Note that in the true variances, $(\sigma_{d1}^{\text{true}})^2$ and $(\sigma_{d2}^{\text{true}})^2$ are needed to form the quantity, Z (Table 12.1, row 13). If they are unavailable, then one must estimate variances from the data, itself. This estimate can be made in either of two ways

$$(\sigma_d^{\text{est}})^2 = \begin{cases} \dfrac{1}{N}\sum_{i=1}^{N}(d_i^{\text{obs}} - \bar{d}^{\text{true}})^2 & \text{if } \bar{d}^{\text{true}} \text{ is known} \\[2ex] \dfrac{1}{N-1}\sum_{i=1}^{N}(d_i^{\text{obs}} - \bar{d}^{\text{est}})^2 & \text{otherwise} \end{cases} \tag{12.15}$$

depending on whether the true mean of the data is known. Both are random variables and so cannot be used to form Z, as it would not be Normally distributed. An estimated variance can be used to create the analogous quantity, t (Table 12.1, row 15), but such a quantity is only approximately t-distributed, because the difference between two t-distributed variables is not exactly t-distributed. The approximation is improved by defining *effective* degrees of freedom, M (as in Table 12.1, row 16). We find in this case that $t^{\text{est}} = 0.376$ and $M = 48$. The probability $P(|t| \geq t^{\text{est}}) = 0.71$, which is much larger than the 0.05 needed to reject the Null Hypothesis.

Question 4. Did the variance change between tests? The estimated variance is 0.876 in the first test and 0.974 in the second. Is it possible that it is getting worse? The Null Hypothesis is that the difference between these estimates is due to random variation. The quantity, F (Table 12.1, row 18), is defined as the ratio of the sum of squares of the two sets of measurements, and is thus proportional to the ratio of their estimated variances. In this case, $F^{\text{est}} = 1.11$ (Table 12.1, row 18). An F that is greater than unity implies that the variance appears to get larger (worse). An F less than unity would mean that the variance appeared to get smaller (better). As F is defined in terms of a ratio, $1/F^{\text{est}}$ is as better as F^{est} is worse. Thus, a two-sided test is needed to assess whether the variance is unchanged (neither better nor worse); that is, one based on the probability, $P(F \leq 1/F^{\text{est}} \text{ or } F \geq F^{\text{est}})$, which in this case is 0.79. Once again, the Null Hypothesis cannot be rejected.

The *MatLab* code for computing $P(F \leq 1/F^{\text{est}} \text{ or } F \geq F^{\text{est}})$ is

```
if(F<1)
    F=1/F;
end
PF = 1 - (fcdf(F,NA,NB)-fcdf(1/F,NA,NB));   (MatLab eda12_07)
```

The function, `fcdf()`, computes the cumulative F-probability distribution. Note that F is replaced with its reciprocal if it is less than unity. Here NA and NB are the degrees of freedom of tests 1 and 2, respectively (both 25, in this case).

We summarize below what we have learned about the instrument:

Question *1*: Is the calibration correct? Answer: *We cannot reject the Null Hypothesis that the difference between the estimated and manufacturer-stated calibration is caused by random variation.*

Question 2: Is the variance within specs? Answer: *We cannot reject the Null Hypothesis that the difference between the estimated and manufacturer-stated variance is caused by random variation.*

Question *3*: Has the calibration changed between the two tests? Answer: *We cannot reject the Null Hypothesis that difference between the two calibrations is caused by random variation.*

Question *4*. Did the variance change between tests? Answer: *We cannot reject the Null Hypothesis that difference between the two estimated variances is caused by random variation.*

Note that in each case, the results are stated with respect to a particular Null Hypothesis.

12.5   Chi-squared test for generalized least squares

In the previous section, we developed a chi-squared test to assess the Null Hypothesis that an estimated (posterior) variance differs from a prescribed (prior) value only because of random variation. As discussed in Section 4..8, the posterior variance can be estimaed from the error of fit (Equation 4.31), such as the fit of a model determined by generalized least squares. Thus, we can use a chi-squared test to assess the following:

Null Hypothesis

The difference of the generalized error from its expected value is due to random variation.

The generalized error E_T is the sum of two terms (Equation 5.15):

$$E_T^{est} = E^{est} + E_p^{est} \tag{12.16}$$

The first term is the estimated error E^{obs} in the data:

$$E^{est} = \sum_{i=1}^{N} \frac{\left(d_i^{obs} - d_i^{pre}\right)^2}{\sigma_d^2} \quad \text{where} \quad \mathbf{d}^{pre} = \mathbf{G}\mathbf{m}^{est} \tag{12.17}$$

and the second term is the estimated error E_p^{est} in the prior information:

$$E_p^{est} = \sum_{i=1}^{K} \frac{\left(h_i^{pri} - h_i^{pre}\right)^2}{\sigma_h^2} \quad \text{where} \quad \mathbf{h}^{pre} = \mathbf{H}\mathbf{m}^{est} \tag{12.18}$$

The data vector $\mathbf{d}$ has length N, the prior information vector $\mathbf{h}$ has length K and the model parameter vector $\mathbf{m}$ has lengtth M. The variance σ_d^2 of the data and the variance σ_h^2 of the prior information are prior variances; that is, they have been determined by a method unrelated to the least squares estimation process. For instance, σ_d^2 could be based on the published precision of a laboratory apparatus (e.g. the manufacturer stated accuracy) and σ_d^2 could be based on a typical range of variation of the particular kind of prior information, as published in the scientific literature (most measurements of this kind are within a known percent of a stated value).

The total error is the sum of squares of Normally-distributed random variables, each with unit variance, and so is chi-squared distributed. The number of degrees of freedom v_T is the number $(N + K)$ of component errors reduced by the number M of model parameters, so $v_T = N + K - M$. The reduction reflects the notion that the data and prior information can be fit exactly when $M = N + K$, yielding a posterior variance of zero. The total error has mean v_T and variance $2v_T$, and the 95% confidence interval is approximately:

$$v_T - 2(2v_T)^{\frac{1}{2}} < E_T^{est} < v_T + 2(2v_T)^{\frac{1}{2}} \tag{12.19}$$

The generalized least squares error is incompatible with the prior variances σ_d^2 and σ_h^2 only when it falls outside this interval; that is, the Null Hypothesis is unlikely in this case. Errors that are to the right of the interval correspond to models that fit the data more poorly than can be expected by random variation alone. A poor fit case can occur when the model is too simple. Too few model parameters are availble to capture the true variability of the data and prior information. Errors to the left of the interval correspond to models that *overfit* the data; that is, the error is smaller than can be expected by random variation alone. The overfit case can occur when the model is too complex. So many model parameters are available that even the noise is being fit.

The individual compatibility of the errors E^{est} and of E_p^{est} with their respective prior variances can also be tested. These errors are chi-squared distributed, but with fewer degrees of error than the total error E_T^{est}. An important issue is how to partition the loss of degrees of freedom associated with the M model parameters between E^{est} and E_p^{est}. The Welch–Satterthwaite approximation spreads it in proportion to the number of component errors, so that E^{est} is assigned $v = v_T N/(N + K)$ degrees of freedom and E_p^{obs} is assigned $v_p = v_T K/(N + K)$ degrees of freedom. The 95% confidence intervals are approximately:

$$v - 2(2v)^{\frac{1}{2}} < E^{est} < v + 2(2v)^{\frac{1}{2}} \text{ and } v_p - 2(2v_p)^{\frac{1}{2}} < E_p^{est} < v_p + 2(2v_p)^{\frac{1}{2}} \tag{12.20}$$

Crib Sheet 12.1 Steps in generalized least squares

Step 1: State the problem in words
How are the data related to the model

Step 2: Organize the problem in standard form
identify the data **d** (length N) the model parameters **m** (length M)
define the data kernel **G** so that $\mathbf{d}^{obs} = \mathbf{Gm}$

Step 3: Examine the data
make plots of the data

Continued

Crib Sheet 12.1—cont'd

Step 4: Establish the accuracy of the data
state a prior variance σ_d^2 based on accuracy of the measurement technique

Step 5: State the prior information in words, for example:
the model parameters are close to a known values, $\mathbf{h}^{pri}$
the mean of the model parameters is close to a known value
the model parameters vary smoothly with space and time

Step 6: Organize the prior information in standard form:
$$\mathbf{h}^{pri} = \mathbf{Hm}$$

Step 7: Establish the accuracy of the prior information
state a prior variance σ_h^2 based on the accuracy of the prior information

Step 8: Estimate model parameter $\mathbf{m}^{est}$ and their covariance $\mathbf{C}_m$
$$\mathbf{m}^{est} = \left[\mathbf{F}^{T}\mathbf{F}\right]^{-1}\mathbf{F}^{T}\mathbf{f}^{obs} \quad \text{and} \quad \mathbf{C}_m = \left[\mathbf{F}^{T}\mathbf{F}\right]^{-1}$$

$$\text{with } \mathbf{F} = \begin{bmatrix} \sigma_d^{-1}\mathbf{G} \\ \sigma_h^{-1}\mathbf{H} \end{bmatrix} \quad \text{and} \quad \mathbf{f}^{obs} = \begin{bmatrix} \sigma_d^{-1}\mathbf{d}^{obs} \\ \sigma_h^{-1}\mathbf{h}^{pri} \end{bmatrix}$$

Step 9: State estimates and their 95% confidence intervals
$$m_i^{true} = m_i^{est} \pm 2\sigma_{mi}\ (95\%) \quad \text{with} \quad \sigma_{mi} = \sqrt{[\mathbf{C}_m]_{ii}}$$

Step 10: Examine the individual errors
$$\mathbf{d}^{pre} = \mathbf{Gm}^{est} \quad \text{and} \quad \mathbf{e} = \mathbf{d}^{obs} - \mathbf{d}^{pre}$$
$$\mathbf{h}^{pre} = \mathbf{Hm}^{est} \quad \text{and} \quad \mathbf{e}_p = \mathbf{h}^{pri} - \mathbf{h}^{pre}$$
plot $\mathbf{e_i}$ vs. i and plot $\mathbf{e}_{pi}$ vs. i
scatter plot of d_i^{pre} vs. d_i^{obs} and scatter plot of h_i^{pre} vs. h_i^{pri}
any unusually large errors?

Step 11: Examine the total error E_T
$$E_T = E + E_p \quad \text{with} \quad E = \sigma_d^{-2}\mathbf{e}^{T}\mathbf{e} \quad \text{and} \quad E_p = \sigma_h^{-2}\mathbf{e}_p^{T}\mathbf{e}_p$$
use a chi-squared test on E_T to assess the likelihood of the Null Hypothesis that E_T
is different than expected only because of random variation

Step 12: Two different models?
use an F-test on the E's of the two models to assess the likelihood
of the Null Hypothesis that the E's are
different from each other only because of random variation

12.6   Testing improvement in fit

Very common is the situation where two alternative models are proposed for a single dataset. Neither fits the data exactly, but one has a smaller total error, E, than the other. Calling the model with the smaller corresponding error the *better-fitting* model is natural. Recall, however, that the error, E, is a random variable. Different realizations of it will vary in size, even when drawn from probability density functions with the exactly the same mean. Thus, when the two errors are similar in size, their difference may be due to random variation and not to one model "really" being better than the other.

The F-test is used to assess the significance in the ratio of the estimated variance of the two fits (Figure 12.4):

$$F_{K_1, K_1} = \frac{(\sigma_d^2)_{\mathrm{model1}}^{\mathrm{est}}}{(\sigma_d^2)_{\mathrm{model2}}^{\mathrm{est}}} = \frac{E_1/K_1}{E_2/K_2} \quad \text{with} \quad K_1 = N_1 - M_1 \quad \text{and} \quad K_2 = N_2 - M_2$$

(12.21)

Here, the first model has M_1 model parameters, N_1 data, and a total error, E_1 and the second model has M_2 model parameters, N_2 data, and a total error, E_2. Note that a model with M parameters ought to be able to fit a dataset with $N = M$ data exactly, so the degrees of freedom are $K = N - M$. If the variances of individual errors, e_i, used to compute the Es are all equal, then they cancel out of the fraction, E_1/E_2. Thus, the statistic, F, does not depend on the variance of the data and one is free to use the unweighted error, $E = \Sigma_i (d_i^{\mathrm{obs}} - d_i^{\mathrm{pre}})^2$.

As an example, we consider the rising temperatures during the first 60 h of the Black Rock Forest dataset (Figure 12.5). A straight line (Figure 12.5A) fits the data fairly well, but a cubic function fits it better, with $F = 1.112$. The Null Hypothesis is that both functions fit the data equally well, and that the difference in error is due to random variation. A two-sided test gives

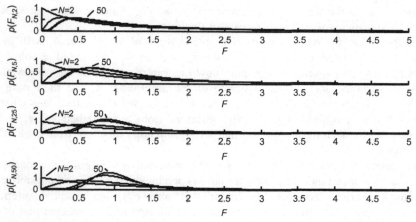

Figure 12.4 F-probability density function, $p(F_{N, M})$, for selected values of M and N. *MatLab* script eda12_04.

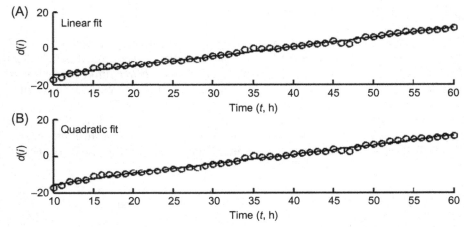

Figure 12.5 Comparison of two fits to a fragment of the Black Rock Forest temperature dataset. (A) Observed data (circles), linear fit (solid line), (B) Observed data (circles), cubic fit. (solid line) The cubic reduces the error by 14% compared to the linear fit. *MatLab* Script eda12_08.

$P(F \leq 1/F^{\text{est}}$ or $F \geq F^{\text{est}}) = 0.71$, which is much larger than 0.05, so the Null Hypothesis that the fits are equally good cannot be rejected.

12.7   Testing the significance of a spectral peak

A complicated time series, meaning one with many short-period fluctuations, usually has a complicated spectrum, with many peaks and troughs. Some peaks may be particularly high-amplitude and stand above other lower-amplitude ones. We would like to know whether the high-amplitude peaks are *significant*. Pinning down just what we mean by significant requires some careful thinking.

Suppose that the data consists of a cosine wave of amplitude, A, with just a little random observational noise superimposed on it. We could employ the rules of error propagation to compute how the variance, σ_d^2, of the observations leads to variance, σ_A^2, in our estimate of the amplitude, A, and then state the confidence intervals for the amplitude as $A \pm 2\sigma_A$. We could then test whether a peak has an amplitude significantly different from a prescribed value, or test whether two peaks have amplitudes that are significantly different from each other, or so forth.

The problem is that this is almost *never* what we mean by the significance of a spectral peak. The much more common scenario is one in which a long and possibly continuous time series is dominated by "noise" that has no obvious sinusoidal patterns at all. Furthermore, the noise is usually some complicated and unmodeled part of the time series itself, and not observational error in the strict sense. We take the spectrum of a portion of the time series—the first hour, say—and detect a spectral peak at frequency, f_0. We want to know if this peak is significant in the sense that, if we were to take the spectra of subsequent hourly segments, they would also have peaks at frequency, f_0. The alternative is that the observed peak arises from some "accident" of the noise pattern in that particular hour of data that is not shared by the other hourly segments.

The Null Hypothesis that corresponds to this case is that the spectral peak can be explained by random variation within a time series that consists of *nothing* but random noise. The easiest case to analyze is a time series that consists of nothing but uncorrelated, Normally distributed random noise with constant variance. The power spectral density of such a time series will have peaks, and the height of these peaks has a probability density function that will allow us to quantify the likelihood that the height of a peak will exceed a specified value. If the likelihood of an observed peak is low, then we have reason to reject the Null Hypothesis and claim that the peak is *significant*, in the sense of being unlikely to have arisen from random fluctuations.

Before starting the analysis, we need to carefully specify whether Fourier coefficients are defined in the frequency range, $(0, f_{ny})$ or $(-f_{ny}, f_{ny})$, because the former have twice the amplitude of the latter. We choose the former, for then plotting power spectral density on the $(0, f_{ny})$ interval, which is the shorter of the two, seems more natural.

Suppose that time series, d_i, of length, N, consists of uncorrelated random noise with zero mean and variance, σ_d^2. Before computing the Fourier transform, the time series is modified by multiplication with a taper, w_i, so that it has elements, $w_i d_i$. The variance of the tapered time series is $N^{-1} \Sigma_{i=1}^N w_i^2 d_i^2$, but this is approximately $f_f \sigma_d^2$, where $f_f = N^{-1} \Sigma_{i=1}^N w_i^2$ is the variance of the taper. This can be seen from the $N = 3$ example:

$$N^2 f_f \sigma_d^2 \rightarrow (w_1^2 + w_2^2 + w_3^2)(d_1^2 + d_2^2 + d_3^2)$$

$$= (w_1^2 d_1^2 + w_2^2 d_2^2 + w_3^2 d_3^2) + (w_1^2 d_2^2 + w_2^2 d_3^2 + w_3^2 d_1^2) + (w_1^2 d_3^2 + w_2^2 d_1^2 + w_3^2 d_2^2)$$

$$\approx 3(w_1^2 d_1^2 + w_2^2 d_2^2 + w_3^2 d_3^2) \rightarrow N \sum_{i=1}^N w_i^2 d_i^2 \tag{12.22}$$

The approximation holds because all the ds have the same statistical properties, so their order has little influence on the sums. This behavior suggests that we normalize the taper to unit variance, $f_f = 1$, so that it has the least effect on the variance of the data (and hence on the overall power in the power spectral density). However, in the discussion, below, we allow f_f to have an arbitrary value.

When the data are uncorrelated and Normally distributed, with uniform variance and zero mean, the coefficients of their Fourier series are also uncorrelated and Normally distributed, with uniform variance and zero mean. This follows from the fact that the Fourier transform is a linear function of the data of the form, $\mathbf{Gm} = \mathbf{d}$, where $\mathbf{m}$ is a vector of the Fourier coefficients, together with the relationship, $[\mathbf{G}^T\mathbf{G}]^{-1} \propto \mathbf{I}$ (Equation 6.14) (except for the first and last frequencies, which we will ignore in this analysis). Each element of the power spectral density, $s^2(t)$, is proportional to the sum of squares of two Fourier coefficients. If we normalize the Fourier coefficients to unit variance, then the power spectral density is chi-squared distributed with $p = 2$ degrees of freedom. The normalization factor, c, is calculated using the relationship between the variance of the time series and the frequency integral of the power spectral density (Equation 6.44), together with the fact that a chi-squared probability density function with two degrees of freedom has mean, 2, and variance, 4:

$$f_f \sigma_d^2 = \int_0^{f_{ny}} s^2(f) df \approx \Delta f \sum_{i=1}^{N_f} s_i^2 = \frac{2N_f \Delta f}{2} \overline{s^2} \quad \text{or} \quad \frac{\overline{s^2}}{c} = 2 \quad \text{with} \quad c = \frac{f_f \sigma_d^2}{2N_f \Delta f}$$

$$(12.23)$$

Here, $N_f = N/2 + 1$. Hence, the power spectral density has mean, $\overline{s^2}$, and variance, $\sigma_{s^2}^2$, given by

$$\overline{s^2} = 2c \quad \text{and} \quad \sigma_{s^2}^2 = 4c^2 \qquad\qquad (12.24)$$

Thus, we need know only the basic parameters that define a random time series—the ones that make up the constant, c—in order to predict the statistical properties of its power spectral density (Figures 12.6 and 12.7). The probability that an element of the power spectral density will exceed a given value, say s_0^2, is $1 - P(s_0^2/c)$, where P is the cumulative chi-squared distribution with two degrees of freedom.

As an example, we consider a length $N = 1024$ time series built up from a 5 Hz cosine wave plus uncorrelated noise, drawn from a Normal probability density function with zero mean and variance, σ_d^2. The amplitude of the cosine is chosen to be small, only $0.25\sigma_d$, so that its presence cannot be detected readily though visual inspection of the time series (Figure 12.8A). Nevertheless, the power spectral density (Figure 12.8B) has a prominent peak at 5 Hz, with height, s_0^2, approximately 10 times the mean level (Figure 12.9). The probability that an element of the power spectral density will have a probability at or below this level can be calculated using *MatLab*'s inverse chi-squared probability distribution:

```
ppeak = chi2cdf(speak/c,p);
```
                                                                    (*MatLab* eda12_11)

Here, `speak` is s_0^2, `c` is the constant defined in Equation 12.23, and p=2 denotes the degrees of freedom. We find that `ppeak=0.99994`, which is to say that the power spectral density is predicted to be less than the level, s_0^2, 99.994% of the time.

At this point, we must be exceedingly careful in stating the Null Hypothesis. If we are specifically looking for a 5-Hz oscillation, then the Null Hypothesis would be that a peak *at* 5-Hz arose solely by random variation. In this case, the probability is $1 - 0.99994 = 0.00006$ or 0.006%, which is very small, indeed. Thus, we can reject the Null Hypothesis with very high confidence. However, this analysis relies on prior knowledge that 5 Hz is special—that a peak occurs there.

Instances will indeed arise when we suspect spectral peaks at specific frequencies—the annual and diurnal periodicities that we discussed in the context of the Black Rock Forest temperature dataset are examples. But another common scenario is one where we have no special knowledge about what frequencies might be associated with peaks.

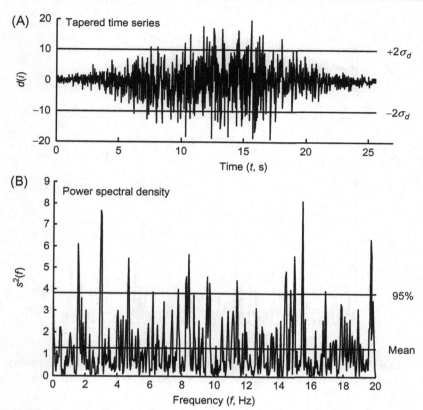

Figure 12.6 (A) Random time series, $d(t)$, after multiplication by Hamming taper. (B) power spectral density, $s^2(f)$, of time series, $d(t)$. *MatLab* scripts eda12_09 and eda12_10.

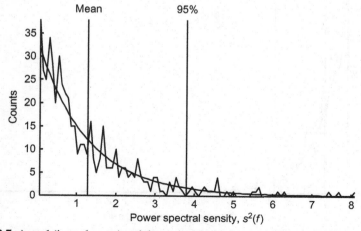

Figure 12.7 Actual (jagged curve) and theoretical (smooth curve) histogram of power spectral density, $s^2(f)$, of the time series shown in Figure 12.6. *MatLab* scripts eda12_09 and eda12_10.

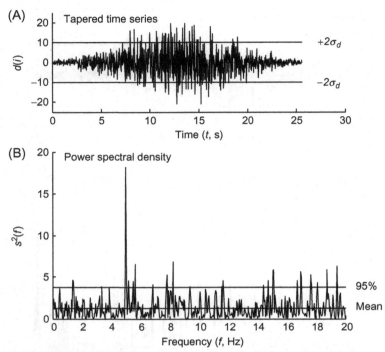

Figure 12.8 (A) Time series, $d(t)$, consisting of the sum of a 5-Hz sinusoidal oscillation plus random noise, after multiplication by Hamming taper. (B) power spectral density, $s^2(f)$, of time series, $d(t)$. *MatLab* scripts eda12_11.

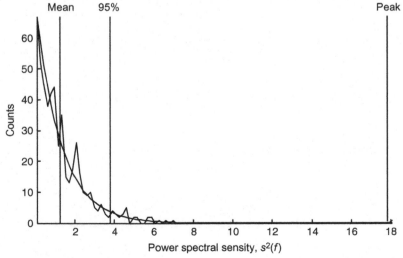

Figure 12.9 Actual (jagged curve) and theoretical (smooth curve) histogram of power spectral density, $s^2(f)$, of the time series shown in Figure 12.8. *MatLab* script eda12_11.

We see a peak *somewhere* in the power spectral density and want to know whether or not it is significant. In this case, the Null Hypothesis is that any peak *at any frequency* in the record arose solely by random variation.

In this example, the power spectral density has $N/2 + 1 = 513$ elements. Thus, there are 513 independent chances for a peak to occur. The probability that a peak of amplitude, s_0^2, occurs somewhere among those 513 possibilities is $(0.99994)^{513} = 0.97$. Thus, there is a 3% chance that a peak arose from random variation—still a small probability, but much larger than the 0.006% that we calculated previously We can still reject the Null Hypothesis at greater than 95% confidence, but with much less confidence than before.

Crib Sheet 12.2 Computing power spectral density

Step 1, Compute time and frequency parameters
number N of data should be even (truncate if necessary)
duration of time series, `T=N*Dt;`
Nyquist (maximum) frequency, `fmax=1/(2*Dt);`
frequency interval, `Df =fmax/(N/2);`
number of non-negative frequencies, `Nf=N/2+1;`
frequency vector, `f=Df*[0:N/2,-N/2+1:-1]';`

Step 2: Pre-process time series, $d(t)$
subtract mean, `d = d-mean(d);`
apply Hamming window function,
`w=0.54-0.46*cos(2*pi*[0:N-1]'/(N-1));`
`dw=w.*d;`

Step 3: Compute Fourier Transform, $\tilde{d}\ (f)$
`dtilde = Dt * fft(dw);`
`dtilde = dtilde[1:Nf];`

Step 4: Compute power spectral density, $s^2(f)$
`s2=(2/T)*abs(dtilde).^2;`

Step 5: Plot power spectral density and look for peaks
`plot( f(1:Nf), s2, 'k-' );`

Step 6: Compute 95% confidence level
Null Hypothesis: time series is uncorrelated random noise
variance of time series, `s2est=std(d);`
power in window function, `ff=sum(w.*w)/N;`
scaling constant, `c = (ff*sd2est)/(2*Nf*Df);`
95% confidence level, `c195 = c*chi2inv(0.95,2);`

Step 7: Plot confidence level and assess significance of peaks
`plot([f(1), f(Nf)], [c195, c195], 'k-');`

12.8   Bootstrap confidence intervals

Many special-purpose statistical tests have been put forward in the literature. Each proposes a statistic appropriate for a specific data analysis scenario and provides a means for testing its significance. However, data analysis scenarios are extremely varied and many have no well-understood tests. Sometimes, model parameters will have a sufficiently complicated relationship to the data that error propagation by normal means will be impractical, making the determination of confidence intervals difficult.

Consider a scenario in which a model parameter, m, is determined through a complicated data analysis procedure. If a large number of *repeated datasets* are available—repeated in the sense of having made all the measurements again at another time—then the problem of determining confidence intervals for the parameter, m, could be approached empirically. The same analysis could be performed on each dataset and a histogram for parameter, m, constructed. With enough repeat datasets, the histogram will approximate the probability density function, $p(m)$. Confidence intervals could then be derived from $p(m)$. While true repeat datasets are seldom available, this method will also work if *approximate* repeat datasets could somehow be constructed from the single, available dataset.

A dataset, $\mathbf{d}$, can be viewed as consisting of N realizations of the probability density function, $p(d)$. The probability density function, $p(d)$, itself can be viewed—loosely—as containing an infinite number of realizations. Suppose that we construct another probability density function, $p'(d)$, by duplicating the N realizations an indefinite number of times and mixing them together (Figure 12.10). As $N \rightarrow \infty$,

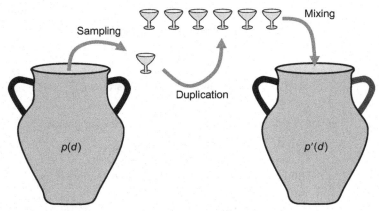

Figure 12.10 A probability density function, $p(d)$, is represented by the large urn at the left and a few of realizations of this function are represented by the small goblet. The contents of the goblet are duplicated indefinitely many times, mixed together, and poured into the large urn at the right, creating a new probability density function, $p'(d)$. Under some circumstances, $p'(d) \approx p(d)$.

$p'(d) \rightarrow p(d)$. As long as N is large enough, $p'(d)$ will be an adequate approximation to $p(d)$.

This scenario suggests that an approximate repeat dataset can be created by randomly *resampling the original dataset with duplications*. If the original data are in a column-vector, $\mathbf{d}^{\text{orig}}$, then a new dataset, $\mathbf{d}$, is constructed by randomly choosing an element from $\mathbf{d}^{\text{orig}}$ each of N times. The two datasets will not be the same, because duplicates from $\mathbf{d}^{\text{orig}}$ are allowed in $\mathbf{d}$. Furthermore, many such resampled datasets can be constructed, all distinct from one another. Identical analyses can then be performed on each resampled dataset, and a histogram of the estimated model parameters assembled. This procedure is called the *bootstrap* method.

We start with a simple case—determining confidence intervals for the slope, b, of a straight line fit to data. We already know how to determine confidence intervals for this linear problem, so it provides a good way to verify the bootstrap results. The probability density function, $p(b)$, of the slope is Normal with variance, σ^2_b, given by a simple formula (Equation 4.29), so 95% confidence is within $\pm 2\sigma_b$ of the mean.

The bootstrap method contains two steps, both within a loop over the number, N_r, of times that the original data are resampled. In the first step, the data, $\mathbf{d}^{\text{orig}}$, and corresponding times, $\mathbf{t}^{\text{orig}}$, are resampled into $\mathbf{d}$ and $\mathbf{t}$. In the second step, standard methodology (least-squares, in this case) is used to estimate the parameter of interest (slope, b, in this case) from $\mathbf{d}$ and $\mathbf{t}$.

```
for p = [1:Nr]
    % resample
    rowindex=unidrnd(N,N,1);
    t = torig(rowindex);
    d = dorig(rowindex);
    % straight line fit
    M=2;
    G=zeros(N,M);
    G(:,1)=1;
    G(:,2)=t;
    mest=(G'*G)\(G'*d);
    slope(p)=mest(2);
end                                           (MatLab eda12_12)
```

The `rowindex` array specifies how the original data are resampled. It is created with the `unidrnd()` function, which returns a column-vector of random integers between 1 and N. The end result is a length-N_r column-vector of slopes. A histogram can then be formed from these slopes and converted into an estimate of the probability density, $p(b)$:

```
Nbins = 100;
[shist, bins]=hist(slope, Nbins);
Db = bins(2)-bins(1);
pbootstrap = shist / (Db*sum(shist));       (MatLab eda12_12)
```

The last line turns the histogram into a properly normalized probability density function, `pbootstrap`, with unit area (Figure 12.11). As expected in this case,

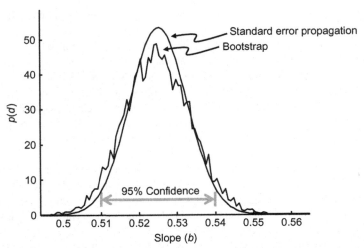

Figure 12.11 Bootstrap method applied to estimating the probability density function, $p(b)$, of slope, b, when a straight line is fit to a fragment of the Black Rock Forest temperature dataset. (Smooth curve) Normal probability density function, with parameters determined by standard error propagation. (Rough curve) Bootstrap estimate. *MatLab* script eda12_12.

the probability density function, $p(b)$, is a good approximation to the Normal probability density function predicted by the standard error propagation formulas. We can use this probability density function to derive estimates of the mean and variance of the slope (see Equations 3.3 and 3.4):

```
mb = Db*sum(bins.*pbootstrap);
vb = Db*sum(((bins-mb).^2).*pbootstrap);  (MatLab eda12_12)
```

Here, mb is the mean and vb is the variance of the slope, b. As the probability density function is approximately Normal, we can state the 95% confidence interval as $mb \pm 2\sqrt{vb}$. However, in other cases, the probability density function may be non-Normal, in which case an explicit calculation of confidence is more accurate:

```
Pbootstrap = Db*cumsum(pbootstrap);
ilo = find(Pbootstrap >= 0.025,1);
ihi = find(Pbootstrap >= 0.975,1);
blo = bins(ilo);
bhi = bins(ihi);                            (MatLab eda12_12)
```

Here, the cumsum() function is used to integrate the probability density function into a probability distribution, Pbootstrap, and the find() function is used to find the values of slope, b, that enclose 95% of the total probability. The 95% confidence interval is then, $b^{lo} < b < b^{hi}$.

In a more realistic example, we return to the varimax factor analysis that we performed on the Atlantic Rock dataset (Figure 8.6). Suppose that the CaO to

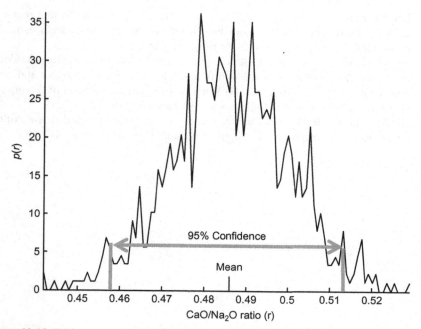

Figure 12.12 Bootstrap method applied to estimating the probability density function, $p(r)$, of a parameter, r, that has a very complicated relationship to the data. Here, the parameter, r, represents the CaO to Na_2O ratio of the second varimax factor of the Atlantic Rock dataset (see Figure 8.6). The mean of the parameter, r, and its 95% confidence intervals are then estimated from $p(r)$. *MatLab* script eda12_13.

Na_2O ratio, r, of factor 2 is of importance to the conclusions drawn from the analysis. The relationship between the data and r is very complex, involving both singular-value decomposition and varimax rotation, so deriving confidence intervals by standard means is impractical. In contrast, bootstrap estimation of $p(r)$, and hence the confidence intervals of r, is completely straightforward (Figure 12.12).

Problems

12.1 The first and twelfth year of the Black Rock Forest temperature dataset are more-or-less complete. After removing hot and cold spikes, calculate the mean of the 10 hottest days of each of these years. Test whether the means are significantly different from one another by following these steps: (A) State the Null Hypothesis. (B) Calculate the t-statistic for this case. (C) Can the Null Hypothesis be rejected?

12.2 Revisit Neuse River prediction error filters that you calculated in Problem 7.2 and analyze the significance of the error reduction for pairs of filters of different length.

12.3 Formally show that the quantity, $Z = (d - \bar{d})/\sigma_d$, has zero mean and unit variance, assuming that d is Normally distributed with mean, $\bar{d}$, and variance, σ_d^2, by transforming the probability density function, $p(d)$ to $p(Z)$.

12.4 Figure 12.6B shows the power spectral density of a random time series. A) Count up the number of peaks that are significant to the 95% level or greater and compare with the expected number. B) What is the significance level of the highest peak? (Note that $N = 1024$ and $c = 0.634$ for this dataset).

12.5 Analyze the significance of the major peaks in the power spectral density of the Neuse River Hydrograph (see Figure 6.11 and *MatLab* script eda06_13).

13 Notes

Note 1.1   On the persistence of *MatLab* variables

MatLab variables accumulate in its *Workspace* and can be accessed not only by the script that created them but also through both the Command Window and the Workspace Window (which has a nice spreadsheet-like matrix viewing tool). This behavior is mostly an asset: You can create a variable in one script and use it in a subsequent script. Further, you can check the values of variables after a script has finished, making sure that they have sensible values. However, this behavior also leads to the following common errors:

(1) you forget to define a variable in one script and the script, instead of reporting an error, uses the value of an identically named variable already in the Workspace;

(2) you accidentally delete a line of code from a script that defines a variable, but the script continues to work because the variable was defined when you ran an earlier version of the script; and

(3) you use a predefined constant such as `pi`, or a built-in function such as `max()`, but its value was reset to an unexpected value by a previous script. (Note that nothing prevents you from defining `pi=2` and `max=4`).

Environmental Data Analysis with MATLAB®. http://dx.doi.org/10.1016/B978-0-12-804488-9.00013-6

Such problems can be detected by deleting all variables from the workspace with a clear all command and then running the script. (You can also delete particular variables with a clear followed by the variable's name, e.g., clear pi). A really common mistake is to overwrite the value of the imaginary unit, i, by using that variable's name for an index counter. Our suggestion is that a clear i be routinely included at the top of any script that expects i to be the imaginary unit.

Note 2.1   On time

Datasets often contain time expressed in calendar (year, month, day) and clock (hour, minute, second) format. This format is not suitable for data analysis and needs to be converted into a format in which time is represented by a single, uniformly increasing variable, say t. The choice of units of the time variable, t, and the definition of the *start time*, $t = 0$, will depend on the needs of the data analysis. In the case of the Black Rock Forest, which consisted of 12.6 years of hourly samples, a time variable that expresses days, starting on January 1 of the first year of the dataset, is a reasonable choice, especially because the diurnal cycle is so strong. However, time in years starting on January 1 of the first year of the dataset might be preferred when examining annual periodicities. In this case, having a start time that allows us to easily recognize the season of a particular time is important.

The conversion of calendar/clock time to a single variable, t, is complicated, because of the different lengths of the months and special cases such as leap years. MatLab provides a *time arithmetic* function, datenum() that expedites this conversion. It takes the calendar date (year, month, day) and time (hour, minute, second) and returns *date number*; that is, the number of days (including fractions of a day) that have elapsed since midnight on January 1, 0000. The time interval between two date numbers can be computed by subtraction. For example, the number of seconds between Feb 11, 2008 03:04:00 and Feb 11, 2008 03:04:01 is

```
86400*(datenum(2008,2,11,4,4,1)-datenum(2008,2,11,4,4,0))
```

which evaluates to 1.0000 s.

Finally, we note a complication, relevant to cases where time accuracy of seconds or better is required, which is related to the existence of *leap seconds*. Leap seconds are analogous to leap years. They are integral-second clock corrections, applied on June 30 and December 31 of each year, that account for small irregularities in the rotation of the earth. However, unlike leap years, which are completely predictable, leap seconds are determined semiannually by the International Earth Rotation and Reference Systems Service (IERS). Hence, time intervals cannot be calculated accurately without an up-to-date table of leap seconds. To make matters worse, while the most widely used time standard, Coordinated Universal Time (UTC), uses leap seconds, several other standards, including the equally widely used Global Positioning System (GPS), do not. Thus, the determination of long time intervals to second-level accuracy is tricky. The time standard used in the dataset must be known and, if that standard uses leap seconds, then they must be properly accounted for by the time arithmetic software. As of the end of 2010, a total of 34 leap seconds have been declared since they were first implemented in 1972. Thus, very long (decade) time intervals can be in error by tens of seconds, if leap seconds are not properly accounted for. The *MatLab* function, datenum(), does not account for leap seconds and hence does not provide second-level accuracy for UTC times.

Note 2.2   On reading complicated text files

MatLab's load() function can read only text files containing a table of numerical values. Some publicly accessible databases, including many sponsored by government agencies, provide data as more complicated text files that are a mixture of numeric and alphabetic values. For instance, the Black Rock Forest temperature dataset, which contains time and temperature, contains lines of text such as:

```
2100-2159 31 Jan 1997    -1.34
2200-2259 31 Jan 1997    -0.958
2300-2400 31 Jan 1997    -0.601
0000-0059 1 Feb 1997     -0.245
0100-0159 1 Feb 1997     -0.217          (file brf_raw.txt)
```

In the first line above, the date of the observation is 31 Jan 1997, the start and end times are 2100-2159, and the observed temperature is -1.34. This data file is one of the simpler ones, as each line has the same format and most of the fields are delimited by tabs or spaces. We occasionally encounter much more complicated cases, in which the number of fields varies from line to line and where adjacent fields are run together without delimiters.

Some of the simpler cases, including the one above, can be reformatted using the *Text Import Wizard* module of *Microsoft's Excel* spreadsheet software. But we know of no universal and easy-to-use software that can reliably handle complicated cases. We resort to writing a custom *MatLab* script for each file. Such a script sequentially processes each line in the file, according to what we perceive to be the rules under which it was written (which are sometimes difficult to discern). The heart of such a script is a for loop that sequentially reads lines from the file:

```
fid = fopen(filename);
for i = [1:N]
    tline = fgetl(fid);
    % now process the line
    ------
end
fclose(fid);                              (MatLab brf_convert)
```

Here, the function, fopen(), *opens* a file so that it can be read. It returns an integer, fid, which is subsequently used to refer to the file. The function, fgetl(), reads one line of characters from the file and puts them into the character string, tline. These characters are then processed in a portion of the script, omitted here, whose purpose is to convert all the data fields into numerical values stored in one of more arrays. Finally, after every line has been read and processed, the file is closed with the `fclose()` function. The processing section of the script can be quite complicated. One *MatLab* function that is extremely useful in this section is `sscanf()`, which can convert a character string into a numerical variable. It is the inverse of the previously discussed `sprintf()` function, and has similar arguments (see Section 2.4 and the *MatLab* Help files). Typically, one first determines the portion of the character string, `tline`, that contains a particular data field (for instance, `tline(6:9)` for the second field, above) and then converts that portion to a numerical value using `sscanf()`.

Data format conversion scripts are tedious to write. They should always be tested very carefully, including by spot-checking data values against the originals. Spot checks should always include data drawn from near the *end* of the file.

Note 3.1   On the rule for error propagation

Suppose that we form M_A model parameters, $\mathbf{m}_A$, from N data, $\mathbf{d}$, using the linear rule $\mathbf{m}_A = \mathbf{M}_A \mathbf{d}$. We have already shown that when $M_A = N$, the covariance matrices are related by the rule, $\mathbf{C}_{MA} = \mathbf{M}_A \mathbf{C}_d \mathbf{M}_A^\mathrm{T}$. To verify this rule for the $M_A < N$ case, first devise $M_B = N - M_A$ complementary model parameters, $\mathbf{m}_B$, such that $\mathbf{m}_B = \mathbf{M}_B \mathbf{d}$. Now concatenate the two sets of model parameters so that their joint matrix equation is square:

$$\begin{bmatrix} \mathbf{m}_A \\ \mathbf{m}_B \end{bmatrix} = \begin{bmatrix} \mathbf{M}_A \\ \mathbf{M}_B \end{bmatrix} \mathbf{d}$$

The normal rule for error propagation now gives

$$\mathbf{C}_m = \begin{bmatrix} \mathbf{M}_A \\ \mathbf{M}_B \end{bmatrix} \mathbf{C}_d \begin{bmatrix} \mathbf{M}_A^\mathrm{T} & \mathbf{M}_B^\mathrm{T} \end{bmatrix} = \begin{bmatrix} \mathbf{M}_A \mathbf{C}_d \mathbf{M}_A^\mathrm{T} & \mathbf{M}_A \mathbf{C}_d \mathbf{M}_B^\mathrm{T} \\ \mathbf{M}_B \mathbf{C}_d \mathbf{M}_A^\mathrm{T} & \mathbf{M}_B \mathbf{C}_d \mathbf{M}_B^\mathrm{T} \end{bmatrix} = \begin{bmatrix} \mathbf{C}_{m_A} & \mathbf{C}_{m_{A,B}} \\ \mathbf{C}_{m_{B,A}} & \mathbf{C}_{m_B} \end{bmatrix}$$

The upper left part of $\mathbf{C}_m$, $\mathbf{C}_{m_A} = \mathbf{M}_A \mathbf{C}_d \mathbf{M}_A^\mathrm{T}$, which comprises all the variances and covariances of the $\mathbf{m}_A$ model parameters, satisfies the normal rule of error proposition and is independent of the choice of $\mathbf{M}_B$. Hence, the rule can be applied to the $M_A < N$ case in which $\mathbf{M}_A$ is rectangular, without concern for the particular choice of complementary model parameters.

Note 3.2   On the `eda_draw()` function

We provide a simple function, `eda_draw()`, for plotting a sequence of square matrices and vectors as grey-shaded images. The function can also place a caption beneath the matrices and vectors and plot a symbol between them. For instance, the command

```
eda_draw(d, 'caption d', '=', G, 'caption G', m, 'caption m');
```
MatLab eda12_01

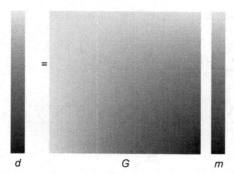

Figure 13.1 Results of call to eda_draw() function. *MatLab* script note03_02.

creates a graphical representation of the equation, $\mathbf{d} = \mathbf{Gm}$ (Figure 13.1). The function accepts vectors, square matrices, and character strings, in any order. A character string starting with the word "caption", as in 'caption d', is plotted beneath the previous vector or matrix (but with the word "caption" removed). Other character strings are plotted to the right of the previous matrix or vector.

Note 4.1   On complex least squares

Least-squares problems with complex quantities occasionally arise (e.g., when the model parameters are the Fourier transform of a function). In this case, all the quantities in $\mathbf{Gm} = \mathbf{d}$ are complex. The correct definition of the total error is

$$E(\mathbf{m}) = \mathbf{e}^{*\mathrm{T}}\mathbf{e} = (\mathbf{d} - \mathbf{Gm})^{*\mathrm{T}}(\mathbf{d} - \mathbf{Gm})$$

where * signifies complex conjugation. This combination of complex conjugate and matrix transpose is called the Hermitian transpose and is denoted $\mathbf{e}^{\mathrm{H}} = \mathbf{e}^{*\mathrm{T}}$. Note that the total error, E, is a nonnegative real number. The least-squares solution is obtained by minimizing E with respect to the real and imaginary parts of $\mathbf{m}$, treating them as independent variables. Writing $\mathbf{m} = \mathbf{m}^R + i\mathbf{m}^I$, we have

$$E(\mathbf{m}) = \sum_{i=1}^{N}\left(d_i^* - \sum_{j=1}^{M}G_{ij}^*m_j^*\right)\left(d_i - \sum_{k=1}^{M}G_{ik}m_k\right) = \sum_{i=1}^{N}d_i^*d_i - \sum_{j=1}^{M}\sum_{i=1}^{N}d_i^*G_{ij}(m_j^R + im_j^I)$$

$$-\sum_{k=1}^{M}\sum_{i=1}^{N}d_iG_{ik}^*(m_j^R - im_j^I) + \sum_{j=1}^{M}\sum_{k=1}^{M}\sum_{i=1}^{N}G_{ij}^*G_{ik}(m_k^R - im_j^I)(m_k^R + im_k^I)$$

Differentiating with respect to the real part of **m** yields

$$\frac{\partial E(\mathbf{m})}{\partial m_p^R} = 0 = -\sum_{j=1}^{M}\sum_{i=1}^{N} d_i^* G_{ij} \frac{\partial m_j^R}{\partial m_p^R} - \sum_{k=1}^{M}\sum_{i=1}^{N} d_i G_{ik}^* \frac{\partial m_j^R}{\partial m_p^R}$$

$$+ \sum_{j=1}^{M}\sum_{k=1}^{M}\sum_{i=1}^{N} G_{ij}^* G_{ik} \frac{\partial m_j^R}{\partial m_p^R}(m_k^R + im_k^I) + \sum_{j=1}^{M}\sum_{k=1}^{M}\sum_{i=1}^{N} G_{ij}^* G_{ik}(m_j^R - im_j^I)\frac{\partial m_k^R}{\partial m_p^R}$$

$$= -\sum_{i=1}^{N} d_i^* G_{ip} - \sum_{i=1}^{N} d_i G_{ip}^* + \sum_{k=1}^{M}\sum_{i=1}^{N} G_{ip}^* G_{ik}(m_k^R + im_k^I) + \sum_{j=1}^{M}\sum_{i=1}^{N} G_{ij}^* G_{ip}(m_j^R - im_j^I)$$

Note that $\partial m_k^R / \partial m_p^R = \delta_{kp}$, as m_k^R and m_p^R are independent variables. Differentiating with respect to the imaginary part of **m** yields

$$\frac{\partial E(\mathbf{m})}{\partial m_p^I} = 0$$

$$= -i\sum_{i=1}^{N} d_i^* G_{ip} + i\sum_{i=1}^{N} d_i G_{ip}^* - i\sum_{k=1}^{M}\sum_{i=1}^{N} G_{ip}^* G_{ik}(m_k^R + im_k^I) + i\sum_{j=1}^{M}\sum_{i=1}^{N} G_{ij}^* G_{ip}(m_j^R - im_j^I)$$

$$= \sum_{i=1}^{N} d_i^* G_{ip} - \sum_{i=1}^{N} d_i G_{ip}^* + \sum_{k=1}^{M}\sum_{i=1}^{N} G_{ip}^* G_{ik}(m_k^R + im_k^I) - \sum_{j=1}^{M}\sum_{i=1}^{N} G_{ij}^* G_{ip}(m_j^R - im_j^I)$$

Finally, adding the two derivative equations yields

$$-2\sum_{i=1}^{N} d_i G_{ip}^* + \sum_{k=1}^{M}\sum_{i=1}^{N} G_{ip}^* G_{ik}(m_k^R + im_k^I) = 0$$

$$\text{or}$$

$$-2\mathbf{G}^H\mathbf{d} + 2[\mathbf{G}^H\mathbf{G}]\,\mathbf{m} = 0$$

The least-squares solution and its covariance are

$$\mathbf{m}^{\text{est}} = [\mathbf{G}^H\mathbf{G}]^{-1}\mathbf{G}^H\mathbf{d} \quad \text{and} \quad \mathbf{C_m} = \sigma_d^2[\mathbf{G}^H\mathbf{G}]^{-1}$$

In *MatLab*, the Hermitian transpose of a complex matrix, G, is denoted with the same symbol as transposition, as in G′, and transposition without complex conjugation is denoted G.′. Thus, no changes need to be made to the *MatLab* formulas to implement complex least squares.

Note 5.1   On the derivation of generalized least squares

Strictly speaking, in Equation (5.4), the probability density function, $p(\mathbf{h})$, can only be said to be proportional to $p(\mathbf{m})$ when the $K \times M$ matrix, $\mathbf{H}$, in the equation, $\mathbf{Hm} = \mathbf{h}$, is square so that $\mathbf{H}^{-1}$ exists. In other cases, the Jacobian determinant is undefined. Nonsquare cases arise whenever only a few pieces of prior information are available. The derivation can be *patched* by imagining that $\mathbf{H}$ is made square by adding $M - K$ rows of complementary information and then assigning them negligible certainty so that they have no effect on the generalized least-squares solution. This patch does not affect the results of the derivation; all the formulas for the generalized least-squares solution and its covariance are unchanged. The underlying issue is that the uniform probability density function, which represents a state of no information, does not exist on an unbounded domain. The best that one can do is a very wide normal probability density function.

Note 5.2   On *MatLab* functions

MatLab provides a way to define functions that perform in exactly the same manner as built-in functions such as `sin()` and `cos()`. As an example, let us define a function, `areaofcircle()`, that computes the area of a circle of radius, r:

```
function a = areaofcircle(r)
% computes area, a, of circle of radius, r.
a = pi * (r^2);
return                              MatLab areaofcircle
```

We place this script in a separate m-file, `areaofcircle.m`. The first line declares the name of the function to be `areaofcircle`, its input to be `r`, and its output to be a. The last line, `return`, denotes the end of the function. The interior lines perform the actual calculation. One of them must set the value of the output variable. The function is called in from the main script as follows:

```
radius=2;
area = areaofcircle(radius);       MatLab eda12_02
```

Note that the variable names in the main script need not agree with the names in the function; the latter act only as *placeholders*.

MatLab functions can take several input variables and return several output variables, as is illustrated in the following example that computes the circumference and area of a rectangle:

```
function [c,a] = CandAofrectangle(l, w)
% computes circumference, c, and area, a, of
% a rectangle of length, l, and width, w.
c = 2*(l+w);
a = l*w;
return                              MatLab CandAofrectangle
```

The function is placed in the m-file, `CandAofrectangle.m`. It is called in from the main script as follows:

```
a=2;
b=4;
[circ, area] = CandAofrectangle(a,b);          MatLab eda12_02
```

Note 5.3   On reorganizing matrices

Owing to the introductory nature of this book, we have intentionally omitted discussion of a group of advanced *MatLab* functions that allow one to reorganize matrices. Nevertheless, we briefly describe some of the key functions here. In *MatLab*, a key feature of a matrix is that its elements can be accessed with a single index, instead of the normal two indices. In this case, the matrix, say A, acts a column-vector containing the elements of A arranged column-wise. Thus, for a 3 × 3 matrix, A(4) is equivalent to A(1,2). The MatLab functions, `sub2ind()` and `ind2sub()`, translate between two "subscripts", i and j, and a vector "index", k, such that A(i,j)= A(k). The `reshape()` function can reorganize any $N \times M$ matrix into a $K \times L$ matrix, as long as $NM = KL$. Thus, for example, a 4 × 4 matrix can be easily converted into equivalent 1 × 16, 2 × 8, 8 × 2, and 16 × 1 matrices. These functions often work to eliminate `for` loops from the matrix-reorganization sections of the scripts. They are demonstrated in *MatLab* script eda12_03.

Note 6.1   On the *MatLab* `atan2()` function

The phase of the Discrete Fourier Transform, $\phi = \tan^{-1}(B/A)$, is defined on the interval, $(-\pi, +\pi)$. In *MatLab*, one should use the function, `atan2(B,A)`, and not the function `atan(B/A)`. The latter version is defined on the wrong interval, $(-\pi/2, +\pi/2)$, and will also fail when $A = 0$.

Note 6.2   On the orthonormality of the discrete Fourier data kernel

The rule, $[\mathbf{G^{*T}G}] = N\,\mathbf{I}$, for the complex version of the Fourier data kernel, $\mathbf{G}$, can be derived as follows. First write down the definition of un-normalized version of the data kernel for the Fourier series:

$$G_{kp} = \exp(2\pi i(k-1)(p-1)/N)$$

Now compute $[\mathbf{G^{*T}G}]$

$$[\mathbf{G^{*T}G}]_{pq} = \sum_{k=1}^{N} G_{kp}^* G_{kq} = \sum_{k=0}^{N-1} \exp(2\pi ik(p-q)/N) = \sum_{k=0}^{N-1} z^k = f(z)$$

with

$$z = \exp(2\pi i(p - q)/N)$$

Now consider the series

$$f(z) = \sum_{k=0}^{N-1} z^k = 1 + z + z^2 + \cdots + z^{N-1}$$

Multiply by z

$$zf(z) = \sum_{k=0}^{N-1} z^{k+1} = z + z^2 + \cdots + z^N$$

and subtract, noting that all but the first and last terms cancel:

$$f(z) - zf(z) = 1 - z^N \quad \text{or} \quad f(z) = \frac{1 - z^N}{1 - z}$$

Now substitute in $z = \exp(2\pi i(p - q)/N)$:

$$f(z) = \frac{1 - \exp(2\pi i(p - q))}{1 - \exp(2\pi i(p - q)/N)}$$

The numerator is zero, as $\exp(2\pi i s) = 1$ for any integer, $s = p - q$. In the case, $p \neq q$, the denominator is nonzero, so $f(z) = 0$. Thus, the off-diagonal elements of $\mathbf{G}^{*T}\mathbf{G}$ are zero. In the case, $p = q$, the denominator is also zero, and we must use l'Hopital's rule to take the limit, $s \rightarrow 0$. This rule requires us to take the derivative of both numerator and denominator before taking the limit:

$$f(z) = \lim_{s \rightarrow 0} \frac{2\pi i \exp(2\pi i s)}{\left(\frac{2\pi i}{N}\right) \exp(2\pi i s/N)} = N$$

The diagonal elements of $\mathbf{G}^{*T}\mathbf{G}$ are all equal to N.

Note 6.3   On the expansion of a function in an orthonormal basis

Suppose that we approximate an arbitrary function $d(t)$ on the interval $t^{min} < t < t^{max}$ as a sum of *basis functions* $g_i(t)$:

$$d(t) \approx d^s(t) \quad \text{with} \quad d^s(t) = \sum_{i=1}^{M} m_i g_i(t)$$

Here, m_i are unknown coefficients. The Fourier series is one such approximation, with sines and cosines as the basis functions:

$$g_i(t) = \begin{cases} \cos(\omega_i t) & i \text{ odd} \\ \sin(\omega_i t) & i \text{ even} \end{cases}$$

For any given sets of coefficients, the quality of the approximation can be measured by defining an error:

$$E = \int_{t^{min}}^{t^{max}} [d(t) - d^s(t)]^2 \, dt$$

This is a generalization of the usual least squares error, and has the properties that $E = 0$ when $d(t) = d^s(t)$. In general, zero error can be achieved only in the $M \rightarrow \infty$ limit. We now take an approach that is very similar to the one used in the derivation of the least squares formula in Section 4.7 and view E as a function of the unknown coefficients and minimize it with respect them by solving $\partial E / \partial m_k = 0$. This procedure leads to an equation for the unknown coefficients:

$$\int_{t^{min}}^{t^{max}} g_j(t) \, d(t) \, dt = \sum_{i=1}^{M} m_i \int_{t^{min}}^{t^{max}} g_j(t) \, g_i(t) \, dt \quad \text{or} \quad b_j = \sum_{i=1}^{M} M_{ij} m_i$$

$$\text{with} \quad b_j = \int_{t^{min}}^{t^{max}} g_j(t) \, d(t) \, dt \quad \text{and} \quad M_{ij} = \int_{t^{min}}^{t^{max}} g_j(t) \, g_i(t) \, dt$$

Solving for the coefficients, we find that $\mathbf{m} = \mathbf{M}^{-1}\mathbf{b}$.

In some cases, the basis functions $g_i(t)$ may be orthonormal, meaning that any pair of them obeys:

$$\int_{t^{min}}^{t^{max}} g_i(t) \, g_j(t) \, dt = \begin{cases} 1 & i = j \\ 0 & i \neq j \end{cases}$$

In this case, $\mathbf{M} = \mathbf{I}$ and the formula for the coefficients simplifies to $\mathbf{m} = \mathbf{b}$. Each coefficient is determined separately; any given basis function has the same coefficient regardless of the number of terms in the summation, or even on the identity of the other basis functions. The coefficients are even the same in the $M \rightarrow \infty$ limit. However, whether $E \rightarrow 0$ in this limit depends on whether the set of basis functions is *complete*, an issue that we do not address further here.

Now suppose that only a discrete version of $d(t)$ is available; that is, we know the time series $d_i = d(t_i)$ (for $i = 1, \ldots, N$). We can approximate the integrals as Riemann sums:

$$M_{ij} \approx \Delta t \sum_{k=1}^{N} g_i(t_k) g_j(t_k) = \Delta t \sum_{k=1}^{N} G_{kj} G_{kj}$$

$$b_j \approx \Delta t \sum_{k=1}^{N} g_j(t_k) d(t_k) \approx \Delta t \sum_{i=1}^{N} G_{kj} d_k$$

where $G_{ij} = g_j(t_i)$. We have achieved a result that is identical to least squares: $\mathbf{m} \approx [\mathbf{G}^T \mathbf{G}]^{-1} \mathbf{G}^T \mathbf{d}$. Furthermore, when the functions are orthonormal, $\mathbf{m}$ can be determined without computing a matrix inverse, since $\Delta t [\mathbf{G}^T \mathbf{G}] \approx \mathbf{I}$ and $\mathbf{m} = \mathbf{G}^T \mathbf{d}$. The estimated coefficients are uncorrelated, since by the usual rules of error propagation, $\mathbf{C}_m = \sigma_d^2 [\mathbf{G}^T \mathbf{G}]^{-1} = \sigma_d^2 \mathbf{I}$, where σ_d^2 is the variance of d_i. These results explain the popularity of series of orthogonal functions.

Note 8.1   On singular value decomposition

The derivation of the singular value decomposition is not quite complete, as we need to demonstrate that the eigenvalues, λ_i, of $\mathbf{S}^T \mathbf{S}$ are all nonnegative so that the singular values of $\mathbf{S}$, which are the square roots of the eigenvalues, are all real. This result can be demonstrated as follows. Consider the minimization problem

$$E(\mathbf{m}) = (\mathbf{d} - \mathbf{Sm})^T (\mathbf{d} - \mathbf{Sm})$$

This is just the least-squares problem with $\mathbf{G} = \mathbf{S}$. Note that $E(\mathbf{m})$ is a nonnegative quantity, irrespective of the value of $\mathbf{m}$; therefore, a point (or points), $\mathbf{m}_0$, of minimum exists, irrespective of the choice of $\mathbf{S}$. In Section 4.9, we showed that in the neighborhood of $\mathbf{m}_0$ the error behaves as

$$E(\mathbf{m}) = E(\mathbf{m}_0) + \Delta\mathbf{m}^T \mathbf{S}^T \mathbf{S} \Delta\mathbf{m} \quad \text{where} \quad \Delta\mathbf{m} = \mathbf{m} - \mathbf{m}_0$$

Now let $\Delta\mathbf{m}$ be proportional to an eigenvector, $\mathbf{v}^{(i)}$, of $\mathbf{S}^T \mathbf{S}$; that is, $\Delta\mathbf{m} = c\mathbf{v}^{(i)}$. Then,

$$E(\mathbf{m}) = E(\mathbf{m}_0) + c^2 \mathbf{v}^{(i)T} \mathbf{S}^T \mathbf{S} \mathbf{v}^{(i)} = E(\mathbf{m}_0) + c^2 \lambda_i$$

Here, we have used the relationship, $\mathbf{S}^T \mathbf{S} \mathbf{v}^{(i)} = \lambda_i \mathbf{v}^{(i)}$. As we increase the constant, c, we move away from the point, $\mathbf{m}_0$, in the direction of the eigenvector. By hypothesis, the error must increase, as $E(\mathbf{m}_0)$ is the point of minimum error. The eigenvalue, λ_i, must be positive or else the error would decrease and $\mathbf{m}_0$ could not be a point of minimum error.

As an aside, we also mention that this derivation demonstrated that the point, $\mathbf{m}_0$, is nonunique if any of the eigenvalues are zero, as the error is unchanged when one moves in the direction of the corresponding eigenvector.

Note 9.1   On coherence

The coherence *can* be interpreted as the zero lag cross-correlation of the band-passed versions of the two time series, $u(t)$ and $v(t)$. However, the band-pass filter, $f(t)$, must have a spectrum, $\tilde{f}(\omega)$, that is *one-sided*; that is, it must be zero for all negative frequencies. This is in contrast to a normal filter, which has a *two-sided* spectrum. Then, the first of the two integrals in Equation (9.32) is zero and no cancelation of imaginary parts occurs. Such a filter, $f(t)$, is necessarily complex, implying that the band-passed time series, $f(t) * u(t)$ and $f(t) * v(t)$, are complex, too. Thus, the interpretation of coherence in terms of the zero-lag cross-correlation still holds, but becomes rather abstract.

Note that the coherence must be calculated with respect to a finite bandwidth. If we were to omit the frequency averaging, then the coherence is unity for all frequencies, regardless of the shapes of the two time series, $u(t)$ and $v(t)$:

$$C_{uv}^2(\omega_0, \Delta\omega) = \frac{\left|\tilde{u}^*(\omega_0)\tilde{v}(\omega_0)\right|^2}{|\tilde{u}(\omega_0)|^2\,|\tilde{v}(\omega_0)|^2} \rightarrow \frac{\tilde{u}^*(\omega_0)\tilde{u}(\omega_0)\tilde{v}^*(\omega_0)\tilde{v}(\omega_0)}{\tilde{u}^*(\omega_0)\tilde{u}(\omega_0)\tilde{v}^*(\omega_0)\tilde{v}(\omega_0)} = 1 \quad \text{as} \quad \Delta\omega \rightarrow 0$$

This rule implies that $C_{uv}^2(\omega_0 = \omega') = 1$ when the two time series are pure sinusoids, regardless of their relative phase. The coherence of $u(t) = \cos(\omega't)$ and $(t) = \sin(\omega't)$, where ω' is an arbitrary frequency of oscillation, is unity. In contrast, $\mathcal{C}_{uv}^2(\omega_0 = \omega') = 0$, as the zero lag cross-correlation of, $u(t)$ and $v(t)$ is

$$\int_{-\infty}^{+\infty} \sin(\omega't)\cos(\omega't)\mathrm{d}t = \frac{1}{2}\int_{-\infty}^{+\infty} \sin(2\omega't)\mathrm{d}t = 0$$

This is the main difference between the two quantities, $C_{uv}^2(\omega_0, \Delta\omega)$ and $\mathcal{C}_{uv}^2(\omega_0, \Delta\omega)$(Menke 2014).

Note 9.2   On Lagrange multipliers

The method of Lagrange multipliers is used to solve constrained minimization problems of the following form: minimize $\Phi(\mathbf{x})$ subject to the constraint $C(\mathbf{x}) = 0$. It can be derived as follows: The constraint equation defines a surface. The solution, say $\mathbf{x}_0$, must lie on this surface. In an unconstrained minimization problem, the gradient vector, $\partial\Phi/\partial x_i$, must be zero at $\mathbf{x}_0$, as Φ must not decrease in any direction away from $\mathbf{x}_0$. In contrast, in the constrained minimization, only the components of the gradient tangent to the surface need be zero, as the solution cannot be moved off the surface to further minimize Φ (Figure 13.2). Thus, the gradient is allowed to have a nonzero component parallel to the surface's normal vector, $\partial C/\partial x_i$. As $\partial\Phi/\partial x_i$ is parallel

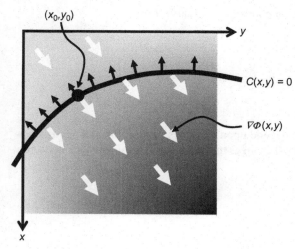

Figure 13.2 Graphical interpretation of the method of Lagrange multipliers, in which the function $\Phi(x, y)$ is minimized subject to the constraint that $C(x, y) = 0$. The solution (bold dot) occurs at the point, (x_0, y_0), on the surface, $C(x, y) = 0$, where the surface normal (black arrows) is parallel to the gradient, $\nabla \Phi(x, y)$ (white arrows). At this point, Φ can only be further minimized by moving it off the surface, which is disallowed by the constraint. *MatLab* script eda12_04.

to $\partial C / \partial x_i$, we can find a linear combination of the two, $\partial \Phi / \partial x_i + \lambda \partial C / \partial x_i$, where λ is a constant, which is zero at $\mathbf{x}_0$. The constrained inversion satisfies the equation, $(\partial / \partial x_i)(\Phi + \lambda C) = 0$, at $\mathbf{x}_0$. Thus, the constrained minimization is equivalent to the unconstrained minimization of $\Phi + \lambda C$, except that the constant, λ, is unknown and needs to be determined as part of the solution process.

Note 11.1   On the chain rule for partial derivatives

Consider a variable $f(x)$ that depends upon a variable x. The notion that a small change Δx in x causes a small change Δf in f is denoted $\Delta f = df/dx \, \Delta x$, where df/dx is the derivative of f with respect to x. Now suppose that $f(x, y)$ depends upon two variables, x and y. The notion that small changes in x and y causes a small change Δf in f is denoted:

$$\Delta f = \frac{\partial f}{\partial x} \Delta x + \frac{\partial f}{\partial y} \Delta y$$

The quantities $\partial f / \partial x$ and $\partial f / \partial y$ are called *partial derivatives*. If another variable $g(x, y)$ also depends upon x and y, then by analogy, small changes in x and y also cause a small change in g:

$$\Delta g = \frac{\partial g}{\partial x} \Delta x + \frac{\partial g}{\partial y} \Delta y$$

These two equations can be written compactly in matrix form:

$$\begin{bmatrix} \Delta f \\ \Delta g \end{bmatrix} = \begin{bmatrix} \dfrac{\partial f}{\partial x} & \dfrac{\partial f}{\partial y} \\ \dfrac{\partial g}{\partial x} & \dfrac{\partial g}{\partial y} \end{bmatrix} \begin{bmatrix} \Delta x \\ \Delta y \end{bmatrix}$$

If two variables $u(f, g)$ and $v(f, g)$ depend upon variables f and g, the analogous equation expressing the notion that small changes in f and g cause small changes in u and v is denoted:

$$\begin{bmatrix} \Delta u \\ \Delta v \end{bmatrix} = \begin{bmatrix} \dfrac{\partial u}{\partial f} & \dfrac{\partial u}{\partial g} \\ \dfrac{\partial v}{\partial f} & \dfrac{\partial v}{\partial g} \end{bmatrix} \begin{bmatrix} \Delta f \\ \Delta g \end{bmatrix}$$

The notion that small changes in x and y causes small changes in f and g which in turn causes small changes in u and v is expressed by substituting one matrix equation onto the other:

$$\begin{bmatrix} \Delta u \\ \Delta v \end{bmatrix} = \begin{bmatrix} \dfrac{\partial u}{\partial f} & \dfrac{\partial u}{\partial g} \\ \dfrac{\partial v}{\partial f} & \dfrac{\partial v}{\partial g} \end{bmatrix} \begin{bmatrix} \dfrac{\partial f}{\partial x} & \dfrac{\partial f}{\partial y} \\ \dfrac{\partial g}{\partial x} & \dfrac{\partial g}{\partial y} \end{bmatrix} \begin{bmatrix} \Delta x \\ \Delta y \end{bmatrix} \quad \text{or} \quad \begin{bmatrix} \Delta u \\ \Delta v \end{bmatrix} = \begin{bmatrix} \dfrac{\partial u}{\partial x} & \dfrac{\partial u}{\partial y} \\ \dfrac{\partial v}{\partial x} & \dfrac{\partial v}{\partial y} \end{bmatrix} \begin{bmatrix} \Delta x \\ \Delta y \end{bmatrix}$$

Equating the matrices yields the *chain rule*:

$$\begin{bmatrix} \dfrac{\partial u}{\partial x} & \dfrac{\partial u}{\partial y} \\ \dfrac{\partial v}{\partial x} & \dfrac{\partial v}{\partial y} \end{bmatrix} = \begin{bmatrix} \dfrac{\partial u}{\partial f} & \dfrac{\partial u}{\partial g} \\ \dfrac{\partial v}{\partial f} & \dfrac{\partial v}{\partial g} \end{bmatrix} \begin{bmatrix} \dfrac{\partial f}{\partial x} & \dfrac{\partial f}{\partial y} \\ \dfrac{\partial g}{\partial x} & \dfrac{\partial g}{\partial y} \end{bmatrix}$$

References

Menke, W., 2014. Coherence Modified for sensitivity to relative phase of real band-limited time series. App. Math 5, 2739–2745. doi:10.4236/am.2014.517261.

Index

Note: Page numbers followed by "*f*" indicate figures, and "*t*" indicate tables.

Printed in the United States
By Bookmasters